CELL SEPARATION METHODS AND APPLICATIONS

CELL SEPARATION METHODS AND APPLICATIONS

edited by

DIETHER RECKTENWALD
AmCell Corporation
Sunnyvale, California

ANDREAS RADBRUCH
Deutsches Rheuma-Forschungszentrum Berlin
Berlin, Germany

MARCEL DEKKER, INC. NEW YORK • BASEL • HONG KONG

Library of Congress Cataloging-in-Publication Data

Cell separation methods and applications / edited by Diether Recktenwald, Andreas Radbruch.
p. cm.
Includes bibliographical references and index.
ISBN 0-8247-9864-3
1. Cell separation. I. Recktenwald, Diether. II. Radbruch, A. (Andreas)
QH585.5.C44C435 1997
571.6—dc21 97-33113
CIP

The publisher offers discounts on this book when ordered in bulk quantities. For more information, write to Special Sales/Professional Marketing at the address below.

This book is printed on acid-free paper.

MARCEL DEKKER, INC.
270 Madison Avenue, New York, New York 10016
http://www.dekker.com

Current printing (last digit):
10 9 8 7 6 5 4 3 2 1

PRINTED IN THE UNITED STATES OF AMERICA

Foreword

It should not come as a surprise that the fields of developmental biology as an intellectual discipline and cell separation as a technological discipline have grown in an interdependent fashion. Just as the cell is the unit of biological organization, cells are organized into groupings from the most mature effector cells back to the most primitive progenitors—the stem cells. Stem cells and progenitor cells have the capacity of being clonogenic precursors for large numbers of progeny; stem cells are the most primitive subset and are defined as the cells that can both self-renew and at the single-cell level give rise to progeny of several different mature cells. Progenitors may be multipotent or oligopotent and can be distinguished from stem cells by their lack of self-renewal capacity. The generation and regeneration of all tissues in the organs in the body depend on the actions of stem and progenitor cells. When we think of generation, we think of those events that occur in embryonic and fetal life to give rise to the organ systems; and when we think of regeneration, we think largely of repair systems, which, in the end, can or should be exploited clinically to replace, wholly or in part, tissue and organ systems that are damaged or are defective.

Identification of the stem and progenitors cells involved in generation or regeneration of a particular organ system or tissue concerns fairly rare populations. When one is dealing with regeneration of tissue or organ systems in a clinical circumstance where diseased or malignant cells may contaminate that organ system, practical regeneration can occur only if the stem/progenitor cells are purified nearly to homogeneity and do not contain contaminating diseased cells. To do this not only in animal models but in humans, one must be thinking of separation technologies that not only are high fidelity in terms of identifying and isolating rare populations, but also are on a scale that is large enough to deliver adequate numbers of stem/progenitor cells for clinically effective and rapid regeneration.

As an immunologist, I must also add that large-scale identification and separation of subsets of cells of the immune system for experiment and clinical treatment involve clonogenic cells of another sort—lymphocytes that respond to antigen by clonal expansion and differentiation into memory or effector cells. Because most methods that allow the culture and expansion of these important potential effector cells also alter their life-span and homing properties to the extent that normal regeneration of the immune system cannot occur, again large-scale high-fidelity separations are required.

Within this exciting volume two masters of cell separation technologies, Diether Recktenwald and Andreas Radbruch, have gathered together the most accomplished researchers at the leading edge of several cell separation technologies. It is fitting that there is input from the commercial arena, as the development of these technologies as research tools and as clinical scale separation devices can only come through entrepreneurial and commercial efforts. These articles collectively provide the technological and scientific basis for advancing cell separation and identification technologies, and I commend them to you as the state of the art just prior to a new era when this marriage of cell, developmental, and cell separation technologies is about to transform medicine and, at the same time, reveal new insights into the cell and molecular biology that allows stem and progenitor cells to be just that.

Irving Weissman, Ph.D.
Department of Pathology
and Developmental Biology
Stanford University
Stanford, California

Preface

The need for ever more powerful methods in cell separation has grown tremendously with the identification of many specialized cell subsets because of rapid progress in cell biology, immunology, and molecular biology. This book on *Cell Separation Methods and Applications* describes all of the important methods for the analytical and preparative isolation of specific populations of biological cells. In Chapter 1, Esser includes basic methods such as lytic removal of cell subsets. These methods were developed several decades ago and are still used, primarily as a prepurification step for the preparation of certain common mammalian hematopoietic cell populations for studies in biochemistry, cell biology, immunology, molecular biology, and clinical research.

The newest developments in methods using endogenous physical cell properties, such as cell size and cell density, are described in two chapters by leaders in the field of centrifugal elutriation (Van Vlasselaer and co-workers) and density gradient separations (Figdor and colleagues).

All of the more specific cell separation methods, based on monoclonal antibodies against cell surface markers, are described by academic and industrial pioneers working on the perfection of the techniques. The methods include polystyrene immunoaffinity devices (Lebkowski and co-workers), biotin avidin immunochromatography (Heimfeld and co-workers), high-density particles (Kenyon and co-workers), complement-based cell lysis (Gee), various immunomagnetic methods for cell sorting (Kantor and co-workers), magnetophoresis (Hausmann and co-workers), and fluorescence-activated cell sorting based on flow cytometry (Hoffman and Houck)

The book also describes the basic technical principles of the respective cell separation methods in detail. Typical examples with performance data are discussed, and the limitations of the methods are outlined. Applications in basic research and medicine and an outlook on future applications are reviewed. For ease of understanding, flowcharts, figures, and tables support the text in all chapters.

Finally, the book includes several chapters on important applications of combinations of cell separation techniques, including research in molecular and cellular biology and genetics (Siebenkotten and co-workers) and the clinical isolation of stem and progenitor cells for cell therapy of cancer and other diseases (Hassan and co-workers).

To aid the researcher with the design of cell separation projects, there is an appendix with cell properties for cell separations and one with CD antigens updated with results from the February 1996 leukocyte antigen workshop in Osaka.

This book will help those working with cellular preparations to select an optimal cell separation approach, and it will provide a detailed understanding of the possibilities and limitations of cell purification for those involved in clinical cellular therapy for transplantation, cancer, and AIDS, or with gene therapy for a variety of genetic defects.

We hope that the methods explained in this book will contribute to major advances in cellular therapy and diagnostics, which, in turn, will lead to refinements in methodology.

We would like to acknowledge those who provided valuable support in the preparation of the book, including Larry Transue, technical typist; and Kathryn Rubenstein and her co-workers, graphics. Without MaryAnn Foote, Ph.D., this book would not exist. Without her energy and skill in organizing the book and keeping it on schedule, the project would not have succeeded.

Diether Recktenwald, Ph.D.
Andreas Radbruch, Ph.D.

Contents

Part V: Flow Cytometry

Part VI: Special Applications

Contributors

Karen Auditore-Hargreaves, Ph.D. CellPro, Incorporated, Bothell, Washington

Mark Benyuenes, M.D. CellPro, Incorporated, Bothell, Washington

Ute Behrens-Jung Miltenyi Biotec GmbH, Bergisch Gladbach, Germany

Willy S. Bont, Ph.D. Tumor Immunology Department, University Hospital Nijmegen, Nijmegen, The Netherlands

Christoph Cremer, Prof. Dr. University of Heidelberg, Heidelberg, Germany

Theo J. M. de Witte, M.D., Ph.D. Hematology Department, University Hospital Nijmegen, Nijmegen, The Netherlands

Michael Emde Research and Development Department, CellPro, Incorporated, Bothell, Washington

Charlotte Esser, Dr. rer. nat. Department of Immunology, Medical Institute of Environmental Hygiene, Düsseldorf, Germany

Carl G. Figdor, Ph.D. Tumor Immunology Department, University Hospital Nijmegen, The Netherlands

Adrian P. Gee, Ph.D. Division of Transplantation Medicine, University of South Carolina School of Medicine, Columbia, South Carolina, and Bone Marrow Transplantation, M.D. Anderson Cancer Center, Houston, Texas

Ian Gibbons Research Department, AmCell Corporation, Sunnyvale, California

Bernhard Goller, Dipl. Ing. Boehringer Mannheim GmbH, Penzberg, Germany

John G. Gribben, M.D., Ph.D., F.R.C.P. Division of Hematologic Malignancies, Dana Farber Cancer Institute, Boston, Massachusetts

K. Gutensohn, M.D. Department of Transfusion Medicine and Transplantation Immunology, Hamburg University Hospital Eppendorf, Hamburg, Germany

Roland Hartig, Dr. Max-Planck-Institute for Cell Biology, Rosenhof, Ladenburg, Germany

Hassan T. Hassan, M.D., Ph.D. Bone Marrow Transplantation Center, Hamburg University Hospital Eppendorf, Hamburg, Germany

Michael Hausmann, Dr. Institute of Applied Physics, University of Heidelberg, Heidelberg, Germany

Shelly Heimfeld, Ph.D. Discovery Department, CellPro, Incorporated, Bothell, Washington

Robert A. Hoffman, Ph.D. Instrumentation Systems Technology Development, Becton Dickinson Immunocytometry Systems, San Jose, California

David W. Houck Engineering Department, Becton Dickinson Immunocytometry Systems, San Jose, California

Richard Huijbens, Eng. Tumor Immunology Department, University Hospital Nijmegen, Nijmegen, The Netherlands

Cindy Jacobs, M.D., Ph.D. Clinical Department, CellPro, Incorporated, Bothell, Washington

Mark Jones, Ph.D. CellPro, Incorporated, Bothell, Washington

Aaron B. Kantor Research Department, AmCell Corporation, Sunnyvale, California

Norma S. Kenyon, Ph.D. Diabetes Research Institute, University of Miami School of Medicine, Miami, Florida

Manfred Kubbies Boehringer Mannheim GmbH, Penzberg, Germany

P. Kuehnl Department of Transfusion Medicine and Transplantation Immunology, Hamburg University Hospital Eppendorf, Hamburg, Germany

Jane S. Lebkowski, Ph.D. Discovery and Product Development Department, RPR Gencell, Santa Clara, California

Hans-Georg Liebich, Prof. Dr. Institute for Veterinary Anatomy I, University of Munich, Munich, Germany

Georg H. Lüers, M.D., Ph.D. Institute of Anatomy and Cell Biology, University of Heidelberg, Heidelberg, Germany

Ian McNiece Amgen, Incorporated, Thousand Oaks, California

Stefan Miltenyi Miltenyi Biotec GmbH, Bergisch Gladbach, Germany

Dewey J. Moody, Ph.D RPR Gencell, Santa Clara, California

Lee M. Nadler, M.D. Division of Tumor Immunology, Dana Farber Cancer Institute, Boston, Massachusetts

Thomas B. Okarma, M.D., Ph.D. RPR Gencell, Santa Clara, California

David A. Okrongly, Ph.D. Research Department, Xytronyx, Inc., San Diego, California

Varghese Palathumpat, Ph.D. Cellular Immunology and Hematology Departments, Activated Cell Therapy, Mountain View, California

Rukmini Pennathur-Das, Ph.D. Discovery Research Department, RPR Gencell, Santa Clara, California

Katja Petry Institut für Genetik, Universität zu Köln, Köln, Germany

Ramila Philip, Ph.D. RPR Gencell, Santa Clara, California

Frank Preijers, Ph.D. Hematology Department, University Hospital Nijmegen, Nijmegen, The Netherlands

Nicole Provost, Ph.D. Research and Development Department, CellPro, Incorporated, Bothell, Washington

Andreas Radbruch, Prof. Dr. Deutsches Rheuma Forschungszentrum, Berlin, Germany

Camillo Ricordi, M.D. Diabetes Research Institute, University of Miami School of Medicine, Miami, Florida

Grant Risdon, Ph.D. Research and Development Departments, CellPro, Incorporated, Bothell, Washington

Paul Ruijs, Eng. Hematology Department, University Hospital Nijmegen, Nijmegen, The Netherlands

Thomas R. Russell, Ph.D. Biological Sciences Department, Coulter Cellular Therapies, Inc., Roxbury, Massachusetts

Armin Saalmüller, Ph.D. Department of Vaccines, Federal Research Centre for Virus Diseases of Animals, Tübingen, Germany

Lisa Schain Molecular Biology Department, RPR Gencell, Santa Clara, California

Jürgen Schmitz, Ph.D. Research and Development Department, Miltenyi Biotec GmbH, Bergisch Gladbach, Germany

Michael H. Shapero, Ph.D. Department of Molecular Immunology, Activated Cell Therapy, Mountain View, California

Gregor Siebenkotten Miltenyi Biotec GmbH, Bergisch Gladbach, Germany

George Strang, Ph.D. Cellular Immunology Department, Activated Cell Therapy, Mountain View, California

Sohel Talib Applied Immune Science Department, RPR Gencell, Santa Clara, California

Joe Tarnowski, Ph.D. Research and Development Department, CellPro, Incorporated, Bothell, Washington

Reinhard Teichmann Department of Surgery, University Clinic of Tübingen, Tübingen, Germany

Peter Van Vlasselaer, Ph.D. Research & Development, Activated Cell Therapy, Mountain View, California

A. R. Zander Bone Marrow Transplantation Center, Hamburg University Hospital Eppendorf, Hamburg, Germany

Robert K. Zwerner, Ph.D. Hematology Research & Development, Coulter Corporation, Miami, Florida

1

Historical and Useful Methods of Preselection and Preparative Scale Sorting

Charlotte Esser
Medical Institute of Environmental Hygiene, Düsseldorf, Germany

I. BACKGROUND

Enormous progress has been achieved in recent decades in cell-sorting techniques. Fluorescence-activated cell sorting (FACS) for multiple parameters and more recently magnetic cell sorting (MACS) allow highly sophisticated and accurate sorting of even the rarest cells from very inhomogeneous cell mixtures. A decisive factor in this progress has been the variety and quality of antibodies (Abs) now available for many surface markers, that distinguish functional cell subsets.

The history of cell separation is long, dating back to the 1960s. Available parameters were exploited as soon as they were identified as being specific for a given cell type. Thus, physical characteristics of cells (e.g., density) or biochemical characteristics (e.g., availability of selectable enzymes) were used. For instance, L-leucine-methyl ester kills the lysosome-rich monocytes and natural killer (NK) cells and can thus be used for their depletion. In 1968, Bøyum published his seminal paper on Ficoll-density gradients for the isolation of lymphocytes from blood. Later, a battery of methods based on antigen (Ag)-Ab reactions were developed (e.g., panning was described in 1978 and complement lysis in the early 1970s). Idiosyncratic characteristics of cells, such as adherence to certain matrices or other cells, were also used. One example is rosetting of T cells with sheep erythrocytes followed by Ficoll-gradient separation. Another good example is the adherence of B cells and macrophages to nylon wool.

Some of the old methods are obsolete, but many are not. Most are simple and reliable routine procedures, and many have the advantage of requiring little equipment. Simple single-parameter sorting can save time and even money where a compromise in the purity of sorted cells is experimentally acceptable. Moreover, for preparative scale sorting (either by FACS or by MACS) of large numbers of cells or very rare cells, a preenrichment is mandatory. Preenrichment is advised to improve electronic gating later, to remove cells from the mixture that may take up the stain unspecifically, or simply to reduce the time required for linear sorting. The latter is also important to reduce stress on the cells. Thus, it is advisable to develop sorting protocols that are optimal in time and efficiency, and best for the cells in question ("best" means high recovery of viable cells that are not changed or compromised in their functionality). No book on cell sorting could be complete without protocols of some preselection methods. Therefore, protocols that I consider most useful, in particular for the immunologist, are provided here.

II. BASIC PROTOCOLS

A. Accessory Techniques

Erythrocyte Lysis

Lysis of erythrocytes is based on their susceptibility to hypotonic shock compared with leukocytes and other nucleated cells. This may be due to differing membrane stability or the fact that mammalian erythrocytes have no nuclei. The phenomenon is exploited for selective lysis at low salt concentrations. Suspensions of spleen cells contain approximately 50% erythrocytes. It may be desirable to remove these cells before further cell sorting to reduce sorting time and to improve electronic gate discrimination. Because of their high abundance (up to 99% in blood), it is always necessary to remove erythrocytes for flow-cytometric analysis of peripheral blood cells.

Several protocols work well for murine spleen cells and blood (Tables 1 and 2). When whole blood was used, distilled water produced better results than hypotonic medium. Commercial lysis solutions have been satisfactory in my hands as well, albeit some of them required two rounds of lysis.

Table 1 Spleen Erythrocyte Lysis

Pellet spleen cell suspension.
Add ACT buffer (140 mM NH_4Cl, 13 mM Tris-HCl, pH 7.2; filter sterilized; 1 mL for each spleen).
Incubate at room temperature for 5 minutes with occasional shaking.
Fill up with medium or physiological buffer (e.g., PBS) and wash cell suspension two to three times.
Expect 5–10 × 10^7 lymphocytes per spleen, depending on age and mouse strain.

Note: The lysis buffer can be buffered by$KHCO_3$ (1 mM) and EDTA (0.1 mM) instead of Tris-HCl.
PBS = phosphate-buffered saline; EDTA = ethylenediaminetetraacetic acid.

Ficoll Gradients

Different cell types can differ in density. This fact can be used to separate cells on single-step or continous-density gradients. Most frequently, nonionic iodinated Ficoll and Percoll are used to prepare gradients of appropriate low viscosity, density, and osmolality. Sucrose or CsCl-based density gradients are not suitable because of their high osmotic potential, which would kill or at least deform cells. Two important areas of use for Ficoll gradients are the isolation of peripheral blood mononuclear cells (PBMCs) and erythrocytes/granulocytes, and separation of living cells (low density) from dead ones (high density) (Fig. 1; Tables 3 and 4). Removal of dead cells becomes necessary after lytic treatment of cells with cytotoxic antibodies before cell

Table 2 Blood Erythrocyte Lysis

Bleed mice.
Immediately add 1 unit of blood to 1 unit of PBS/10 mM EDTA to inhibit clotting.
Add 10 units of distilled water.
Incubate 15–20 seconds (no longer).
Fill with 1.1 units of 10 × PBS to bring back to isotonicity.
Wash cell suspension two or three times.
Expect approximately 10^4 leukocytes/μL blood (i.e., 10^3 lymphocytes).

Note: For larger samples of human blood, use 1 unit 10 mM Tris HCl/5 mM $MgCl_2$/300 mM sucrose/1% Triton X-100 (pH 7.5) per volume of anticoagulated pelleted blood, centrifuge, and take up in physiological buffer.
PBS = Phosphate-buffered saline; EDTA = ethylenediaminetetraacetic acid.

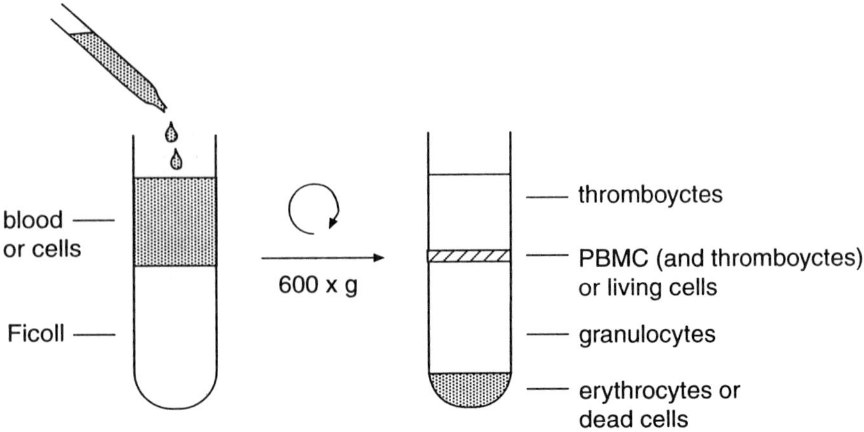

Figure 1 One-step gradient separation of cells. Diluted blood or a mixture of dead and living cells (italics) is layered on top of a Ficoll cushion of appropriate density. After centrifugation, peripheral blood mononuclear cells (PBMCs) and thrombocytes form an interface on top of the gradient, whereas erythrocytes and granulocytes have passed through the Ficoll and are found at the bottom of the tube. Likewise, living cells form the interface and dead cells form the pellet. Appropriate Ficoll density must be chosen for the cells to be separated (e.g., human or rodent cells).

Table 3 Single-Step Gradient to Remove Dead Cells

Determine number of living and dead cells by trypan blue exclusion in a counting chamber. Calculate the number and size of gradients necessary: in a 15-mL tube, 2×10^7 living cells can be recollected; in a 50-mL tube, 2×10^8 cells.
Fill 15- (or 50-)mL tubes with 5 (or 10) mL Ficoll-Paque[a] ($\rho = 1.36$ for murine cells).
Take up cells in a small volume (e.g., 5 mL for the small gradient, 10 mL for the larger one).
Layer cells slowly onto the Ficoll-Paque surface.
Centrifuge gradients at $600 \times g$, 20°C for 20 minutes, with brakes switched off.
Collect cells from the interface with a Pasteur pipette.
Dilute collected cells with PBS, spin down at $120 \times g$, 10 minutes. Repeat washing twice. Count and determine cell viability and yield.

[a]This solution can also be prepared in the lab by mixing 70 mL 14% (w/v) Ficoll 400 with 30 mL sodium metrizoate. Sterilize by filtration, store protected from light.

Note: It is important to prepare and run the gradients at room temperature because cell density is temperature dependent. Ficollation is stressful for cells and should not be considered a routine for tissue culturing.

PBS = Phosphate-buffered saline.

Table 4 Isolation of Human Peripheral Blood Mononuclear Cells on a Single-Step Gradient

Fill tubes with Ficoll-Hypaque ($\rho = 1.077$): one gradient in a 50-mL tube requires 15 mL of whole blood.
Dilute blood 1:2 with PBS or saline.
Carefully layer blood on top of Ficoll cushion.
Centrifuge at $600 \times g$, 20°C for 30 minutes, with brakes switched off.
PBMCs and thrombocytes will be at the interface, granulocytes will enter the Ficoll, and erythrocytes will pass through.
Collect PBMCs from the interface.
Wash with cold PBS ($120 \times g$, 10 minutes) several times until supernatant is clear (to remove Ficoll residue and thrombocytes).

Note: The erythrocytes can also be collected from the same gradient. Obtain blood from patients who have fasted, otherwise the lipids in the serum make it very difficult to see the PBMC layer. PBS = Phosphate-buffered saline; PBMCs = peripheral blood mononuclear cells.

sorting. Also, density separation has been described to isolate a whole variety of cells and organelles.

Cotton-Wool Columns to Remove Dead Cells and Tissue Debris

A tissue homogenate contains debris and dead cells that must be removed before further handling and sorting. Dead cells are autofluorescent, tend to take up Ab nonspecifically, and stick to other cells. Such phenomena interfere with high-quality flow-cytometric analysis and cell sorting. Tissue debris might clog the nozzle in a flow-cytometer. Cotton-wool columns are an easy tool to avoid these problems (Fig. 2 and Table 5). Dead cells stick to cotton for some unknown reason, and tissue debris is held back physically as well. However, cotton-wool columns are not intended to remove high percentages of dead cells. In this case, Ficoll gradients are the method of choice.

B. Simple and Useful Bulk Selections

Panning

The "panning" method is based on the specific adherence of cells to Ab-coated polystyrene plastic dishes. Antibodies specific for the cell type to be depleted are coated onto plastic, and then the cell mixture is added and cells bind via their surface markers to the Abs (Fig. 3). Antigen density on the

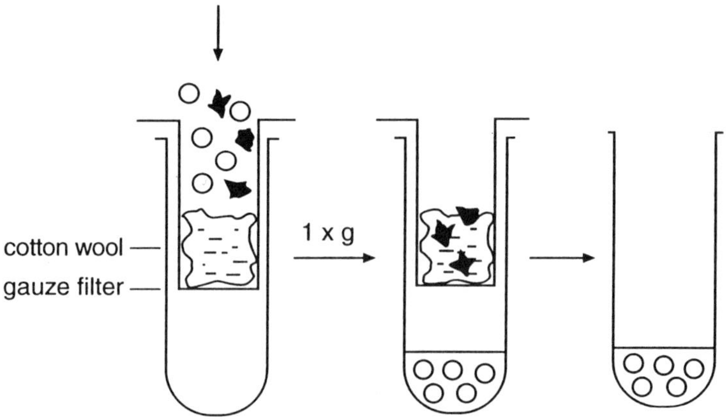

Figure 2 Pure cotton wool, which is lightly stuffed into a syringe closed with nylon gauze at the lower end, is equilibrated with buffer of medium. The cell suspension with dead cells and debris is loaded on top and allowed to pass through. The column is washed with several volumes of buffer or medium, until no further cells elute. The cotton wool is discarded, and the column can be refilled and reused after autoclaving.

Table 5 Cotton-Wool Columns to Remove Dead Cells and Tissue Debris

Prepare a stock of single-use sterilized cotton-wool columns for convenience: cut off the bottom of 5- or 10-mL syringes with a scalpel, briefly melt the rim on a Bunsen burner, and press onto a precut piece of nylon gauze (mesh-size 80 μm) to close the syringe end.
Fill column lightly with 1–2 mL of pure cotton-wool.
Close upper end of column with aluminum foil and autoclave (e.g., after bundling several columns in a glass beaker).
Hang column in tube (e.g., 15-mL tube).
Prewet cotton-wool with buffer (e.g., PBS); discard wash eluent.
Add cell suspension onto column and wash with several volumes of buffer until no more cells elute (check in microscope if necessary).
Discard column, count cells, spin down, and take up in an appropriate volume of buffer or medium.

Note: The columns can be recycled. Discard cotton-wool, clean column first with cell-compatible detergent (e.g., 7 ×) and then with several rinses of deionized water. Refill with fresh cotton-wool. PBS = Phosphate-buffered saline.

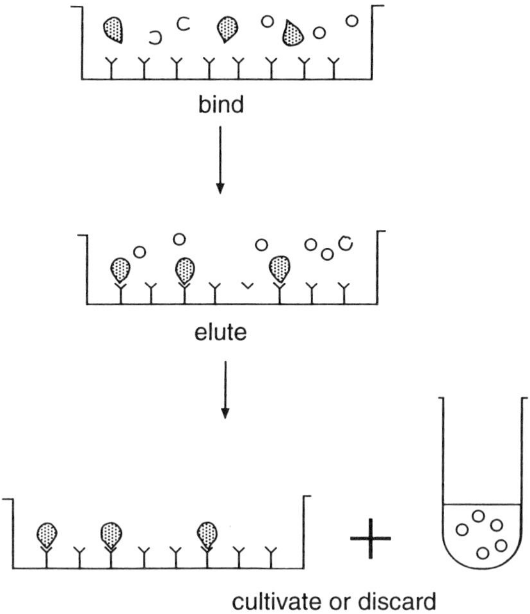

Figure 3 Schematic drawing of "panning" procedure. A cell mixture is added to plastic dishes with an antibody (Ab) coat. Cells bearing surface markers for the Abs stick to the dishes; unbound cells are washed off and collected. Either of the populations can be used further.

surface must be sufficient for the bonds of cell-to-Ag coat to withstand the physical shearing forces of the procedure. Either the Ag^+ or the Ag^- cells can be used further. However, positive selection of cells is only possible if no nontarget cells bind specifically to the polystyrene plastic. Unbound cells (Ag^-) are washed off, and bound cells (Ag^+) can be scraped off the plastic dish with ice-cold buffer or cultivated in the dish directly. Purity, yield, and viability of cells separated by this method are reasonably satisfying, and because it is simple to perform and requires only minimal equipment, this method is still useful. It has been used by many laboratories to separate T cells and B cells (Table 6).

Table 6 Panning Protocol

Calculate number of culture dishes needed: approximately 10^3 positive cells/mm^2 can be bound (e.g., 2×10^7 per 15-cm-diameter round culture dish).
Add Ab solution to polystyrene tissue culture flask or dish, just covering them. Incubate overnight at 4°C or at least for 1 hour at 37°C.
Block uncoated sites on culture dishes by adding PBS/1% BSA.
Incubate overnight at 4°C or for 1 hour at 37°C; discard blocking solution.
Prepare a single cell suspension (important to avoid false negative/positive from cell clumps and doublets) in either PBS/BSA or medium buffered with HEPES; aim for a suspension volume large enough to cover the culture dish (e.g., 15 mL for 15-cm-diameter dish).
Add cells onto Ab-coated dishes and incubate for 30 minutes at 4–6°C (refrigerator); swirl dishes gently after approximately 15 minutes.
Remove unbound cells with a pipette and wash dish two or three times with ice-cold PBS/BSA by tipping the dish to the side, slowly flooding the surface of the dish with buffer, swirling gently, and removing buffer; the dish should contain a monolayer of bound cells. Analyze quality of separation flow-cytometrically or in the fluorescence microscope. The unbound cells can be subjected to a second round of panning (of equal or different specificity) if desired.
Bound cells can be either cultivated directly in the dish or recovered by scraping them off carefully with a rubber scraper. Alternatively, add culture medium containing FCS and incubate for 1 hour at 37°C. Cells can then be pipetted off rather easily.
Count cells to calculate recovery rate and viability.
Expect viability and recovery of 80–95% and a purity of > 95%, depending on cell type and frequency in the mixture.

Note: We used freshly prepared 2 μg/mL dilutions in PBS for panning splenic B cells. Do not filter-sterilize such low concentrations to avoid loss of Ab. We have successfully used polyclonal antisera and mAbs. It may be necessary to determine the Ab concentration for specific pannings experimentally. Occasionally charges of plastic do not work for panning.

Ab = Antibody; mAb = monoclonal antibody; BSA = bovine serum albumin; PBS = phosphate-buffered saline; FCS = fetal calf serum; HEPES = N-2-hydroxyethylpiperazine-N′-2-ethanesulfonic acid.

T-Cell Separation on Nylon Wool

For some unknown reason, B cells and macrophages adhere to nylon wool. This phenomenon can be exploited for a quick and simple purification of T cells from spleens or lymph nodes (Table 7). Similar to the method of cotton-wool columns, cells are added onto nylon wool in a column and eluted (see Fig. 2). The method is easy to perform and does not require antibodies or

Table 7 T-Cell Separation on Nylon Wool

Stuff 10-mL syringe with 0.7–1 g nylon wool; fit a piece of tubing to the syringe outlet and close with a clamp. Alternatively use a three-way stop cock. Sterilize syringe by autoclaving wrapped in aluminum foil or bundled into a closed glass beaker.

Fix the column over a recipient tube and wash with 15 mL warm medium/5% FCS as protein source. Remove air bubbles with a pipette. Fill column with medium, close at the bottom, and incubate for at least 30 minutes at 37°C. Again fix column over tube (in the laminar flow for sterile sorting) and wash with an additional 10 mL warm medium. Finally, leave enough medium on column to cover nylon wool. Close column.

Add single spleen-cell suspension containing up to 2.5×10^8 cells in 1.5 mL warm medium to the column and allow slowly to enter nylon matrix. Close the bottom of the syringe column and add 0.5. mL medium more.

Incubate column at 37°C for 45 minutes.

Elute column slowly (approximately 2 mL/min) with 20 mL prewarmed medium. Do not use excessive eluent to avoid eluting contaminating cells from the nylon wool. Check eluted cells flow cytometrically or in the fluorescence microscope. Count cells to determine recovery and viability.

Expect a recovery rate of 70–85% T cells (i.e., 25% of total spleen cells) and a purity of about 85% splenic T cells.

Note: Cell capacity of such a column is approximately 1.5×10^8 cells (i.e., sufficient for separation of cells from two mouse spleens). Columns can be adapted; rule of thumb is 1 g nylon wool for 1.5×10^8 bound cells.
FCS = Fetal calf serum.

complement. Yields and purity are satisfactory for many analytical purposes, although they do not achieve the levels of panning or complement lysis, let alone FACS. The method is particularly useful when the cells to be used for separation are already covered with Abs (e.g., from mice treated in vivo). This would preclude separation by panning or complement.

Complement Lysis

"Complement" is the term used for a group of serum factors capable of destroying cells marked with complement-activating Abs. Upon binding to the Ab, complement factors are converted is a cascade, eventually resulting in a complex that can punch holes in the cell membrane. Thus, complement lysis (Table 8) is used strictly for elimination of unwanted cell populations from a

Table 8 Complement Lysis

Prepare a single cell suspension, remove erythrocytes if present.
Pellet cells and take up in antibody solution at 10^7 cells/mL.
Stain cells for 30 minutes at 4°C, wash cells once with medium/FCS.
Dilute low-tox complement in medium/FCS, filter-sterilize.
Pellet the stained cells and take up in prewarmed complement at 10^7 cells/mL.
Incubate for 30 minutes at 37°C.
Wash two times with $10 \times$ volume of medium/FCS at $120 \times g$.
Count cells with Trypan blue exclusion. Determine cell recovery and viability. Check success of complement lysis flow cytometrically or in the fluorescence microscope.
Expect 80%–90% purity of the remaining cells. Repeat complement lysis for higher enrichment rates, or use panning.

Note: Optimal Ab concentration must be determined experimentally. Rule of thumb: 1–10 μg/mL for mAbs, 1:2 or 1:4 for culture supernatants. Dilute in medium/FCS and filter-sterilize on low-protein binding filters.
The dilution of each complement batch must be determined experimentally. Also, test the toxicity of each batch by running a complement-only (no Ab) control with cells at least once. Rule of thumb: dilute1:6 to 1:12. Do not freeze-thaw complement several times.
FCS = Fetal calf serum.

mixture, so its applicability is more limited than other simple bulk separations. As with panning, a defined surface Ag on the cells to be eliminated is necessary. For instance, complement lysis is often used to eliminate T cells with anti–Thy-1 or B cells with anti–major histocompatability class (MHC) II Abs. The major problem with complement lysis is finding a good, low-toxicity batch of serum complement and titrated antisera Abs of a complement-fixing immunoglobulin isotype. Note that complement will kill all immunoglobulin-covered cells, which may be present due to specific experimental conditions (e.g., in vivo treatment with Abs).

III. CONCLUSIONS

A rudimentary listing of often-used protocols is supplied in the tables in this chapter, and Table 9 lists many other cell-isolation methods that make use of physical, immunological, or biochemical characteristics. The reader can make use of this information if necessary. It is hoped that this information will be useful to researchers in the field of cell sorting and separation.

Table 9 Methods of Cell Separation and Isolation

Name	Principle	Area of use	Ref.
I. Separation by physical parameters			
Ficoll density[a]	Buoyancy	Leukocyte separation from whole blood; removal of dead cells	1–3
Percoll density lysis	Buoyancy	Separation of lymphocyte subsets	3
Cotton wool column[a]	Idiosyncratic adhesion	Removal of dead cells and debris	
Nylon wool[a]	Idiosyncratic adhesion	Isolation of T cells	4–6
Sephadex G-10	Idiosyncratic adhesion	Removal of macrophages and Ag-presenting cells	7,8
II. Cell separation/isolation by immunological parameters			
Complement lysis[a]	Specific lysis of Ab-coated cells	Depletion of cells bearing a distinct surface marker	2,9–11
Panning[a]	Specific adherence of cells to Ab-coated plastics	Enrichment or depletion cells bearing a distinct surface marker	2,12–14
Rosetting	Change of buoyancy by cell-cell contact across surface molecules	Isolation of T cells	6,15–17
Hapten-derivatized nylon fibers	Ag-covered nylon	Isolation of lymphocyte subsets	18,19
Immunosorbents	Chromatography	Isolation of lymphocyte subsets	15,20
Avidin columns	Ag-binding	Isolation of functional T cells	21,22
Silica particles	Ag-binding	Isolation of B cells	23
III. Cell separation/isolation using biological characteristics			
Fe-Phagocytosis		Isolation of macrophages	24
Adhesion to plastic		Isolation of macrophages	2,25,26
Erythrocyte lysis[a]	Osmotic shock	Removal of erythrocytes	20,27–29
Fe-magnetism	Magnetism	Isolation of erythrocytes	30
IV. Cell separation/isolation by biochemical characteristics			
LME	Enzyme	Removal of monocytes	31
Azaguanin	Enzyme	Selection of hybridomas	32,33
G 148/GANC	Antibiotic resistance	Selection of gene-targeted cells	34

Table 9 Continued

Name	Principle	Area of use	Ref.
"Suicide" of secreting cells	Complement lysis	Selection of mutants and isotype switch variants	11,35
V. Other methods described in the text of the book			
FACS, MACS, elutriation, immunotoxins, cell-affinity chromatography			

[a]See text and tables for detailed protocol.
Ag = Antigen; FACS = fluorescence-activated cell sorting; MACS = magnetic cell sorting.

REFERENCES

1. Bøyum A. Isolation of mononuclear cells and granulocytes from human blood. *Scand J Clin Lab Invest* 1968; 21:77.
2. *Current Protocols in Immunology.* John Wiley and Sons, Ltd., 1994.
3. Sharpe PT. Centrifugation. In: Burdon RH, van Knippenberg PH, eds. *Methods of Cell Separation.* Amsterdam: Elsevier; 1988:18.
4. Julius MH, Simpson E, Herzenberg LA. A rapid method for the isolation of functional thymus-derived murine lymphocytes. *Eur J Immunol* 1973; 3:645.
5. Eisen SA, Wedner HJ, Porker CW. Isolation of pure human peripheral blood T-lymphocytes using nylon wool columns. *Immunol Commun* 1972; 1:571.
6. Mendes NF, Tolnau M, Silveira N, Gilbertsen R, Metzgar R. Technical aspects of the rosette tests used to detect human complement receptor (B) and sheep erythrocyte-binding (T) lymphocytes. *J Immunol* 1973; 111:860.
7. Ly LA, Mishell RI. Separation of mouse spleen cells by passage through columns of Sephadex G-10. *J Immunol Meth* 1974; 5:239.
8. Hathcock KS, Singer A, Hodes RJ. Passage over Sephadex G-10 columns. In: Hershkovitz HH, Holden HT, Eelanti JA, Ghaffer A, eds. *Manual of Macrophage Methodology: Collection, Characterization, and Function.* New York: Marcel Dekker, 1981:127.
9. Miller JF, Sprent J. Thymus-derived cells in mouse thoracic duct lymph. *Nature New Biol* 1971; 230:267.
10. Miller JF, Sprent J, Basten A, Warner NL. Selective cytotoxicity of anti-kappa serum for B-lymphocytes. *Nature New Biol* 1972; 237:18.

11. Köhler G, Shulman MJ. Immunoglobulin M mutants. *Eur J Immunol* 1980; 10:467.
12. Mage MG, McHugh LL, Rothstein TL. Mouse lymphocytes with and without surface immunoglobulin: preparative scale separation in polystyrene tissue culture dishes coated with specifically purified anti-immunoglobulin. *J Immunol Meth* 1977; 15:47.
13. Wysocki LJ, Sato VL. "Panning" for lymphocytes: a method for cell selection. *Proc Natl Acad Sci USA* 1978; 75:2844.
14. Small M, Majumdar AS, Liebermann M, Weissman I. Isolation of $CD3^-$, $CD4^-$, $CD8^-$, $IL\text{-}2R^+$ thymocyte precursors by panning. *J Immunol Meth* 1994; 167:103.
15. Chess L, Schlossman S. Methods for separation of unique human lymphocyte subpopulations. In: *Manual of Clinical Immunology*. Washington, DC: American Society for Microbiology, 1976:77.
16. Weiner MS, Bianco C, Nussenzweig V. Enhanced binding of neuraminidase treated sheep erythrocytes to human T-lymphocytes. *Blood* 1973; 42:939.
17. Indiviri F, Huddlestone J, Pellegrino MA, Ferroni S. Isolation of human T lymphocytes: comparison between wool filtration and rosetting with neuraminidase (VCN) and 2-aminoethylisothiouronium (AET)-treated sheep red blood cells. *J Immunol Meth* 1980; 34:107.
18. Kiefer H. Binding and release of lymphocytes by hapten-derivatized nylon fibers. *Eur J Immunol* 1973; 3:181.
19. Krawinkel U, Rajewsky K. Specific enrichment for antigen-binding receptors from sensitized murine lymphocytes. *Eur J Immunol* 1976; 6:529.
20. Boyle W. An extension of the ^{51}Cr release assay for the estimation of mouse cytotoxins. *Transplantation* 1964; 6:761.
21. Berenson RJ, Bensinger WI, Kalamsz D. Positive selection of viable cell populations using avidin-biotin immunoadsorption. *J Immunol Meth* 1986; 91:11.
22. Davis LS, McIlraith MI, Paecheo T, et al. Assessment of a positive selection technique using an avidin column to isolate human peripheral blood T cell subsets. *J Immunol Meth* 1994; 175:247.
23. Seeger M, Sepulveda C, Vallejos P, Aguillon JC, Ferreira A. A simple method for negative and positive selection of murine and human IgM-bearing lymphocytes based on the use of antibody-coated silica microparticles. *J Immunol Meth* 1993; 162:109.
24. Lundgren G, Zukoski CF, Möller G. Differential effects of human granulocytes and lymphocytes on human fibroblasts in vitro. *Clin Exp Immunol* 1968; 3:817.
25. Edelson PJ, Cohn ZA. Purification and cultivation of monocytes and macrophages. In: Bloom BA, Davids JR, eds. *In Vitro Methods in Cell-Mediated and Tumor Immunity*. San Diego: Academic Press, 1976:333.
26. Steinman RM, Kaplan G, Witmer MD, Cohn ZA. Identification of a novel cell type in peripheral lymphoid organs of mice. V. Purification of spleen dendritic cells, new surface markers, and maintenance in vitro. *J Exp Med* 1979; 149:1.
27. Webster GA, Pockley AG. Effect of red cell lysis protocols on the expression of

rat peripheral blood lymphocyte subset and activation antigens. *J Immunol Meth* 1993; 163:115.

28. Terstappen LW, Meiners H, Loken MR. A rapid sample preparation technique for flow cytometric analysis of immunofluorescence allowing absolute enumeration of cell subpopulations. *J Immunol Meth* 1989; 123:103.
29. Mansour I, Bourin P, Rouger P, Doinel C. A rapid technique for lymphocyte preparation prior to two-colour immunofluorescence analysis of lymphocyte subsets using flow-cytometry. *J Immunol Meth* 1990; 127:61.
30. Melville D, Paul F, Roath S. Direct magnetic separation of red cells from whole blood. *Nature* 1975; 255:706.
31. Thiele DL, Kurosaka M, Lipsky PE. Phenotype of the accessory cell necessary for mitogen-stimulated T and B cell responses in human peripheral blood: delineation by its sensitivity to the lysosomotropic agent, L-leucine methyl ester. *J Immunol* 1983; 131:2282.
32. Köhler G, Milstein C. Derivation of specific antibody-producing tissue culture and tumor lines by cell fusion. *Eur J Immunol* 1976; 6:292.
33. Köhler G, Milstein C. Continuous cultures of fused cells secreting antibody of predefined specificity. *Nature* 1975; 265:495.
34. Mansour SL, Thomas KR, Capecchi MR. Disruption of the proto-oncogene int-2 in mouse embryo-derived stem cells: a general strategy for targeting mutations to non-selectable genes. *Nature* 1988; 336:348.
35. Faguet GB, Agee JF. A simple technique for the rapid enrichment of class and subclass hybridoma switch variants. *J Immunol Meth* 1993; 165:217.

2

New Approaches in Density Gradient Separation Using Colloidal Silica Gradients in the Processing of Human Hematopoietic Progenitor Cells

Peter Van Vlasselaer, Varghese C. Palathumpat, George Strang, and Michael H. Shapero
Activated Cell Therapy, Mountain View, California

I. INTRODUCTION

Physiological cell mixtures are complex in composition, and the assessment of the biological function of a particular cell type in a pool of mixed cells is difficult. This observation has fueled the development of various cell-separation technologies (1,2). In many cell-separation protocols, the target cells are initially enriched from a complex cell mixture by buoyant density centrifugation. This makes further purification, by techniques using specific cell binding agents such as antibodies (Abs) or lectins in combination with magnetic beads (3,4), solid phase-adsorption (5,6), or fluorescence-activated cell sorting (FACS) (7,8), easier and more cost effective. Over the years, many different density gradient materials such as serum albumin, Ficoll, metrizamide, dextran, methylcellulose, salts, and colloidal silica have been used for this purpose (9–11). The literature on this subject is extensive. Hence, this chapter is focused on the use of colloidal silica for the separation of human hematopoietic progenitor cells for transplantation. It has become well established that a convenient cell surface marker for the subset of cells that harbor hematopoietic progenitor cells is recognized by the antibody to the CD34 surface antigen (Ag) (12,13).

II. THEORETICAL PRINCIPLES OF SINGLE-STEP GRADIENT CENTRIFUGATION

Cells behave in suspension as small spheroid particles that sediment at a rate proportional to their size and the difference between their density and that of the surrounding medium. The sedimentation of spherical particles in solution occurs according to Stokes law, whether it be at $1 \times g$ sedimentation or during centrifugation. According to Stokes law, the rate of sedimentation (v) of a sphere of diameter d in a gravitational field g equals:

$$v = \frac{d^2(\rho p - \rho l)g}{18\eta}$$

where ρp and ρl are the densities of the spherical particle and the liquid, respectively, and η the viscosity of the liquid. Depending on the magnitude of the centrifugal force and the density of the medium, particles will separate on the basis of either their density, their size, or their density and size combined. The separation of particles during centrifugation at high g force over an increasingly dense medium occurs on the basis of density and is called isopycnic separation (14). Velocity sedimentation, which occurs at low g force, refers to a separation on the basis of the size of the particle (15). This chapter focuses on the separation of cells in a solution with a defined density (single-step gradient) using high g force during which cells separate based on their size and their density. Since the size, morphology, and density of cells are variable and are determined by the prior history or the stage of differentiation of cells, Stokes law predicts only in part how a particular cell type can be separated on a single-step gradient. An empirical approach is required to determine the density of a particular cell type.

III. SINGLE-STEP GRADIENTS OF COLLOIDAL SILICA

A. Physicochemical Properties of Colloidal Silica

Apart from the inherent variations in the density of cells, the limitations of density separation often relate to the physicochemical characteristics of the density medium itself. At the density required for optimal resolution, many media are not physiological. This forces the scientist to compromise between the degree of purification and maintenance of biological integrity and viability of the target cells. There are a number of characteristics of an ideal density gradient material. Such a material:

Allows the formulation of solutions within a range of densities, which suffices to separate most of the biological particles of interest.
Has a low ionic strength in its undiluted form, allowing the formulation of solutions with a wide variety of densities without affecting the osmolality or the pH of the final product.
Is isotonic at its final density.
Is isosmotic throughout the gradient.
Has a low viscosity.
Does not dissolve or penetrate biological membranes.
Is chemically inert and shows low interaction with storage containers or filtration membranes.
Is stable upon storage for prolonged periods of time.
Does not alter its physicochemical characteristics or chemical nature upon final sterilization by heat or ionizing irradiation.
Has no adverse effects on biological and nonbiological assays.
Can be easily removed from purified materials and cells.

From the different density-gradient materials mentioned above, derivatized colloidal silica are unique since they fulfill almost all these criteria. Colloidal silica were introduced as cell separation media in the late 1950s and early 1960s (16,17). These studies showed that a certain degree of derivatization was required since unmodified colloidal silica are unstable in solution and are toxic at physiological pH. In an attempt to overcome these adverse effects, polymer supplements such as dextran, polyvinylpyrrolidone (PVP), and polyethylene glycol (PEG) were added to the colloidal silica (18,19). Although this abrogated the toxic nature and greatly contributed to the stability of the colloidal silica in solution, this approach is not practical since high concentrations of these additives were required. These additives increased the viscosity and osmolality of the medium, reducing the number of density solutions that could be formulated within an isotonic range. In contrast, when the colloidal silica were covalently coated with PVP (20,21), the stability was increased and the toxic nature of colloidal silica was greatly reduced. Colloidal silica coated with PVP is currently known as Percoll (22), which is completely nontoxic to cells and can be formulated into a wide variety of isotonic solutions over a broad range of densities (1.0–1.3 g/mL) due to its low osmolality (23,24).

Moreover, the viscosity of Percoll is low, which allows very rapid banding of cells or particles during centrifugation. Whereas undiluted Percoll is stable and can be sterilized by autoclaving, it becomes sensitive to heat and ionizing irradiation (gamma and E-beam irradiation) once diluted with a salt solution. Even at low temperatures of 37°C, diluted Percoll tends to aggregate and precipitate, which can interfere with general cell culture.

Recently, we have used a colloidal silica derivatized through silanization. In its undiluted form, silanized colloidal silica has characteristics comparable to Percoll's osmolality, density, and viscosity. Furthermore, it can be formulated into isotonic solutions over a broad range of densities. In contrast to Percoll, silanized colloidal silica diluted in salt solution, retains its physicochemical characteristics during autoclaving or ionizing irradiation. Consequently, silanized colloidal silica lends itself better to Good Manufacturing Procedures (GMP) formulation and final sterilization than Percoll. Similar to Percoll, this silanized colloidal silica is nontoxic both in vitro and in vivo. A variety of cells, including hematopoietic progenitor cells, can be mixed and incubated in the silanized colloidal silica without losing their functional capabilities. Furthermore, silanized colloidal silica did not induce a pyrogenic effect nor result in acute toxicity when injected in laboratory animals at maximal doses as defined by the United States Pharmacopoeia (USP) (data not shown).

B. Use of Colloidal Silica in Preclinical and Clinical Research

Based on the advantageous characteristics outlined above, Percoll was used in a variety of preclinical and clinical applications ranging from the purification of hematopoietic progenitor cells (25,26), and the separation of lymphocytes (27,28), natural killer (NK) cells (29,30), suppressor cells (31,32), monocytes (33,34), and osteoblasts (35,36) to the fractionation of sperm (37,38). The application of Percoll in the clinic is best illustrated by its use in the fractionation of bone marrow and peripheral blood for stem cell transplantation (39–41). Bone marrow and peripheral blood are processed before cryopreservation and infusion to the patient to reduce the total cell volume and the associated amount of cryopreservatives in the transplant and to facilitate the purging of allogeneic T cells and tumor cells (39,42,43). This chapter will discuss the use of colloidal silica in the processing of hematopoietic progenitor cells before transplantation.

C. Technical Considerations for the Use of Colloidal Silica for Processing of Hematopoietic Progenitor Cells

The yield of hematopoietic progenitor cells recovered from bone marrow and peripheral blood varied widely between different clinical centers when Percoll was used (44–48). Across these different studies, the recovery of colony-forming units (CFU) varied between 35 and 90% for bone marrow and 9 and 90% for peripheral blood. A more recent study showed that the median recovery of CD34 cells from the blood of patients with non-Hodgkin's lymphoma mobilized with recombinant human granulocyte colony-stimulating factor (rHuG-CSF) was 81% (range, 10–100%) (39). This illustrates not only the variability among patient samples, but also the inconsistency in reproducibility in cell separation that is the most fundamental hurdle in density gradient separation. In our laboratory the biggest challenge in performing reproducible cell separations was related to the methods of formulation, the storage conditions of silanized colloidal silica, and the accuracy of recovering cells from the interface after centrifugation. Colloidal silica is provided by the manufacturer as a raw material that requires further dilution to obtain a desired density, pH, and osmolality. The formulation is performed using in-house protocols that are not standardized between clinical centers and rarely are adequately quality controlled. Furthermore, the manufacturer and literature instruct the user to formulate Percoll solutions by measuring the raw materials on the basis of volumes rather than weight. In our experience, formulation of small (100 mL)- to medium (1 L)-scale batches of colloidal silica based on volumetric measures, using standard laboratory equipment such as pipettes and regular fluid transfer pipettes, results in significant batch-to-batch variations. Variations in density as great as 0.001 g/mL were observed between different batches formulated by volumetric means. This variation is undesirable for the separation of particular cell types, especially hematopoietic progenitor cells, whose density varies within a 0.001 g/mL range, as will be discussed. Especially for small to medium batches, it is advisable to formulate colloidal silica solutions based on weight measurements and to control the physicochemical specifications of the final product. Another feature of the "home-brew" approach that is often underestimated is the condition and the choice of container used for storage of the final product. Many laboratories store Percoll solutions for prolonged periods in wide-mouth flasks with a large headspace of air. The amount of water that can evaporate in this space is sufficient to change the specifications of the final material. This is especially important for the separation of cell types with a narrow density distribution, such as hematopoietic progenitor cells. In this context, it is advisable to store premade solutions in narrow-mouth containers in volumes that are rapidly used. Under these conditions, isotonic solutions of silanized colloidal silica were stored in our facility for more than a year without observing changes in density ($\leq$ 0.0005 g/mL), osmolality ($\leq$ 10 mOsm), or pH ($\leq$ 0.5).

Lastly, one of the most technically challenging aspects of density-gradient separation is the layering and removal of cells on and from the interface of the density material. To accommodate this, we developed a centrifugation device that allows the recovery of cells from the interface by decantation rather than by pipetting. Figure 1 shows a schematic picture of this device and its use. Briefly, a 50-mL centrifugation tube is equipped with a funnel-shaped insert with a central orifice. During centrifugation, cells of high density migrate through the density solution and roll down the slope of the insert into the lower compartment where they collect in the pellet. After centrifugation, the cells at the interface are collected by inverting the centrifugation tube. During inversion, an air-lock establishes at the orifice due to the surface tension of the density material. This air-lock is sufficiently strong to retain the cells in the lower compartment of the tube and to prevent the mixing of the cells from the pellet with those from the interface during collection.

IV. SEPARATION OF HEMATOPOIETIC PROGENITOR CELLS

The rationale for processing hematopoietic progenitor cells was briefly addressed. The goal of our experiments in this context was to define the physicochemical characteristics of a single-step density gradient that could reduce the total cell volume of a progenitor preparation ("debulk") while maximizing the yield of CD34 cells.

A. The Density of CD34 Cells Varies Between Different Tissues

The density of CD34 cells from different tissues was defined using isotonic solutions of silanized colloidal silica with densities ranging from 1.0550 to 1.0720 g/mL. Samples of bone marrow, cord blood, and peripheral blood harvested by apheresis from rHuG-CSF–treated patients with non-Hodgkin's lymphoma and breast cancer were centrifuged over single-step gradients of different density. The yield of total nucleated cells, red blood cells, and CD34 cells from the interface was calculated in comparison with the number of cells

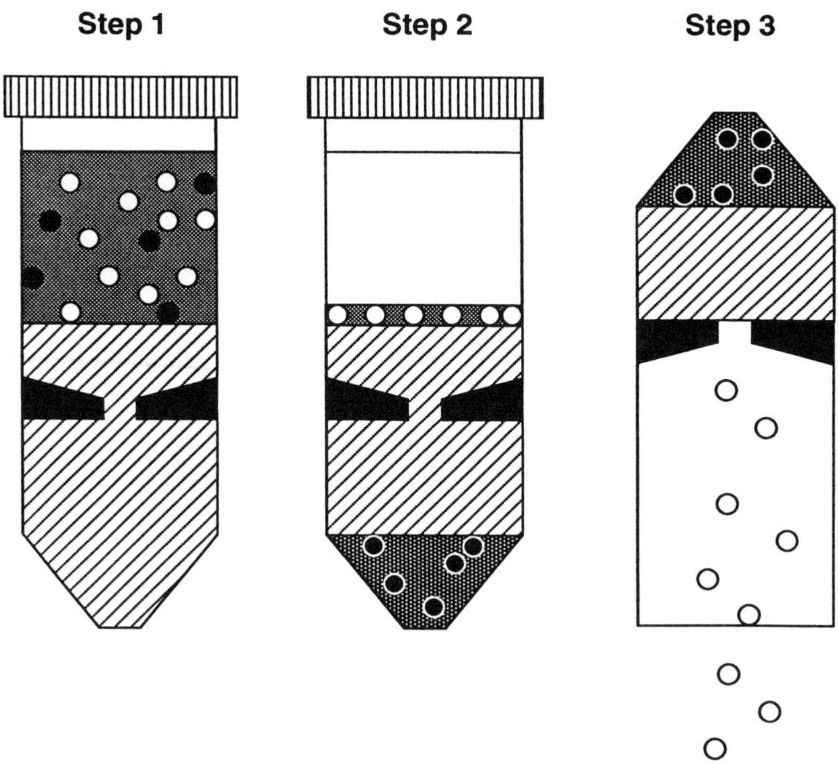

Figure 1 Principle of single-step density gradient centrifugation on silanized colloidal silica in ACT tubes. A sample containing cells of different density (●,○) is loaded onto the colloidal silica single-step gradient (▨) (step 1). During centrifugation, the heavy cells (●) migrate through the density material and pass through the orifice of the insert to end up in the pellet. The low-density cells (○) remain on top of the density material, where they form an interface (step 2). The low-density cells are collected from the interface by decanting the tube (step 3). An air lock is formed in the orifice of the insert and allows the collection of the low-density cells while keeping the high-density cells secluded in the pellet.

in the unprocessed sample. The data in Figure 2 show the distribution of total cells, red cells, and CD34 cells in the interface from the three tissues. In all cases ≥ 99% of the red cells migrate to the pellet at densities equal to or higher than 1.0720 g/mL. At higher concentrations, the amount of red cells at the interface gradually increases (data not shown). The number of nucleated cells

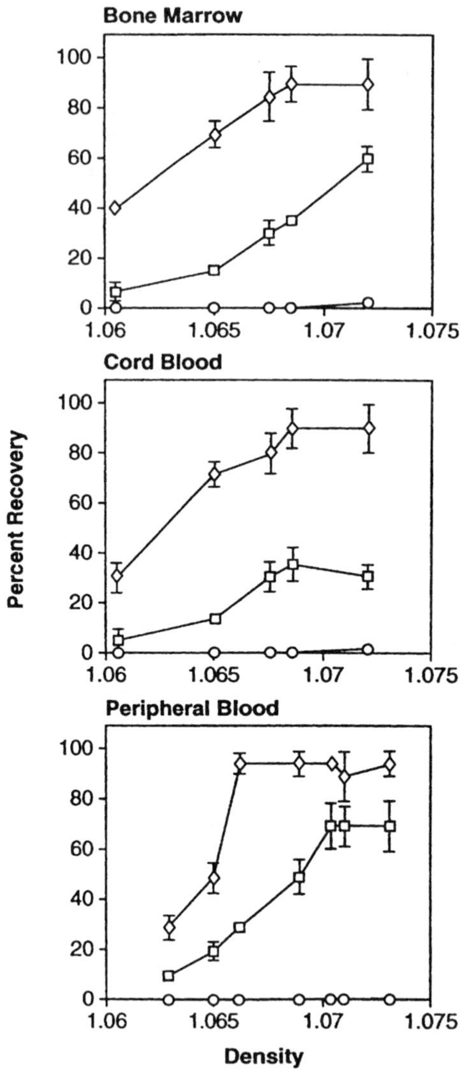

Figure 2 Density distribution of nucleated cells, red blood cells, and CD34 cells from bone marrow, cord blood, and peripheral blood. See text for details of sample preparation. Nucleated cells, red blood cells, and CD34 cells are represented by the symbols (□), (○), and (◇), respectively.

collected from the interface remains fairly constant at densities equal to or higher than 1.0720 g/mL for bone marrow, 1.0720 g/mL for cord blood, and 1.0675 g/mL for peripheral blood. Below this density, the number of cells recovered from the interface decreases rapidly. The reduction of the number of cells at the interface does not occur within the same density range as that observed for other nucleated cells. Consequently, there appears to be a density window at which a single-step density-gradient centrifugation reduces the total cell number by approximately 70% of the initial number while the yield of CD34 cells remains approximately 90%. For apheresis products, a density of 1.0605 ± 0.0005 g/mL is most effective, and for both bone marrow and cord blood a density of 1.0685 ± 0.0005 g/mL is preferred. In both cases, significant numbers of CD34 cells migrated to the pellet when the density of the solution was reduced ≥ 0.0005 g/mL. This narrow 0.0005 g/mL variance of the density specification represents a challenge for the reproducible formulation of density solutions for hematopoietic progenitor cell processing. This emphasizes, once more, the issues related to formulation and storage of colloidal silica, described above. It is not clear why single-step gradients of different density are required to obtain equal enrichment of CD34 cells from different tissues. Most likely, this difference is not associated with differential functional potential since processed cells from the three different tissues were used with success for in vivo stem cell transplantation (39,40,49,50).

B. Effect of the Collection Procedure on the Efficiency of Processing CD34 Cells

Apart from the tissue source, the density of hematopoietic progenitor cells may be determined by the processing procedure that differs for peripheral blood, bone marrow, and cord blood. The effect of manipulation on the density of nucleated cells is further discussed below.

Timing Between Harvest and Processing of CD34 Cells

Many research laboratories process stem cell preparations several hours after collection because of delays associated with shipment. In an attempt to define the best conditions to process hematopoietic progenitor cells by single-step density centrifugation, fresh apheresis samples (approximately 2 hours old) from rHuG-CSF–mobilized patients with non-Hodgkin's lymphoma were centrifuged either immediately on a 1.0605 ± 0.0005 g/mL isotonic silanized colloidal silica solution or after incubation for several hours in isotonic saline. The goal of this study was to define the timing within which single-step density-gradient centrifugation significantly reduces the number of nucleated cells while retaining a high yield of CD34 cells with clonogenic potential as measured by the CFUs. Upon storage, cells appear to become less dense, as illustrated by the increased number of nucleated cells at the interface with the aged samples (Fig. 3). This phenomenon is especially noticeable with samples that were stored for 3 hours and longer. Although the age of the sample does not affect the yield of the CD34 cells or CFUs, older samples may be less enriched for CD34 cells because of increased accumulation of mononuclear cells at the interface. Furthermore, this result indicates that storage of hematopoietic progenitor cells in the presence of silanized colloidal silica does not interfere with their clonogenic potential.

Centrifugation Conditions

The resolution of cell separation on colloidal silica is poor at temperatures below room temperature because of the increase of its viscosity and density at lower temperatures. In addition to temperature, the duration and the speed of centrifugation are also crucial for reproducible processing of cells, especially during the processing of samples with cell concentrations of 10^7 cells/mL and higher. In a study comparing Percoll with Ficoll (Fig. 4),

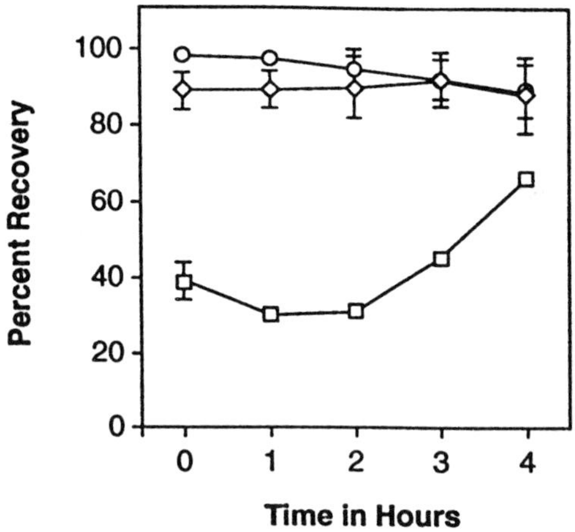

Figure 3 Effect of the sample age on the resolution of cell separation during single-step density-gradient centrifugation. See text for details of sample preparation. Nucleated cells, CFU, and CD34 cells are represented by the symbols (□), (○), and (◇), respectively.

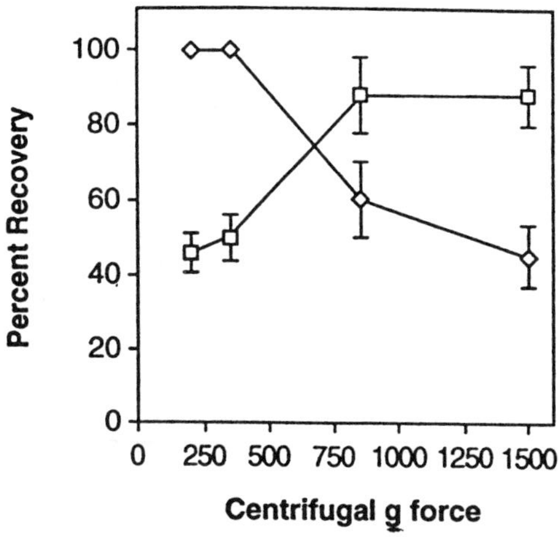

Figure 4 Effect of centrifugation time and force on the resolution of CD34 cell separation on 1.0605 g/mL single-step density gradient centrifugation. Blood was treated as in Figure 2 and was centrifuged either on the single-step density gradient of colloidal silica (□) or on Ficoll (◇). The cells were centrifuged for 30 minutes at ambient temperature at different *g* forces. The cells from the interface were collected and counted, stained with anti-CD45 and anti-CD34 mAbs and the nuclear stain LDS751. The number of CD34 cells is determined using a FACSCAN equipped with a LYSYS II program.

it was observed that a 30-minute centrifugation of at least $550 \times g$ is essential to obtain acceptable cell-separation resolution. Optimal debulking of nucleated cells and CD34 enrichment are obtained at a centrifugal force of $850 \times g$. Whereas the number of CD34 cells collected from the interface with colloidal silica remains constant, increased centrifugal force on Ficoll resulted in loss of CD34 cells to the pellet. During centrifugation on a single-step density gradient, a variety of physicochemical events take place at the interface between the sample and the density solution, resulting in some mixing and diffusion between the two phases. The sedimentation of cells produces convections that cause mixing of the cell suspension with the density gradient material. Consequently, the difference in CD34 recovery may possibly be due to the slight hypertonic character of Ficoll compared with the isotonic Percoll. Cells at the interface of Ficoll gradually shrink and become denser and finally separate to the pellet, a phenomenon that occurs at centrifugal forces exceeding $550 \times g$.

C. Cells That Copurify with CD34 Cells

With the increased use of peripheral blood as the source of hematopoietic progenitor cells for transplantation and adoptive immunotherapy, there is interest in both the bystander cells that copurify with CD34 cells during processing and the effect these cells may have on hematopoietic engraftment, antitumor effect, and overall disease outcome (51–53). Both suppressor cells and NK cells have attracted interest, since they have repeatedly been implicated in regulating immunological phenomena such as graft-versus-host disease (GVHD) and graft-versus-leukemia (GVL) after allogeneic and autologous transplantation, respectively (31,54–61). Apheresis samples from patients with non-Hodgkin's lymphoma who were mobilized with rHuG-CSF were processed on a discontinuous density gradients ranging from 1.0520 to 1.0640 g/mL. Suppressor cell and NK cell activity was determined at the different interfaces by means of the mixed lymphocyte culture and chromium release assay, respectively. The results in Figure 5 show that more than 90% of suppressor cell activity is associated with cells that separate in a density range that is equal to or less than 1.0605 ± 0.0005 g/mL. The same is true for the distribution of NK cell activity (Fig. 6). CD34 cell preparations obtained by processing apheresis samples on single-step gradients in a density window of 1.0605 ± 0.0005 g/mL contain not only the majority of the CD34 cells, but also cells that may play a role in overall posttransplant disease progression.

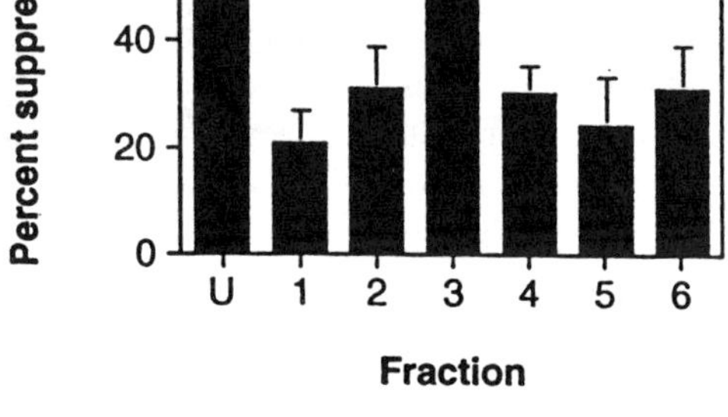

Figure 5 Density distribution of cells with suppressor activity from apheresed blood samples.

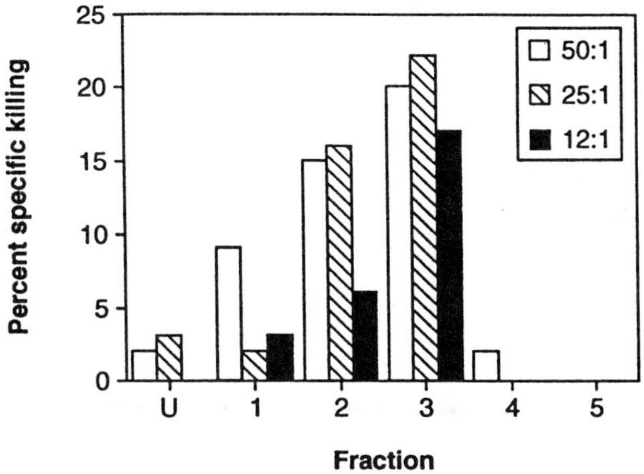

Figure 6 Density distribution of NK cells from apheresed blood samples.

Moreover, it is of interest to note that this suppressor activity acts across histocompatibility barriers since the cells used for the mixed lymphocyte culture and the alleged suppressor cells from the apheresis product are derived from donors with different HLA type.

D. Cells That Do Not Copurify with CD34 Cells

The contribution of T cells to GVHD after allotransplantation is well documented (62–64). Consequently, the number of T cells in the graft pose a major concern to the transplant physician. Our experience shows that the isolation of hematopoietic progenitor cells from peripheral blood using single-step gradients of appropriate density results in a reduction of T cells. Experiments were performed with apheresis samples from rHuG-CSF-treated normal donors. The results in Figure 7 show that centrifugation on a 1.0605 ± 0.0005 g/mL single-step density gradient results in a 90% reduction of the total number of T cells. This 1-log reduction brings the total T-cell number in apheresis products to the level of T cells present in normal, unprocessed bone marrow. The same experiments were performed in the automatic cell processing system, the COBE 2991, and produced the same results. The data of this study are reported elsewhere (65).

Similarly, $CD20^+$ B cells were reduced by the same magnitude. This may explain why in some cases density centrifugation may suffice to purge non-

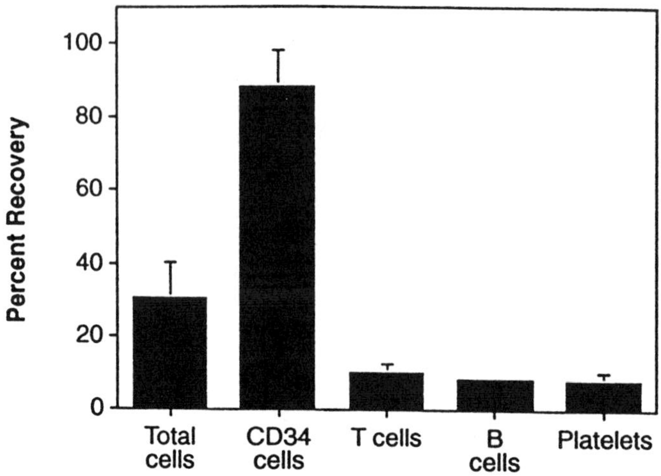

Figure 7 Depletion of T cells, B cells, and platelets by single-step density gradient centrifugation. Cells from apheresed blood of normal rHuG-CSF treated donors were centrifuged on a 1.0605 g/mL colloidal silica single-step gradient. The cells were centrifuged for 30 minutes at room temperature at 850 × *g*. The cells from the interface were counted and stained with anti-CD20, anti-CD3, anti-CD45, and anti-CD34 mAbs and the nuclear stain LDS751. The number of T cells, B cells, CD34 cells, and platelets is determined using a FACSCAN equipped with a LYSYS II program.

Hodgkin's lymphoma tumor cells to a polymerase chain reaction (PCR)–negative signal for the (14;18) translocation (39). The potential to reduce T cells by single-step density centrifugation differs from tissue to tissue. Whereas the T-cell purging works well with autologous and allogeneic peripheral blood samples, T-cell depletions of only 30% were observed during the processing of bone marrow on a 1.0685 g/mL single-step density gradient (data not shown). It should also be noted that the processing of CD34 preparations on single-step density gradients of colloidal silica reduced the platelet content by 90%, which may be a significant advantage for further processing efficiency.

E. Advantages Associated with CD34 Cell Processing

As described earlier, the reduction of T cells in allogeneic transplants may be advantageous for the reduction of GVHD, whereas the reduction of malignant B cells in autologous transplants would be of obvious benefit to the patient, since reinfused tumor cells may contribute to disease relapse (66). Apart from these obvious biological benefits, the reduction of cell volume also reduces clinical complications related to the infusion of large volumes of cryopreserved hematopoietic progenitor cells. Dependent on the origin of tissue, the volume of cells varies; as much as 160 mL may be collected for cord blood, > 300 mL for apheresis samples, and up to 1.5 L for bone marrow collections. The total cell number in these products ranges from as few as 10^9 cells to as many as 10^{12} cells. The transfusion of large volumes of hematopoietic progenitor cells may induce fluid overload and pulmonary edema in the patient (67,68). In addition, due to the lysis of red blood cells, peripheral blood and bone marrow transplants contain free hemoglobin that can induce hemoglobinuria and kidney damage upon transfusion (69). In addition, the infusion of material containing the cryopreservative dimethylsulfoxide (DMSO) frequently induces nausea, vomiting, abdominal cramping, flushing, and dyspnea (70,71). Consequently, volume reduction of the transplant is desirable.

It was mentioned earlier that the rapidity of hematopoietic reconstitution is determined by the number of viable CD34 cells infused/kg patient weight. In this context, we examined the effect of cell processing on the recovery of CD34 cells during a freeze-thawing procedure. The data in Figure 8 show that the recovery of both the total cell number and, more importantly, the number of CD34 cells is greater in the processed than in the unprocessed samples. This may be explained by cell clumping, which was more prominent during thawing of unprocessed samples. In addition, the processed samples have a reduced number of platelets that are notorious for inducing cell clumping (data shown above).

V. DENSITY-ADJUSTED CELL SORTING

Single-step density-gradient centrifugation can be used to reduce the total volume of nucleated cells and to deplete specific cell types, while enriching CD34 cells, NK cells, and suppressor cells. The magnitude of this simultaneous depletion and enrichment process is, however, limited and is not sufficient to obtain pure cell preparations. In this context, we have developed an extension to single-step density-gradient centrifugation that can be used for both positive and negative selection of particular cell types. This novel cell separation system is based on the adjustment of the density of low-density cells with heavy carriers and is referred to as density-adjusted cell sorting (DACS).

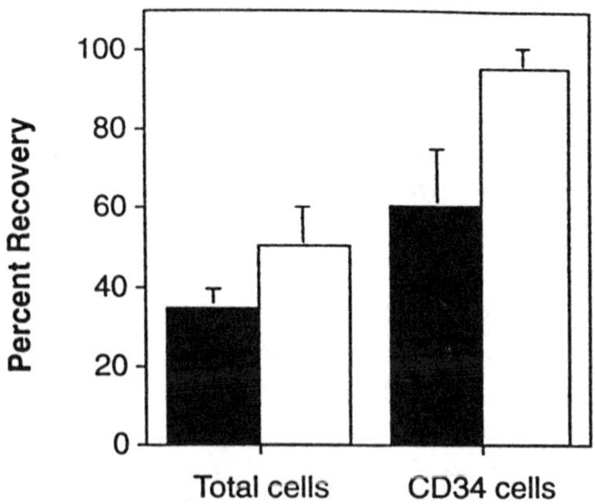

Figure 8 Effect of cell processing on the recovery of total cells and CD34 cells from processed (□) and unprocessed (■) samples after freezing and thawing. Cells from apheresed blood of rHuG-CSF–treated NHL patients were either centrifuged on a 1.0605 g/mL colloidal silica single-step gradient and then frozen in liquid nitrogen or frozen without processing. The cells were rapidly thawed in a 37°C water bath and stained with anti-CD45 and anti-CD34 mAbs and the nuclear stain LDS751. The percent cell recovery is determined by dividing the number of total cells and CD34 cells recovered after freezing and thawing by the number present in the unfractionated sample.

A. Principle

The DACS technology combines two cell-separation principles in one step. First, cells of interest are enriched using a single-step density-gradient of colloidal silica solution with a defined density. Second, low-density unwanted cells, which would otherwise copurify with the desired cells, are positively depleted during centrifugation by increasing their density by binding them to high-density DACS beads. Figure 9 represents a graphic illustration of the principle of DACS. In step 1, the cells are incubated with murine monoclonal antibodies (mAbs) directed to the unwanted cell population. In step 2, these cells are incubated with DACS beads coated with goat antimouse immunoglobulins. After this incubation, the cell-bead mixture is loaded (step 3) onto the single-step density-gradient and centrifuged. During centrifugation, both the high-density cells and the low-density unwanted cells with adjusted density migrate (step 4) to the bottom of the tube where they concentrate in the pellet. The desired cells with low density remain at the interface and are collected

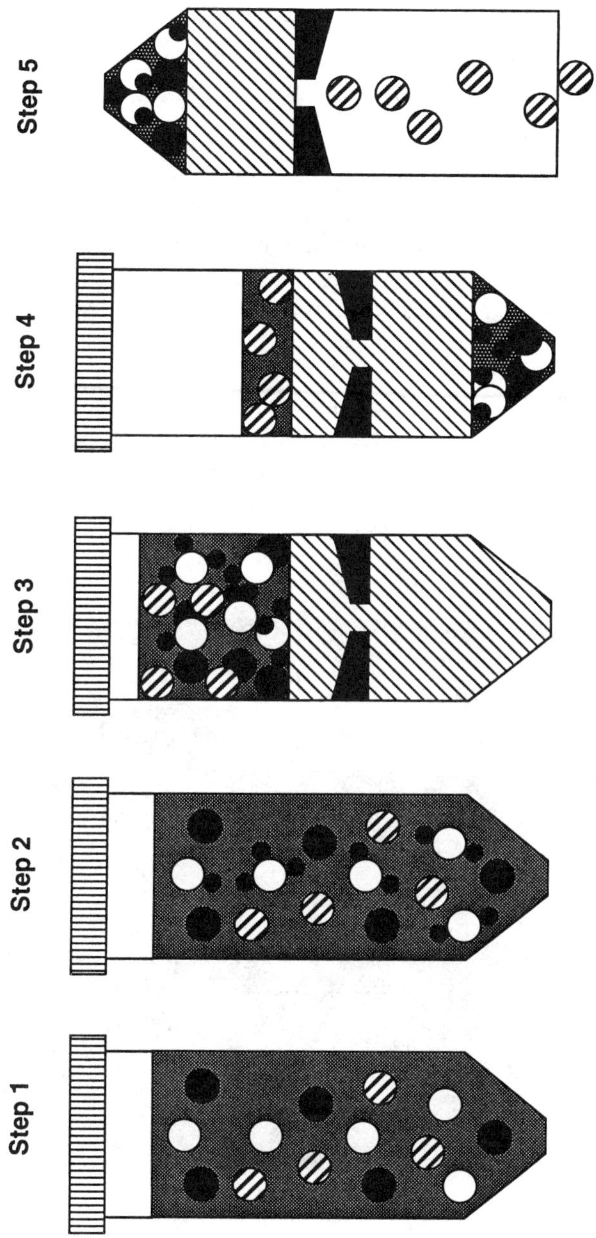

Figure 9 A mixed population of low-density cells which include desired cells (◍) and unwanted cells (○) and high-density cells (●) are mixed with murine mAbs specific for the unwanted cells (step 1). The cells are centrifuged, resuspended in medium, and mixed with DACS beads coated with goat antimouse IgG (●) (step 2). The beads bind to the mAb-coated cells and form bead-cell complexes. The bead and cell mixture is layered onto the single-step density gradient (step 3) and then centrifuged (step 4). High-density cells and bead-cell complexes collect in the lower chamber of the test tube. The desired cells are collected from the interface by decanting the tube (step 5).

(step 5). As shown above, the desired cells can be decanted instead of being collected with a pipette, due to the particular shape of the centrifugation tube.

B. Bead Characteristics

The usefulness of microspheres for cell separation depends on their surface characteristics, uniformity in size, and potential to bind immunoglobulins. Figure 10 shows a scanning electron photomicrograph (SEM) image of anti-CD3-coated DACS beads bound to T cells. This image shows a representative field of beads and illustrates the uniformity in both size and shape of the DACS beads. In addition, their surface appears to be smooth and devoid of obvious porous areas or cracks. The DACS beads were further analyzed by flow cytometry using a FACSCAN, which allows statistical and quantitative analysis of the uniformity, autofluorescence, and ability of the beads to bind mouse immunoglobulin. Scatter analysis shows that DACS beads are composed of two main populations, R1 and R2, which represent 94% and 5% of the total bead population, respectively (Fig. 11A). The R1 population repre-

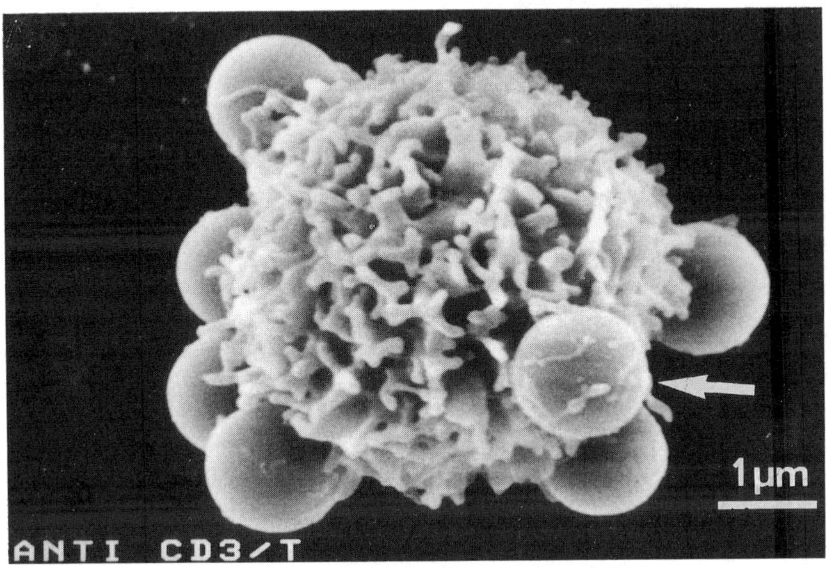

Figure 10 Scanning electron photomicrograph showing a T cell coated with anti-CD3 mAbs conjugated to DACS beads.

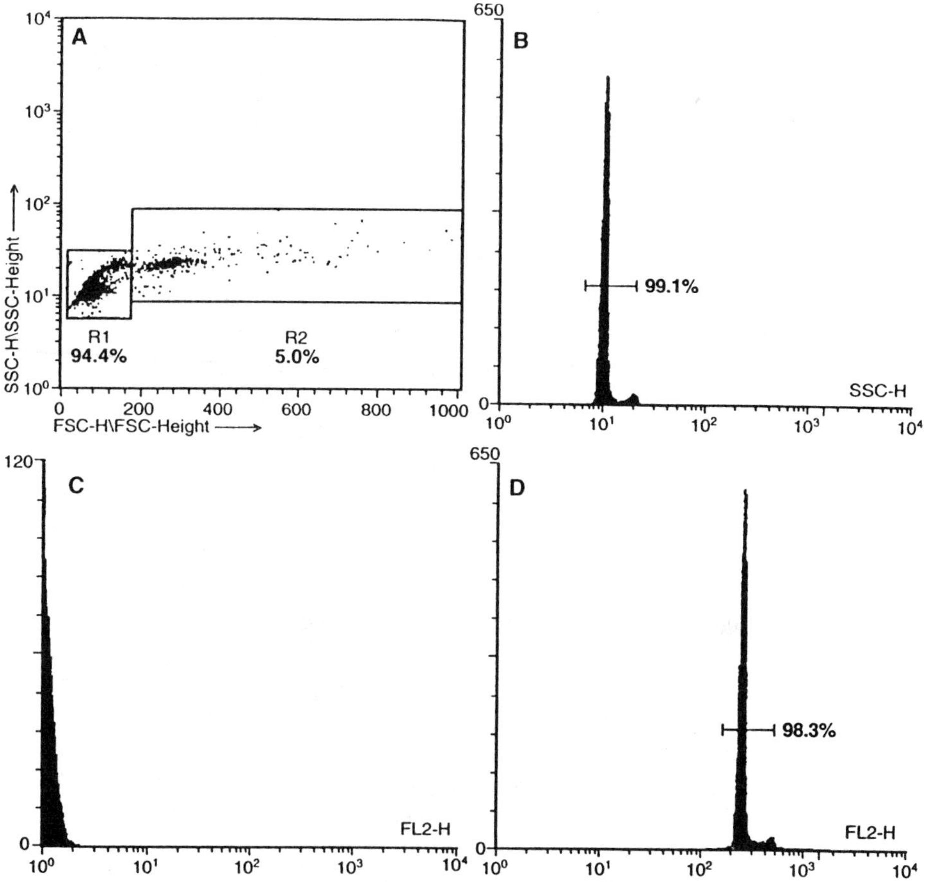

Figure 11 FACS analysis of DACS beads. (A) Forward (FSC) versus sideward (SSC) scatter profile of goat-antimouse AB-coated DACS beads. R1 and R2 represent the singlets and aggregates, respectively. (B) Statistical analysis of the bead population based on SSC shows that approximately 99% of the DACS beads occur as singlets. (C) Demonstration of the low autofluorescence of DACS beads. (D) Indication that 98% of goat-antimouse coated DACS beads uniformly bind to phycoerythrin-conjugated mouse IgG.

sents the single beads, whereas R2 represents the doublets and clumps of multiple beads. DACS beads represent a fairly homogeneous bead population (Fig. 11B). Fluorescence analysis shows that the DACS beads have a very low autofluorescence (Fig. 11C) and have a uniform binding pattern for mouse IgG as illustrated by the single, narrow peak in the second fluorescence channel (FL2) (Fig. 11D).

C. Tumor Cell and T-Lymphocyte Purging

The clinical advantages associated with tumor cell and T-cell depletion have been discussed briefly. In this context, DACS beads were used in combination with a single-step density-gradient of the appropriate density to deplete T cells, B cells, and malignant cells from apheresis samples obtained from rHuG-CSF–mobilized cancer patients. Apheresis samples from rHuG-CSF–mobilized patients with non-Hodgkin's lymphoma were either used as such or were first spiked with either SU-DHL4 lymphoma cells or SK-BR3 breast cancer cells. The cells were then incubated with a cocktail of mAbs directed against T cells (anti-CD3), B cells, or B-lymphoma cells (cocktail of anti-CD9, anti-CD10, anti-CD19, and anti-CD20), or breast cancer cells (anti-Her2NEU). After incubation with DACS beads, the cells were layered onto a 1.0605 ± 0.0005 g/mL single-step density-gradient of colloidal silica and centrifuged for 30 minutes at $850 \times g$. The cells from the interface were decanted and stained with fluoro-chrome-conjugated antiCD45 and antiCD34 mAbs and analyzed in the FACS. The number of CD34 cells was determined in both the unfractionated and the fractionated sample. The results, shown in Figure 12, indicate that a single round of DACS depleted the total number of nucleated cells by approximately 70%, whereas 70% of the $CD34^+$ cells and 80% of the CFUs were retained at the interface. At the same time, the initial number of T cells, SU-DHL4, and SK-BR3 cells were depleted by 2–3 logs. Transplant literature shows that the rapidity of engraftment of CD34 cells after transplantation is directly related to the number infused per kilogram of patient weight (72–74). Prompt recovery of platelets and neutrophils after transplantation occurs when the number of CD34 cells infused exceeds 2.5×10^6 CD34 cells/kg. This implies that the yield of CD34 cell–processing procedures is of crucial importance. A retrospective study was performed in our laboratory in which the nonspecific loss of $CD34^+$ cells from peripheral blood stem cell samples after DACS was determined. In 31 experiments using apheresed blood from a combination of rHuG-CSF–mobilized patients with breast cancer, non-Hodgkin's lymphoma, and Hodgkin's disease, the mean nonspecific loss of

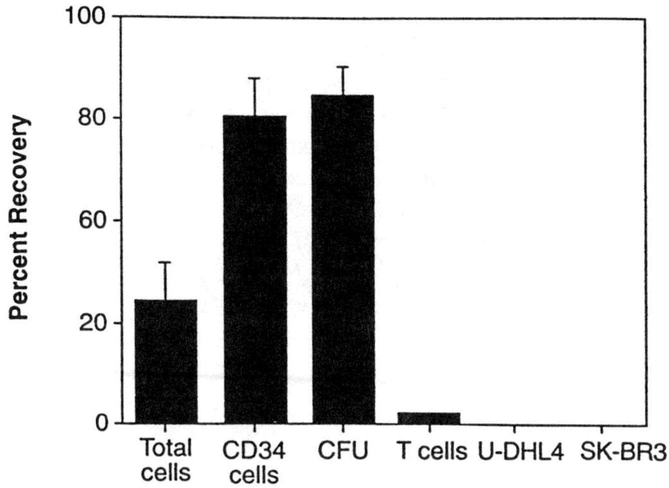

Figure 12 Tumor and T-cell depletion using DACS. Cells from either normal buffy coats or apheresis samples of rHuG-CSF–treated NHL patients were spiked with the lymphoma tumor line SU-DHL4 or the breast cancer line SK-BR3 and centrifuged on a 1.0605 g/mL single-step density gradient of colloidal silica. The cells from the interface were collected and the number of T cells and SK-BR3 cells was determined by FACS analysis using fluorochrome conjugated anti-CD3 and anti-Her2-neu mAbs, respectively. The number of SU-DHL4 cells was determined by quantitative PCR detection of the (14;18) translocation in the cell line.

cells was 24 ± 19%, while the mean recovery of $CD34^+$ cells was 75 ± 15% (Fig. 12).

VI. CONCLUSION

Clinical studies have shown that the use of hematopoietic progenitor cells for transplantation as rescue therapy for patients receiving myeloablative chemo- and radiation therapy can salvage patients with relapsed non-Hodgkin's lymphoma, Hodgkin's disease, and acute leukemia (75–77). Recently, sophisticated mobilization regimens have allowed the collection of sufficient hematopoietic progenitor cells for treatment from one or two aphereses (39). In many instances, subsequent cell processing is used to deplete T lymphocytes and tumor cells before allogeneic and autologous transplantation, respectively. Depending on the technique, cell volume reduction makes purging more effective and cost efficient. Apart from that, volume reduction reduces elements in the transplant with adverse effects for the patient, such as cryopreservatives and free hemoglobin. This chapter described single-step density-gradient centrifugation as an effective and reproducible processing method that reduces the total cell volume of peripheral blood and bone marrow by approximately 70%, while maintaining close to 90% of CD34 cells. Consequently, this procedure reduces the amount of cryopreservatives by 70% and may be cost saving in the context of purging. In addition, cells processed on colloidal silica showed an improved recovery of CD34 cells following cryopreservation in comparison with unprocessed samples.

REFERENCES

1. Silvestri F, Wunder E, Sovalat H, Henon P, Serke S. Positive selection of CD34+ cells: a short review of the immunoabsorption methods currently available for experimental and clinical use. *J Hematother* 1993; 2(4):473–481.
2. Bøyum A. Separation of lymphocytes, lymphocyte subgroups, and monocytes: a review. *Lymphology* 1977; 10(2):71–76.
3. Miltenyi S, Muller W, Weichel W, Radbruch A. High gradient magnetic cell separation with MACS. *Cytometry* 1990; 11(2):231–238.
4. Strauss LC, Trischmann TM, Rowley SD, Wiley JM, Civin CI. Selection of normal human haemopoietic stem cells for bone marrow transplantation using immunomagnetic microspheres and CD34 antibody. *Am J Pediatr Hematol Oncol* 1991; 13(2):217–221.
5. Stray KM, Corpuz S, Kalamasz D, Colter M, Berenson R, Heimfeld S. Purging tumor cells from bone marrow or peripheral blood using avidin-biotin immunoabsorption. In: Gee A, Santos G, Worthington-White D, eds. *Advances in Bone Marrow Purging and Processing. Fourth International Symposium.* New York: Wiley-Liss, 1994:97–103.
6. Wysocki L, Sato V. "Panning" for lymphocytes: a method for cell selection. *Proc Nat Acad Sci USA* 1978; 75(6):2844–2848.
7. Spangrude G, Heimfeld S, Weissman I. Purification and characterization of mouse hematopoietic stem cells. *Science* 1968; 241:58–62.
8. Van Vlasselaer P, Falla N, Snoeck H, Mathieu E. Characterization and purification of osteogenic cells from murine bone marrow by two-color cell sorting using anti-Sca-1 monoconal antibody and wheat germ agglutinin. *Blood* 1994; 84(3):753–763.
9. Bøyum A. Separation of white blood cells. *Nature* 1964; 204:793.
10. Skoog W, Beck W. Studies of the fibrogen, dextran and phytohemagglutinin methods of isolating leukocytes. *Blood* 1956; 11:436.

11. Richter A. Erythrocyte aggregation, lethal dose, and molecular structure of hydrophilic colloids. *Acta Physiol Scand Suppl* 1963; 59:130.
12. Andrews RG, Singer JW, Beynstein ID. Monoclonal antibody 12.8 recognizes a 115 Kd molecule present on both unipotent and multipotent hematopoietic colony forming cells and their precursors. *Blood* 1986; 67(3):842–845.
13. Watt SM, Karhi K, Gatter K, et al. Distribution and epitope analysis of the cell membrane glycoprotein (HPCA-1) associated with human hematopoietic progenitor cells. *Leukemia* 1987; 1(5):417–426.
14. Pertoft H, Laurent T. Isopycnic separation of cells and cell organelles by centrifugation in modified colloidal silica gradients. In: Catsimpoolous, ed. *Methods of Cell Separation*, Vol. 1. New York: Plenum Press, 1977:25.
15. Bøyum A. Isolation of mononuclear cells and granulocytes from human blood. *Scand J Clin Lab Invest Suppl* 1968; 97:77–89.
16. Mateyko G, Kopac M. Cytophysical studies on living normal and neoplastic cells. *Ann NY Acad Sci* 1963; 105:185.
17. Pertoft H. Gradient centrifugation in colloidal silica-polysaccharide media. *Biochem Biophys Acta* 1966; 126(3):594–596.
18. Pertoft H, Back O, Lindahl-Kiessling K. Separation of various blood cells in colloidal silica-polyvinylpyrrolidone gradients. *Exp Cell Res* 1968; 50(2):355–368.
19. Pertoft H. The separation of rat liver cells in colloidal silica-polyethylene glycol gradients. *Exp Cell Res* 1969; 57(2)338–350.
20. Wolff D. The separation of cells and subcellular particles by colloidal-silica density centrifugation. In: Prescott D, ed. *Methods in Cell Biology*, Vol. 10. New York: Academic Press, 1975:85–104.
21. Olofsson T, Gartner I, Olsson I. Separation of human bone marrow cells in density gradients of polyvinylpyrrolidone-coated silica gel (Percoll). *Scand T Haematol* 1980; 24(3):254–262.
22. Pertoft H, Laurent T. The use of gradients of colloidal silica for the separation of cells and subcellular particles. In: Gerritsen T, ed. *Modern Separation Methods of Macromolecules and Particles*, Vol. 2. New York: John Wiley & Sons, 1968:71–90.
23. Pertoft H, Laurent TC, Laas T, Kagedal L. Density gradients prepared from colloidal silica particles coated by polyvinylpyrrolidone (Percoll). *Anal Biochem* 1978; 88(1):271–282.
24. Pertoft H, Rubin K, Kjellen L, Laurent T, Klingebom B. The viability of cells grown or centrifuged in a new density gradient medium, Percoll™. *Exp Cell Res* 1977; 110(2):449–457.
25. Martin H, Hibbin J. Purification of haemopoietic progenitor cells from patients with chronic granulocytic leukemia using Percoll density gradients and elutriation. *Br J Haematol* 1986; 63(1):187–198.
26. Brandt J, Baird N, Lu L, Srour E, Hoffman R. Characterization of a human hematopoietic progenitor cell capable of forming blast cell containing colonies in vitro. *J Clin Invest* 1988; 82(3):1017–1027.

27. Gutierrez C, Bernabe R, Vega J, Kreisler M. Purification of human T and B cells by a discontinuous density gradient of Percoll. *J Immunol Meth* 1979; 29(1):57–63.
28. Kurnick J, Ostberg L, Stegagno M, Kimura AK, Orn A, Sjoberg O. A rapid method for the separation of functional lymphoid cell populations of human and animal origin on PVP-silica (Percoll) density gradients. *Scand J Immunol* 1979; 10(6):563–573.
29. Merluzzi VJ, Trail PA, Last-Barney K. Differential expression of lymphokine-activated killer cells and natural killer cells in adoptive transfer experiments utilizing fractionated bone marrow. *J Immunol* 1986; 137:2425–2427.
30. Timonen T, Saksela E. Isolation of human NK cells by density gradient centrifugation. *J Immunol Meth* 1980; 36(3–4):285–291.
31. Schmidt-Wolf IG, Dejbaklısh-Jones S, Ginzton N, Greenberg P, Strober S. T-cell subsets and suppressor cells in human bone marrow. *Blood* 1992; 80(12):3242–3250.
32. Tarkkanen J, Saksela E. Umbilical-cord-blood-derived suppressor cells of the human natural killer cell activity are inhibited by interferon. *Scand J Immunol (Norway)* 1982; 15(2):149–157.
33. Pertoft H, Johnsson A, Warmegard B, Seljelid R. Separation of human monocytes on density gradients of Percoll®. *J Immunol Meth* 1980; 33(3):221–229.
34. Feige U, Overwien B, Sorg C. Purification of human blood monocytes by hypotonic density gradient centrifugation in Percoll. *J Immunol Meth* 1982; 54(3):309–315.
35. Budenz R, Bernard G. Characterization of bone marrow osteoprogenitor cell lines. *J Cell Biol* 1979; 83:32.
36. Falla N, Van Vlasselaer P, Bierkens J, Borremans B, Schoeters G, Van Gorp U. Characterization of a 5-fluorouracil-enriched osteoprogenitor population of the murine bone marrow. *Blood* 1993; 82(12):3580–3591.
37. Arcidiacono A, Walt H, Campana A, Balerna M. The use of Percoll gradients for the preparation of subpopulations of human spermatozoa. *Int J Andrology* 1983; 6(5):433–455.
38. Kaneko S, Oshio S, Kobanawa K, Kobayashi T, Mohri H, Iizuka R. Purification of human sperm by a discontinuous Percoll density gradient with an innercolumn. *Biol Reprod* 1986; 35(4):1059–1063.
39. Negrin RS, Kusnierz-Glaz CR, Still BJ, et al. Transplantation of enriched and purged peripheral blood progenitor cells from a single apheresis product in patients with non-Hodgkin's lymphoma (NHL). *Blood* 1994; 84(10, suppl 1):396a.
40. Ekert H, Ellis WM, Georgiou GM, Robertson DM, Tiedemann K, Waters KD. Marrow function reconstitution by fraction 3 of Percoll-density gradient-separated cells. *Transplantation* 1986; 42(1):58–68.
41. Schriber J,R Dejbakhsh-Jones S, Kusnierz-Glaz CR, et al. Enrichment of bone marrow and blood progenitor ($CD34^+$) cells by density gradients with sufficient yields for transplantation. *Exp Hematol* 1995; 23(9):1024–1029.

42. Ayello J, Hesdorffer C, Reiss R. A semiautomated technique for volume reduction of stem cell suspensions for autotransplantation. *J Hematother* 1995; 4:545–549.
43. Gee A. Bone marrow purging and processing—a review of ancillary effects. In: Gross S, Gee A, Worthington-White D, eds. *Bone Marrow and Purging Processing* New York: Wiley-Liss, 1990:507–521.
44. Wells JR, Opelz G, Cline MJ. Characterization of functionally distinct lymphoid and myeloid cells from human blood and bone marrow I. Separation by a buoyant density gradient. *J Immunol Meth* (*Netherlands*) 1977; 18(1–2):79–93
45. Gilmore MJ, Prentice HG, Blacklock HA, Janossy G Hoffbrand AV. A technique for rapid isolation of bone marrow mononuclear cells using Ficoll-Metrizoate and the IBM 2991 blood cell processor. *Br J Haematol* 1982; 50(4):619–626.
46. Martin H, Neumann M, Fache I, Fliedner TM, Pfliger H. Gradient separation of granulocytic progenitor cells (CFUc) from human blood mononuclear leukocytes. *Exp Hematol* 1985; 13(2):79–86.
47. Lasky C, Zanjani ED. Size and density characterization of human committed and multipotent hematopoietic progenitors. *Exp Hematol* 1985; 13(7):680–684.
48. Roberton DM, Georgiou GM, Tiedemann K, et al. Immunological reconstitution in a patient with severe combined immune deficiency using non-sibling bone marrow depleted of T cells with HuLy-m1. *Aust NZ J Med* 1986; 16(3):385–388.
49. Wagner JE, Broxmeyer HE, Byrd RL, et al. Transplantation of umbilical cord blood after myeloablative therapy: analysis of engraftment. *Blood* 1992; 79(7):1874–1881.
50. Broxmeyer HE, Douglas GW, Hangoc G, et al. Human umbilical cord blood as a potential source of transplantable hematopoietic stem/progenitor cells. *Proc Natl Acad Sci USA* 1989; 86(10):3828–3832.
51. Riddell SR, Greenberg PD. Cellular adoptive immunotherapy after bone marrow transplantation. *Cancer Treat Res* 1995; 76:337–369.
52. Kessinger A. Do autologous peripheral blood cell transplants provide more than hematopoietic recovery? *Stem Cells (DAYT)* 1995; 13(4):351–354.
53. Baxevanis CN, Papamichail M. Characterization of the anti-tumor immune response in human cancers and strategies for immunotherapy. *Crit Rev Oncol Hematol* (*Ireland*) 1994; 16(3):157–159.
54. Strober S, Cheng L, Zeng D, et al. Double negative ($CD4^{-}CD8^{-}y\beta+$) T cells which promote tolerance induction and regulate autoimmunity. *Immunol Rev* 1996; 149:217.
55. Palathumpat Y, Dejbakhsh-Jones S, Holm B, Strober S. Different subsets of T cells in the adult mouse bone marrow and spleen induce or suppress acute graft versus host disease. *J Immunol* 1992; 149(3):808–817.
56. Warner JF, Dennert G. Effects of a cloned cell line with NK activity on bone marrow transplants, tumor development and metastasis in vivo. *Nature* 1982; 300(5887):31–34.
57. Kiessling R. Antitumor effector cells: characteristics and basic mechanisms. Introduction. *Immunol Ser* 1994; 61:37–39.

58. Whiteside T, Herberman R. Role of human natural killer cells in health and disease. *Clin Diag Lab Immunol* 1994; 1(2):125–133.
59. Cervantes F, Miller JS, McGlave PB. Cytolytic activity by lymphokines (LAK activity) in chronic myeloid leukemia. *Med Clin (Barc) Spain* 1995; 105(2):50–53.
60. Miller JS Verfaille C, McGlave P. Expansion and activation of human natural killer cells for autologous therapy. *J Hemathother* 1994; 3(1):71–74.
61. Jiang YZ, Cullis JO, Kanfer EJ, Goldman JM, Barrett AJ. T cell and NK cell mediated graft-versus-leukemia reactivity following donor buffy coat transfusion to treat relapse after marrow transplantation for chronic myeloid leukemia. *Bone Marrow Transplant* 1993; 11(2):133–138.
62. Poynton CH. T cell depletion in bone marrow transplantation. *Bone Marrow Transplant* 1988; 3(4):265–279.
63. O'Reilly R, Collins N, Kernan N, et al. Transplantation of marrow-depleted T cells by soybean lectin agglutinin and E-rosette depletion: major histocompatability complex-related graft resistance in leukemia transplant recipients. *Transplant Proc* 1985; 17:455.
64. Champlin R, Gajewski J, Feig S, et al. Selective depletion of CD8 positive lymphocytes for prevention of graft-versus-host disease following allogeneic bone marrow transplantation. *Transplant Proc* 1989; 21(1Pt 3):2947–2948.
65. Przepiorka D, Van Vlasselaer P, Huynh L, et al. Rapid debulking and CD34 enrichment of filgrastim-mobilized peripheral blood stem cells by automated density gradient centrifugation in a closed system. *J Hematother* 1996;5:497–502.
66. Brenner MK, Rill DR, Moen RC, et al. Gene-marking to trace origin of relapse after autologous bone-marrow transplantation. *Lancet* 1993; 341(8837):85–86.
67. Strooncek DF, Fautsh SK, Lasky LC, et al. Adverse reactions in patients transfused with cryopreserved marrow. *Transfusion* 1991; 31(6):521–526.
68. Rapoport AP, Rowe JM, Packman CH, Ginsberg SJ. Cardiac arrest after autologous marrow infusion. *Bone Marrow Transplant (England)* 1991; 7(5):401–403.
69. Smith DM, Weisenberger Dd, Bierman P Kessinger A, Yaughan WP, Armitage JO. Acute renal failure associated with autologous bone marrow transplantation. *Bone Marrow Transplant (England)* 1987; 2(2):195–201.
70. Davis J, Rowley SD, Santos GW. Toxicity of autologous bone marrow graft infusion. In: Gross S, Gee AP, Worthington-White DA, eds. *Bone Marrow Purging and Processing.* New York: Wiley-Liss, 1990:531.
71. Davis J, Rowley SD, Santos GW. Toxicity of autologous bone marrow graft infusion. *Prog Clin Biol Res* 1990; 333:531–540.
72. Haas R, Mohle R, Fruhauf S, et al. Patients characteristics associated with successful mobilizing and autografting of peripheral blood progenitor cells in malignant lymphomas. *Blood* 1994; 83(2):3787–3794.
73. Schwartzberg L, Birch R, Blanco R, et al. Rapid and sustained hematopoietic reconstitution by peripheral blood stem cell infusion alone following high-dose chemotherapy. *Bone Marrow Transplant* 1993; 11(5):369–374.
74. Zimmerman T, Lee W, Bender J. CD34 may be used to determine the adequacy

of a stem cell harvest for hematologic recovery following high dose chemotherapy (HDC). Third International Symposium on Peripheral Blood Stem Cell Autografts, Bordeaux, France, 1993.

75. Philip T, Armitage J, Spitzer G. High-dose therapy and autologous bone marrow transplantation after failure of conventional chemotherapy in adults with intermediate-grade or high-grade non-Hodgkin's lymphoma. *N Engl J Med* 1987; 316(24): 493–498.
76. Jagannath S, Armitage JO, Dicke KA, et al. Progenitor factors for response and survival after high-dose cyclophosphamide, carmustine, and etoposide and autologous bone marrow transplantation for relapsed Hodgkin's disease. *J Clin Oncol* 1989; 7(2):179–185.
77. Clift RA, Buckner CD, Appelbaum FR, et al. Allogeneic marrow transplantation during untreated first relapse of acute myeloid leukemia. *J Clin Oncol* 1992; 10(11):1723–1729.

3

Centrifugal Elutriation: A Powerful Separation Technique in Cell Biology, Immunology, and Hematology

Carl G. Figdor, Frank Preijers, Richard Huijbens, Paul Ruijs, Theo J. M. de Witte, and Willy S. Bont
University Hospital Nijmegen, Nijmegen, The Netherlands

I. INTRODUCTION

Centrifugal elutriation, also called counterflow centrifugation elutriation, or counter-streaming centrifugation, was initially described by Lindahl in 1948 (1). It was used until 1968, when the Elutriator rotor became commercially available (2). During the past decade it has evolved into a separation system particularly suited to processing large numbers of cells, ranging from small (> 1 μm) particles, such as bacteria, to large (40 μm) tumor cells, immature erythrocytes, and megakaryocytes. Earlier elutriation systems were completely closed, thus allowing clinical applications. The design of new, larger separation chambers or parallel usage of separation chambers has stimulated the use of centrifugal elutriation for clinical purposes.

Compared with other separation techniques, centrifugal elutriation provides a rapid, nondestructive, and reproducible method of large-scale cell separation; for example, 2×10^9 to 2×10^{10} mononuclear cells, depending on the equipment used, can be separated in a single separation chamber in a single run. In addition, it allows separation at low temperatures in a medium of choice, preventing activation of cells (monocytes). Centrifugal elutriation does not require any pretreatment of the sample and can be applied to separate cell populations that differ only slightly in sedimentation velocity (see Sec. II). A major disadvantage that thus far has prevented centrifugal elutriation from becoming a major separation technique is that it requires well-trained and dedicated personnel. If centrifugal elutriation can be further automated and made user friendly, this will undoubtedly extend the number of applications.

In this chapter we shall discuss the principles of centrifugal elutriation and recent developments that have improved the technique. In addition, we shall discuss a number of applications for which centrifugal elutriation is frequently used in research laboratories and in clinical practice.

II. BASIC PRINCIPLES OF ELUTRIATION

Centrifugal elutriation is based on differences in the sedimentation velocity of the heterogeneous particles to be separated. A suspension of cells is loaded into a separation chamber (Fig. 1). The separation chamber is mounted onto a rotor

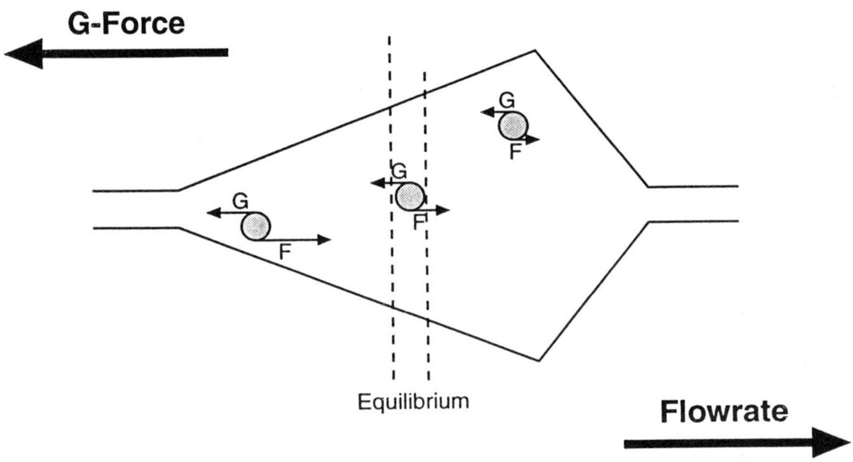

Figure 1 Cells entering an elutriation chamber are subject to two opposing forces: the G-force, because of centrifugation, and the flow rate. Only when the resulting force equals zero will the cells remain in the elutriation chamber. Depending on the cell type and the hydrodynamic characteristics of the chamber, this is near the inlet or near the outlet of the separation chamber.

in a centrifuge. During rotation, the medium in which the cells are suspended flows through the chamber in a direction opposite to the G-force (Fig. 1). A rotating seal allows fluid to be pumped through the center of the rotor and through the internal tubing into the separation chamber. At the end of this chamber the flow reconverges into a small tube, leaves the rotor through the rotating seal, and can be collected. The design of the separation chamber forces a cell that enters the chamber to rest, relative to the rotor, at a point where the outward-directed centrifugal or G-forces are balanced by the inward-directed fluid dynamics, due to an appropriate rotor speed and flow rate, respectively (1,3–5). Therefore, the cells depicted in Figure 1 will ultimately settle in the small equilibrium boundary. Obviously the rotor speed, which determines the G-force, and the flow rate must be chosen such that the cells are not pelleted (too high a G-force) or flushed through (too high a flow rate). If this condition is fulfilled, a suspension of cells that differ in sedimentation velocity (volume and density) redistributes itself after injection into the chamber (Fig. 2) according to the physical property of each cell. Differences in volume are more likely to be used than differences in density of cells (1,3–5). When cells have the same volume, differences in density can be exploited (6). Immediately after loading (Fig. 2, top), the cells will redistribute according to their sedimentation velocity (Fig. 2, middle). Subsequently, cells can be fractionated (stepwise) by increasing the flow rate or decreasing the G-force by reducing the rotor speed (Fig. 2, bottom). This will result in isolation of the cells with the lowest sedimentation velocity, because they will enter the converging outlet of the chamber, with increasing flow dynamics. Thus fractions will be obtained that differ in cells with increasing sedimentation velocity (which usually correlates with increases in size). A typical centrifugal elutriation set-up consists of a centrifuge equipped with an elutriator rotor, a pump to generate a flow, a sample injection device, a unit to cool the elutriation fluid, and a sample collector. A second elutriator to reduce the sample volume (see Sec. III.C) or a flow cytometer to monitor the sample flow by light scatter analysis (see Sec. III) is optional.

Inherent to centrifugal elutriation techniques are several factors that might influence the resolution capacity of the cell separation. The separation process is influenced by several parameters. Although the differences in sedimentation velocity of the cells in the sample largely determine the elutriation profile, we previously demonstrated that the highest resolution is obtained at the highest possible rotor speed (6). Therefore, elutriation at a constant high speed increases the flow rate and would give better results than elutriation at a constant flow rate but reduced rotor speed. The flow rate that can be used is limited, not only because a high flow rate results in large sample volumes

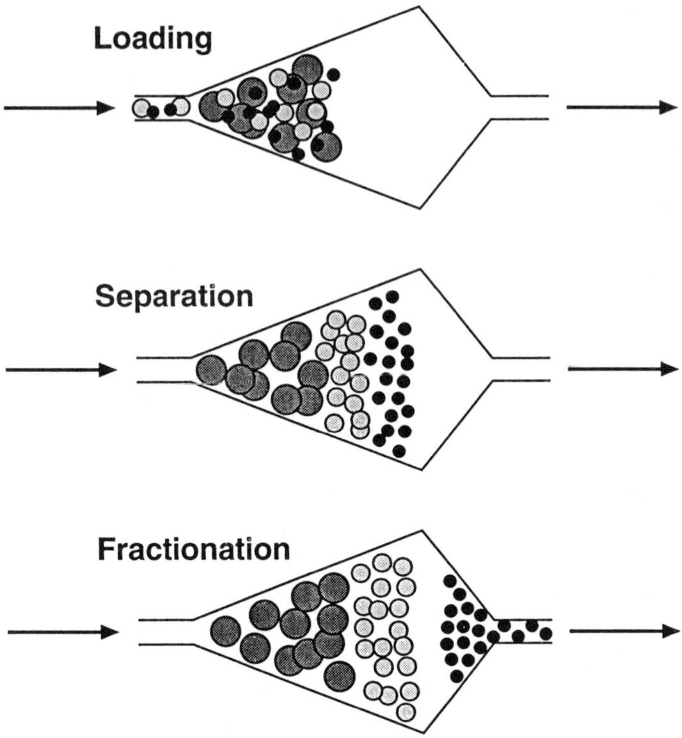

Figure 2 When a mixture of cells is injected (top), the cells will rapidly reorient according to the sedimentation velocity (middle). Cells are fractionated by a stepwise increase in the flow rate or decrease in the rotor speed.

leading to cell loss by concentration procedures, but also because a high flow rate is associated with a high jet stream in the separation chamber, which disturbs the elutriation process. Furthermore, the highest possible speed is limited, depending on the elutriation system used, to approximately 5000 rpm. For these reasons it is advantageous to elutriate at a constant flow rate and fractionate by reducing the rotor speed (7–9).

Other parameters that affect the elutriation process are the temperature of the elutriation fluid (10–13), the density of the elutriation fluid (14), the shape of the elutriation chamber (15), and the number of cells that are loaded.

III. RECENT DEVELOPMENTS

A. Flow Systems to Maintain a Constant Flow Rate

Usually a pump is used to generate a flow. A disadvantage of a pump, in particular a roller pump, is the pulse, which disturbs the elutriation process. Therefore, a small container filled with air is frequently connected to the system to suppress the pulsation (5,9). Another difficulty is the chance of leakage when the flow is obstructed somewhere in the system. Pressure sensors can be used to monitor the system throughout the elutriation process. More recently, gear pumps have been used to obtain a pulse-free flow (14). We have applied hydrostatic forces or air pressure to generate a flow (6,8,16,17). Flow thus generated is pulse-free and simply stops when clogging of the lines occurs (e.g., air bubbles, cell aggregates).

B. Elutriation Chambers

Many investigators use the standard Beckman (Fullerton, CA) separation chamber; however, chambers that differ in shape and size have also been used successfully (5,15). Sanderson et al. (5) developed a chamber with a different hydrodynamic profile, aiming at better separation of small numbers of cells (10^5–10^7). Furthermore, a disposable chamber has been made that fits in a Curamee 3000 centrifuge and which is of special interest for clinical applications (9,18,19).

Most chambers have a separation volume of approximately 3–4 mL, limiting the number of cells that can be loaded. Lutz et al. (14) have altered the design to allow a fivefold higher load of latex particles without a major increase in volume. We have recently constructed a chamber with a volume that is twice as large as the standard Beckman chamber. Particularly in the field of stem-cell transplantation, larger separation chambers able to handle larger cell numbers are needed. Some investigators developed elutriation chambers with larger volumes (up to 40 mL) (14,20,21). A potential problem with larger separation chambers is the formation of deposits of cells around the inlet of the chamber, which affect the purity of the fractions. Alternatively, four-chamber rotors have been built in which separations have been carried out by using four separation chambers in parallel (9). This significantly shortened the time required to remove lymphocytes from bone marrow (9).

C. Reduction of Sample Volume

Centrifugal elutriation results in a relatively large volume of the fractions. When a flow rate of 18 mL/min is used, the fractions are usually 150 mL (16,21). When larger separation chambers are used, the volume of the fractions increases. We previously designed an elutriation system (22) composed of two elutriator rotors in line (Fig. 3). Separation is carried out by the first rotor, and fractions are collected by the rotor spinning at a higher speed. Fractions are collected by reversing the flow in the second rotor (22). Thus, we were able to reduce the volume of fractions to 25 mL. Recently, we have coupled a standard chamber with a separation chamber twice as large, both mounted on the same rotor. Cells elutriated from the standard chamber are, because of the different hydrodynamic profile, concentrated in the second, larger chamber (unpublished results). This has the advantage that fractions are concentrated using only one rotor.

D. Closed Centrifugal Elutriation Systems

In most elutriation systems, a rotating seal is used to connect the inlet and outlet separation chamber in the rotor with the nonrotating tubing. Major problems of rotating seals in general are leakage and cross-contamination of elutriation fluid into and out of the rotor; importantly, for clinical applications, rotating seals cannot be a completely closed system, leading to the potential risk of infection. Therefore, we designed a completely closed disposable elutriation system containing inlet and outlet tubings, each connected to, in total, four elutriation chambers. This system does not use a rotating seal and, therefore, is ideally suited for clinical applications. This development has certainly increased the safe clinical application of centrifugal elutriation. This has been illustrated by more than 400 separate procedures using bone marrow and peripheral blood (after $CD34^+$ mobilization) for transplantation.

E. Flow Cytometry to Monitor and Automate Centrifugal Elutriation

As discussed in Section II, the separation profile of elutriation is dependent not only on the composition of the sample and the volume, shape, and density of the cells, but also on the cell load, the density and viscosity of the medium, and of course the rotor speed, flow rate, and temperature applied. Particularly, differences in cell load and in composition of the sample necessitate adaptations in the separation protocol each time. However, adaptations in rotor speed

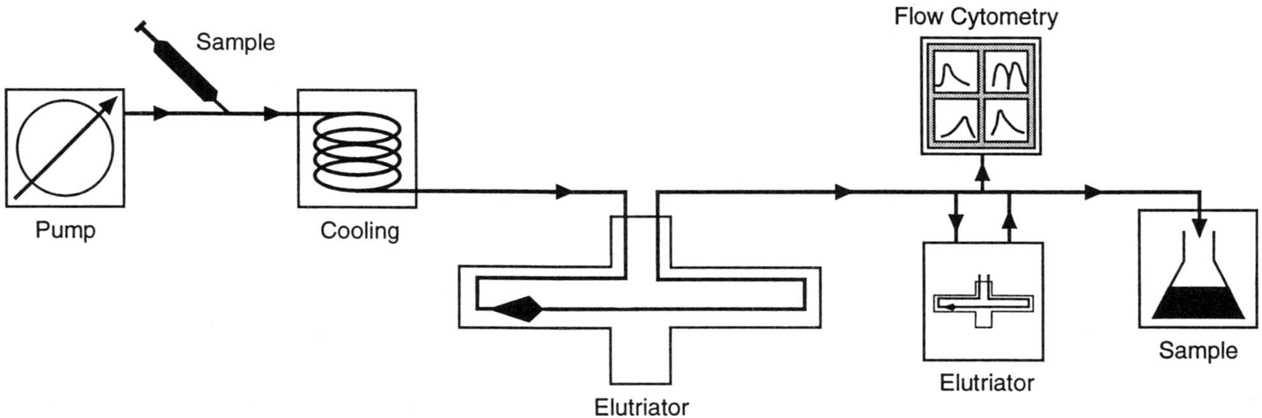

Figure 3 A centrifugal elutriation system consists of an elutriator rotor, a pump to generate a flow, a sample injection unit, a cooling unit to control the temperature of the elutriation fluid, and a sample collection unit. Optional are a flow cytometer to monitor the elutriation process and a second elutriator to concentrate the fractionated cells.

or flow rate can only be made after analysis of the previously elutriated fraction. This stimulates the use of detection systems that allow constant monitoring of elutriated fractions. De Mulder et al. (23) used a one-parameter (forward scatter), inline flow cytometer. Sloot et al. used three-parameter flow cytometry (forward, side, and back scatter) to monitor the cell-separation process (24,25). The latter also developed algorithms for the automated analysis of the fractions (26), which may result in the development of computer-assisted centrifugal elutriation (CACE).

F. Computer-Assisted Centrifugal Elutriation

The ultimate goal is a "turnkey" CACE system, not only capable of monitoring the separation process and analyzing the content of the fractions, but also capable of isolating specific cell populations based on given scatter parameters. A feedback loop from the monitor will control the rotor speed, altering the separation parameters.

IV. APPLICATIONS IN CELL BIOLOGY

A. Isolation of Hematopoietic Cells

Centrifugal elutriation is used frequently to separate large numbers of hematopoietic cells from peripheral blood or bone marrow. The technique is ideally suited to isolate cell populations from mononuclear cell suspensions obtained after density centrifugation (8,18,30–33) or by apheresis (23) (Fig. 4). Usually this results in highly purified lymphocyte (> 98%) and monocyte (> 90%) collection (9,27–29). Because centrifugal elutriation can be done at low temperatures, this technique is particularly suited for the isolation of monocytes, which are easily activated at room temperature (30–33).

The technique has been used to isolate polymorphonuclear cells for functional studies, including basophils and eosinophils, which comprise only a small fraction of the total polymorphonuclear population (34–36). Red blood cell populations have been separated to study cell aging (37).

Bone marrow of human (38–41) or murine (42,43) origin has been separated to enrich for hematopoietic progenitor cells ($CD34^+$) for functional in vitro and in vivo studies. In addition, bone marrow and peripheral blood separations were done in preclinical and clinical studies to deplete T cells (18,44), to remove contaminating malignant cells from bone marrow (45), and to enrich $CD34^+$ cells from peripheral blood [mobilized by recombinant human

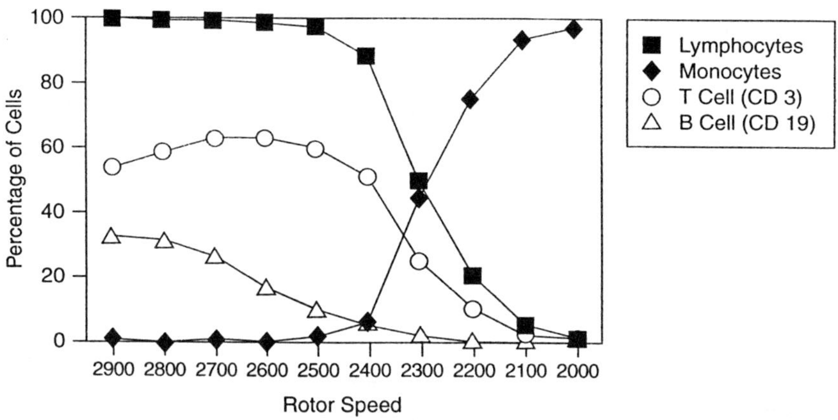

Figure 4 Counterflow centrifugal elutriation fractions of peripheral blood mononuclear cells analyzed for the percentage of lymphocytes, monocytes, T cells, and B cells. The percentage of lymphocytes and monocytes was determined by morphological analysis. The percentage of lymphocyte subpopulations of T cells ($CD3^+$ expression) and B cells ($CD19^+$ expression) was determined by flow cytometric immunofluorescence. The number of nucleated cells per fraction was comparable.

granulocyte colony-stimulating factor (rHuG-CSF) or recombinant human granulocyte macrophage colony stimulating factor (rHuGM-CSF) (46)]. Megakaryocytes and their precursors have been isolated from bone marrow or peripheral blood by centrifugal elutriation (47,48).

B. Separation of Cells from Solid Tissue

Centrifugal elutriation can isolate cells from soft solid tissues. Human and mouse hepatocytes and Kupffer cells have been isolated from liver (49–51). Mouse spleen cell suspensions were generated to isolate dendritic cells (7), and human thymus was separated to isolate thymocytes that differed in their stage of differentiation (16). The technique also has been used to isolate epithelial cells from lung (52) and keratinocytes from skin (33). Centrifugal elutriation has been used to separate spermatocytes and spermatids to study spermatogenesis (52–56). In general, single-cell suspensions are a prerequisite for optimal results, otherwise cell clumps will clog the channels and the separation chamber.

C. Separation of Cells According to the Cell Cycle

Cell-cycle synchronization is important to study cell cycle–dependent processes and their kinetics (57–60). Centrifugal elutriation has been demonstrated to be well suited to separate cells including mammalian cells, yeast (8), and bacteria (61–63). Many investigators have synchronized cancer cells by centrifugal elutriation to study the effects of cytostatic drugs (64–68).

V. CLINICAL APPLICATIONS

A. Bone Marrow Transplantation

At present, centrifugal elutriation is most commonly used for separation of bone marrow intended for transplantation. In allogeneic bone marrow transplantation at least 2×10^8 unseparated bone marrow cells/kg should be given to the patient. To transplant an adult patient, approximately 2×10^9 nucleated cells must be processed. Gradient centrifugation is often applied to reduce the volume and number of nonrelevant cells resulting in a remaining mononuclear cell population of approximately 8×10^9 cells to be separated. Elutriation has proven to be very effective to concentrate bone marrow progenitors and/or to deplete T cells from bone marrow suspensions before transplantation (18,69,70). The development of a completely closed, disposable elutriation system has improved its clinical applicability (9). This set-up, which contains four disposable elutriation chambers, allows the separation of material for one transplantion in one run. T-cell dosing is possible to manipulate graft-versus-host disease (GVHD); $CD34^+$ cells are enriched in the separation.

Besides bone marrow, peripheral blood is increasingly used as a source of $CD34^+$ cells, mobilized from the marrow by a 4-day treatment with rHuG-CSF or rHuGM-CSF. We have currently performed a pilot study on the feasibility of the use of centrifugal elutriation to enrich sufficient $CD34^+$ cells with the simultaneous reduction of T cells. Mononuclear cells obtained by apheresis were separated in two elutriation runs on two consecutive days to collect 4×10^6 $CD34^+$ cells/kg body weight (unpublished). Since apheresis material contains many red blood cells, the cell suspension should be centrifuged over density gradients to reduce the red blood cell number. Alternatively, a larger separation chamber can be used to isolate the progenitor cells. Currently, a new separation chamber is being tested that has a twofold capacity compared with the standard chamber. This would reduce the time required to separate enough material for a peripheral blood transplantation using one run.

Bone marrow can be processed by elutriation to remove potential contaminating tumor cells in the marrow compartment, such as leukemia and neuroblastoma cells (45,71,72). Preliminary studies have demonstrated that $CD34^+$ cells isolated by centrifugal elutriation are efficiently separated from tumor cells.

B. Immunotherapy with Mononuclear Cells

The use of leukocytes is relatively new therapy. Cancer patients have been treated with interferon-γ–activated macrophages isolated from mononuclear cell populations by centrifugal elutriation (73). We have developed a clinical protocol in which patients with melanoma are treated with dendritic cells derived from monocytes isolated by centrifugal elutriation and subsequently loaded with melanoma-associated antigens (Ags). Highly purified (> 90% to 95%) monocytes (49,100) are grown in the presence of rHuGM-CSF and recombinant human interleukin-1 (rHuIL-4) for 5–7 days (30,74). During this time, they mature into dendritic cells particularly good at presenting Ags to T cells (30). These cells are loaded with antigens and administered subcutaneously. We previously demonstrated that the dendritic cells generated as described above have excellent Ag-presenting capacity (74).

IV. FUTURE PERSPECTIVES

Centrifugal elutriation can be used for a wide range of applications. The power of the elutriation technique is its ability to process large numbers of cells without affecting their function. Because it is possible to elutriate a fixed number of T cells from stem cell grafts, capturing progenitor cells, centrifugal elutriation is a technique of choice in stem cell transplantation.

Elutriation has been applied by a limited number of research groups. Wider use of this technique is still unlikely due to the labor involved, as well as the skill and devotion required of investigators. If we succeed in automating a large part of the separation process as suggested in Section III.E, its user-friendliness will increase enormously and may allow more investigators to use it. The development of a completely closed elutriation system is a first major step in that direction, particularly with respect to the clinical applications. We expect that with the enhanced immunotherapeutical use of leukocytes to treat malignancies, the application of centrifugal elutriation to isolate the various leukocyte populations will also increase.

Thus far, centrifugal elutriation mainly has been applied for preparative cell separations. Only in a few cases has it been used as an analytical instrument, mainly because of the lack of suitable separation chambers specifically designed to separate small numbers of cells. It will be a challenge to design such separation chambers.

REFERENCES

1. Lindahl PE. Principles of a counter streaming centrifuge for thc separation of particles of different size. *Nature* 1948; 161:648–649.
2. McEwen CR, Stallard RW, Juhos ET. Separation of biological particles by centrifugal elutriation. *Anal Biochem* 1968; 23(3):369–377.
3. Lindahl PE. On counterstreaming centrifugation in the separation of cells and cell fragments. *Biochim Biophys Acta* 1956; 21:411–415.
4. Pretlow II TG, Pretlow TP. Centrifugal elutriation (counterstreaming and centrifugation) of cells. *Cell Biophys* 1979; 1(2):195–210.
5. Sanderson RJ, Bird KE, Palmer NF, Brenman J. Design principles for a counterflow centrifugation cell separation chamber. *Anal Biochem* 1976; 71(2):615–622.
6. Figdor CG, Leemans JM, Bont WS, de Vries JE. Theory and practice of centrifugal elutriation (CE). Factors influencing the separation of human blood cells. *Cell Biophys* 1983; 5(2):105–118.
7. Ossevoort MA, Tocs RE, De Bruijn ML, Melief CJ, Figdor CG, Kast WM. A rapid isolation procedure for dendritic cells from mouse spleen by centrifugal elutriation. *J Immunol Methods* 1992; 155(1):101–111.
8. Figdor CG, Bont WS, De Vries JE Rapid isolation of mononuclear cells from buffy coats prepared by a new blood cell separator. *J Immunol Methods* 1982; 55(2):221–229.
9. Plas A, de Witte T, Wessels H, Haanen CA. New multichamber counterflow centrifugation rotor with high-separation capacity and versatile potentials. *Exp Hematol* 1988; 16(5):355–359.
10. Abrahamsen TG, Carter CS, Read EJ, et al. Stimulatory effect of counterflow centrifugal elutriation in large-scale separation of peripheral blood monocytes can be reversed by storing the cells at 37 degrees. *J Clin Apheresis* 1991; 6(1):8–53.
11. Grant WD, Morrison M. Resolution of cells by centrifugation elutriation. *Anal Biochem* 1979; 98(1):112–115.
12. Meistrich ML, Hunter NM. Temperature regulation during centrifugation elutriation and its effect on cell separation. *Cell Biophys* 1981; 3(2):127–140.
13. Keng PC, Li CK, Wheeler KT. Characterisation of the separation properties of the Beckman elutriater system. *Cell Biophys* 1981; 3(1):41–56.
14. Lutz MP, Gaedicke G, Hartman W. Large-scale cell separation by centrifugal elutriation. *Anal Biochem* 1992; 200(2):376–380.
15. Meistrich ML. Experimental factors involved in separation by centrifugation

elutriation. In: Pretlow II TG, Pretlow TP, eds. *Cell Separation: Methods and Selected Applications.* London: Academic Press, 1983:33–61.

16. Figdor CG, Vyth-Dreese FA, Bont WS, Spits H, de Vries JE. Isolation of human thymocytes differing in maturation state and function by centrifugal elutriation. *Thymus* 1982; 4(4):243–256.
17. Figdor CG, Bont WS, Touw I, de Roos J, Roosnek EE, de Vries JE. Isolation of functionally different human monocytes by counterflow centrifugation elutriation. *Blood* 1982; 60(1):46–53.
18. Preijers F, Ruijs P, Schattenberg A, de Witte T. T cell depletion from allogenceic bone marrow by counterflow centrifugation is not associated with a substantial loss of $CD34^+$ cells. *Prog Clin Biol Res* 1994; 389:339–344.
19. Bar BM, de Witte T, Schattenberg A, Boezeman J, Hoogenhout J. Favourable outcome of patients older than 40 years of age after transplantation with marrow grafts depleted of lymphocytes by counterflow centrifugation. *Br J Haematol* 1990; 74(1):53–60.
20. Faradji A, Bohbot A, Schmitt Goguel M, et a1. Large scale isolation of human blood monocytes by continuous flow centrifugation leukapheresis and counterflow centrifugation elutriation for adoptive cellular immunotherapy in cancer patients. *J Immunol Methods* 1994; 174(1–2):297–309.
21. Figdor CG, Bont WS, De Vries JE, Van Es Wl. Isolation of large numbers of highly purified lymphocytes and monocytes with a modified centrifugal elutriation technique. *J Immunol Methods* 1981; 40(3):275–288.
22. Figdor CG, Van Es WL, Leemans JM, Bont WS. A centrifugal elutriation system of separating small numbers of cells. *J Immunol Methods* 1984; 68(1):73–87.
23. deMulder PHM, Wessels JM, Rosengrand DA, et al. Monocyte purification with counterflow centrifugation monitored by continuous flow cytometry. *J Immunol Methods* 1981; 47(1):31.
24. Sloot PM, Van der Donk, EH Figdor CG. Computer-assisted centrituga1 elutriation. II. Multiparametric statistical analysis. *Comput Methods Programs Biomed* 1988; 27(1):37–46.
25. Sloot PM, Carels MJ, Tensen P, Figdor CG. Computer assisted centrifugal elutriation. I. Detection system and data acquisition equipment. *Comput Methods Programs Biomed* 1987; 24(3):179–188.
26. Sloot PMA. *Elastic Light Scattering from Leukocytes.* Amsterdam: 1988. (Thesis.)
27. de Vries JE, Vyth Dreese FA, Figdor CG, Spits H, Leemans JM, Bont WS. Induction of phenotypic differentiation, interleukin-2 production, and PHA responsiveness of "immature" human thymocytes by interleukin-1 and phorbol ester. *J Immunol* 1983; 131(1):201–206.
28. te Velde AA, Huijbens R, de Vries JE, Figdor CG. IL-4 decreases Fc gamma R membrane expression and Fc gamma R mediated cytotoxic activity of human monocytes. *J Immunol* 1990; 144(8):3046–3051.
29. van den Bosch G, Preijers F, Vreudenhil A, Hendriks J, Mass F, De Wite J.

Granulocyte macrophage colony-stimulaitng factor (GM-CSF) counteracts the inhibiting effect of monocytes on natural killer (NK) cells. *Clin Exp Immunol* 1995; 101(3):515–520.

30. te Velde AA, Klomp JP, Yard BA, de Vries JE, Figdor CG. Modulation of phenotypic and functional properties of human peripheral blood monocytes by IL-4. *J Immunol* 1988; 140(5):1548–1554.
31. Keizer GD, Te Velde AA, Schwarting R, Figdor CG, De Vries JE. Role of p150,95 in adhesion, migration, chemotaxis and phagocytosis of human monocytes. *Eur J Immunol* 1987; 17(9):1317–1322.
32. te Velde AA, Keizer GD, Figdor CG. Differential function of LFA-I family molecules (CD11 and CD18) in adhesion of human monocytes to melanoma and endothelial cells. *Immunology* 1987; 61(3):261–267.
33. de Waal Malefyt R, Abrams J. Bennet B, Figdor CG, de Vries JE. Interleukin 10 (IL-10) inhibits cytokine synthesis by human monocytes: an autoregulatory role of IL-10 produced by monocytes. *J Exp Med* 1991; 174(5):1209–1220.
34. Whyte MK, Meagher LC, MacDermot J, Haslett C. Impairment of function in aging neutrophils is associated with apoptosis. *J Immunol* 1993; 150(11):5124–5134.
35. Beeken W, Northwood I, Beliveau C, Gump D. Phagocytes in cell suspensions of human colon mucosa. *Gut* 1987; 28(8):976–980.
36. de Boer M, Roos D. Metabolic comparison between basophils and other leukocytes from human blood. *J Immunol* 1986; 136(9):3447–3454.
37. Bosch FH, Werre JM, Schipper L, et al. Determinants of red blood cell deformability in relation to cell age. *Eur J Haematol* 1994; 52(1):35–41.
38. Keller JR, Oritz M, Ruscetti FW. Steel factor (c-kit ligand) promotes the survival of hematopoietic stem/progenitor cells in the absence of cell division. *Blood* 1995; 86(5):1757–1764.
39. Wagner JE, Collins D, Fuller S, et al. Isolation of small, primitive human hematopoietic stem cells: distribution of cell surface cytokine receptors and growth in SCID-Hu mice. *Blood* 1995; 86(2):512–523.
40. Grimm J, Zeller W, Zader AR. Separation and characterization of mobilized and unmobilized peripheral blood progenitor cells by counterflow centrifugal elutriation. *Exp Hematol* 1995; 23(6):535–544.
41. de Witte T, Plas A, Koekman E, et al. Separation of human bone marrow by counterflow centrifugation monitored by DNA-flowcytometry. *Br J Haematol* 1984; 58(2):249–258.
42. Orlic D, Anderson S, Biesecker LG, Sorrentino BP, Bodine DM. Pluripotent hematopoietic stem cells contain high levels of mRNA for c-kit, GATA-2, p45 NF-E2 and c-myb and low levels or no mRNA for c-fms and the receptors for granulocyte colony-stimulating factor and interleukins 5 and 7. *Proc Natl Acad Sci USA* 1995; 92(10):4601–4605.
43. Orlic D, Anderson S, Bodine DM. Biological properties of subpopulatious of

pluripotent hematopoietic stem cells enriched by elutriation and flow cytometry. *Blood Cells* 1994; 20(1):107–117.

44. Brauer RB, Baldwin WM, Wang D, et al. Hepatic and extrahepatic biosynthesis of complement factor C6 in the rat. *J Immunol* 1994; 153:3168–3176.
45. Slaper-Cortenbach IC, de Vries-van Rossen A, Huijbens RJ, van Leeuwen EF, Figdor CG. Neuroblastoma purging by immunorosette depletion. *Prog Clin Bio Res* 1992; 377:147–162.
46. de Witte T, Gratwohl A, Van Der Lely N, et al. Recombinant human granulocyte-macrophage colony-stimulating factor accelerates neutrophil and monocyte recovery after allogeneic T cell depleted bone marrow transplantation. *Blood* 1992; 79(5):1359–1365.
47. Nichol JL, Hornkohl AC, Choi ES, et al. Enrichment and characterization of peripheral blood-derived megakaryocyte progenitors that mature in short-term liquid culture. *Stem Cells* 1994; 12(5):494–505.
48. van Pampus EC, Huijgens PC, Zevenbergen A, Twaalfbhoven H, van Kamp GJ, Langenhuijsen MM. Influence of aspirin on human megakaryoctye prostaglandin synthesis. *Eur J Haematol* 1993; 50(5):264–268.
49. Amouzadeh HR, Pohl LR. Processing of endoplasmlc reticulum luminal antigens associated with halothane hepatitis in rat hepatocytes. *Hepatology* 1995; 22(3):936–943.
50. Heuff G, van de Loosdrecht AA, Betjies MG, Beelen RH, Meijer S. Isolation and purification of large quantities of fresh human Kupffer cells, which are cytotoxic against colon carcinoma. *Hepatology* 1995; 21(3):740–745.
51. Janousek J, Strmen E, Gervais F. Purification of murine Kupffer cells by centrifugal elutriation. *J Immunol Methods* 1993; 164(1):109–117.
52. Belinsky SA, Lechner JF, Johnson NF. An improved method for the isolation of type II and Clara cells from mice. *In Vitro Cell Dev Biol Anim* 1995; 31(5):361–366.
53. Boix J, Roca J. Separation of rooster spermatogenic nuclei by means of centrifugal elutriation. *Cytometry* 1993; 14(4):465–467.
54. Meistrich ML, Trostle Weige PK, Van Beek ME. Separation of specific stages of spermatids from vitamin A-synchronized rat testes for assessment of nucleoprotein changes during spermiogenesis. *Biol Reprod* 1994; 51(1):334–344.
55. Fujisawa M, Bardin CW, Morris PL. A germ cell factor(s) modulates preproenkephalin gene expression in rat Sertoli cells. *Mol Cell Endocrinol* 1992; 84(1):79–88.
56. Pineau C, Le Magueresse B, Courtens JL, Jegou B. Study in vitro of the phagocytic function of Sertoli cells in the rat. *Cell Tissue Res* 1991; 264(3):589–598.
57. Brown KD, Coulson RM, Yen TJ, Cleveland DW. Cyclin-like accumulation and loss of the putative kinetochore motor CENP-E results from coupling continuous synthesis with specific degradation at the end of mitosis. *J Cell Biol* 1994; 125(6):1303–1312.

58. Hengstschlager M, Knofler M, Mullner EW, Ogris E, Wintersberger E, Wawra E. Different regulation of thymidine kinase during the cell cycle of normal versus DNA tumor virus-transformed cells. *J Biol Chem* 1994; 269(19):13836–13842.
59. van de Loosdrecht A, Ossenkoppele GJ, Beelen RH, Broekhoven MG, Langenhuijsen MM. Cell cycle specific effects of tumor necrosis factor alpha in monocyte mediated leukemic cell death and thc role of beta 2-integrins. *Cancer Res* 1993; 53(18):4399–4407.
60. Chiang Y, Schneiderman M H, Vishwanatha JK. Annexin II expression is regulated during mammalian cell cycle. *Cancer Res* 1993; 53(24):6017–6021.
61. De Stefano JA, Foy JM, Sullivan DW, et al. Fractionation of *Pneumocystis carinii* developmental stages by counterflow centrifugal elutriation and sequential filtrations. *Parasitol Res* 1994; 80(1):1–9.
62. Woldringh CL, Huls PG, Vischcr NO. Volume growth of daughter and parent cells during the cell cycle of *Saccharomyces cerevisiae* a/alpha as determined by image cytometry. *J Bacteriol* 1993; 175(10):3174–3181.
63. Figdor CG, Olijhoek AJM, Klencke S, Nanninga N, Bont WS. Isolation of small cells from an exponential growing culture of *Escherichia coli* by centrifugal elutriation. *FEMS Microbiol Lett* 1981; 10:349–352.
64. Lukas J, Bartkova J, Welcker M, et al. Cyclin D2 is a moderately oscillating nucleoprotein required for G1 phase progression in specific cell types. *Oncogene* 1995; 10(11):2125–2134.
65. Davis MA, Tang HY, Maybaum J, Lawrence TS. Dependence of fluorodeoxyuridine-mediated radiosensitization on S phase progression. *Int J Radiat Biol* 1995; 67(5):509–517.
66. Van Der Lely N, Minderman H, Wessels H, et al. Detection of incorporated iododeoxyuridine in colonies by immunoperoxidase staining: a novel method to measure the proportion of cycling colony-forming cells. *Exp Hematol* 1995; 23(3):236–243.
67. Zeng XR, Hao H, Jiang Y, Lee MY. Regulation of human DNA polymerase delta during the cell cycle. *J Biol Chem* 1994; 269(39):24027–24033.
68. Minderman H, Linssen P, Wessels J, Haanen C. Cell cycle related uptake, retention and toxicity of idarubicin, daunorubicin and duxorubicin. *Anticancer Res* 1993; 13(4):1161–1165.
69. de Witte T, Hoogenhout J, de Pauw B, et al. Depletion of donor lymphocytes by counterflow centrifugation successfully prevents acute graft-versus-host disease in matched allogeneic marrow transplantation. *Blood* 1986; 67(5):1302–1308.
70. Schattenberg A, de Witte T, Preijers F, et al. Allogeneic bone marrow transplantation for leukemia with marrow grafts depleted of lymphocytes by counterflow centrifugation. *Blood* 1990; 75(6):1356–1363.
71. Gazitt Y, Reading CC, Hoffman R, et al. Purified CD34+ Lin-Thy+ stem cells do not contain clonal myeloma cells. *Blood* 1995; 86(1);381–389.
72. Slaper-Cortenbach IC, Wingaarden-du Bois MJ, Admiraal LG, Tetteroo, PA.

Figdor CG, van Leeuwen EF. The use of immunorosettes for purging of bone marrow in childhood cancer. *Prog Clin Biol Res* 1990; 333:337–344.

73. Andreesen R, Scheibenbogen C, Brugger W, et al. Adoptive transfer of tumor cytoxic macrophages generated in vitro from circulating blood monocytes: a new approach to cancer immunotherapy. *Cancer Res* 1990; 50(25):7450–7456.
74. Bakker A, Marland G, de Boer A, et al. Generation of antimelanoma cytotoxic T lymphocytes from healthy donors after presentation of melanoma-associated antigen derived epitopes by dendritic cells in vitro. *Cancer Res* 1995; 55(22):5330–5334.

4

Isolation, Activation, Expansion, and Gene Transduction of Cell-Based Therapeutics Using Polystyrene Immunoaffinity Devices

Jane S. Lebkowski, Dewey J. Moody, Ramila Philip, Lisa Schain, Sohel Talib, Rukmini Pennathur-Das, and Thomas B. Okarma
RPR Gencell, Santa Clara, California

David A. Okrongly
Xytronyx, Inc., San Diego, California

I. INTRODUCTION

Many different immunological approaches have been pursued in both preclinical and clinical investigations to enhance immune-mediated clearance of tumor or infected cells (1). In several studies, native antibodies (Abs) or conjugated monoclonal antibodies (mAbs) targeting particular tumor or cell types have been investigated with limited success (2–8). In addition, cytokines, including α-interferon (α-IFN) (9–12), γ-interferon (γ-IFN) (13), and interleukin (IL)-2 (14–18) have been administered in numerous animal models and clinical trials. Because of the pleiotropic activities of many cytokines and growth factors, systemic administration of these agents has been associated with substantial dose-limiting toxicities and minimal efficacy. Most growth factors and other immune modulators are naturally designed to exert their activity in the local microenvironment of the target cell. Therefore, to control the distribution and hence activity of these immune modulators, we chose to purify specific immune cell types and to expose them ex vivo to single or multiple factors. We postulated that subsequent infusion of the activated autologous immune cells could potentiate a therapeutic response without the numerous toxicities associated with direct growth-factor administration.

To purify specific cell types in sufficient quantity for clinical application, we developed methods to covalently immobilize proteins onto polystyrene surfaces. Further, we developed procedures to terminally sterilize the protein-coated polystyrene surfaces to provide a sterile platform for cell manipulation. Taking advantage of the multiple molded and extruded forms of polystyrene, we have produced devices known as RPR CELLectors® that can be used to manipulate up to 3×10^{10} peripheral blood mononuclear cells (PBMC). Using covalently attached mAbs or lectins, these devices can be used to specifically deplete or positively select cells bearing the target antigen (Ag). Cells captured by these devices can be cultured, activated, and genetically modified while still immobilized on the modified surface. This chapter describes several specific cell manipulations that can be performed on these surfaces. Moreover, new embodiments of this technology that now permit the automation of these cell-processing schemes are outlined. Finally, the use of these cell separation devices in several clinical trials in the areas of bone marrow transplantation, solid tumor therapies, and acquired immunodeficiency syndrome (AIDS) treatments are discussed.

II. DESCRIPTION OF THE RPR IMMUNOAFFINITY CELL SELECTION DEVICES

The RPR CELLector is a sterile cell separation device containing covalently immobilized proteins such as antibodies or lectins for manipulation of cells in numerous research and clinical applications. Derivatization of the polystyrene surfaces, covalent attachment of the proteins, and terminal sterilization of the devices is performed as detailed in Lebkowski et al., 1988 (19). Four polystyrene CELLector™devices have been developed to accommodate different size cell loads and different cell processing needs. The three smallest devices, the 25, 150, and 624 cm^2 RPR CELLectors, are prepared using standard tissue-culture plasticware (Corning, NY or Nunc, Naperville, IL) that is derivatized for covalent protein attachment instead of being surface charged for cell adherence. The largest RPR CELLector incorporates eleven parallel 12 cm × 12 cm plates, (providing 3000 cm_2 surface area), which are coated on each side with the desired antibodies. The newest device, the MacroCELLector, was engineered to meet clinicians' flexible needs and replace the limited abilities of the CELLector devices. The MacroCELLector was engineered to enable practical and widespread cell manipulation for cell and gene therapy applications and utilizes uniquely designed technology. The MacroCELLector includes extruded 80 μm polystyrene fibers woven into very high surface area pads yielding 100 cm^2 of polystyrene/ cm^3 of volume. These pads are derivatized, coated, and housed in custom designed acrylic chambers for easy manipulation. Current MacroCELLectors designs incorporate 65,000 cm^2 of fiber in a two-chamber device and will deplete T-cells from up to 3×10^{10} peripheral blood mononuclear cells or mobilized peripheral blood progenitor cell preparation in less than 3 hours. Specific device configurations and use protocols have been developed based on the particular requirements and applications for the processed cells. Positive selection and culture of cells have been performed with the 25 cm^2, 150 cm^2, and 624 cm^2 RPR CELLectors. Systems are now being adapted to optimize culture in the MacroCELLector fiber pads.

III. METHODS FOR THE ISOLATION OF SPECIFIC CELL TYPES

A. Human CD34$^+$ Hematopoietic Progenitor Cells

Human CD34$^+$ hematopoietic progenitor cells are isolated from bone marrow using a modified procedure of Lebkowski et al. (19) To enrich for the CD34$^+$ cells, bone marrow mononuclear cells are depleted of cells binding to the lectin soybean agglutinin (SBA) using RPR CELLector SBA devices (20). Approximately 2×10^7, 1.2×10^8, and 4×10^9 total white blood cells are loaded into the 25, 150, and 3000 cm^2 RPR CELLectors SBA, respectively, for optimum depletion of SBA-binding cells. After a 30- to 60-minute room temperature incubation, the nonadherent, SBA$^-$ cell fraction is removed and used for the further purification of CD34$^+$ cells. The adherent SBA$^+$ cells can also be released by incubation of the devices at 37°C for 0.5–1 hour with Dulbecco's Modified Eagle Medium (DMEM) containing 200 mM N-acetyl galactosamine. For the isolation of CD34$^+$ cells, the SBA$^-$ cells are loaded onto a second polystyrene RPR CELLector containing the CD34 mAb ICH3. Approximately 2×10^7 or 4×10^9 SBA$^-$ cells are loaded onto a single 25 or 624 cm^2 RPR CELLector CD34, respectively. After a 1-hour incubation, the nonadherent, CD34$^-$ cells are removed. The adherent, CD34$^+$ cells are either used directly on the cell separation devices for culture or gene transduction or can be released for further use during a 2-hour incubation at 37°C in serum-free medium (Stempro-34, Life Technologies, Gaithersburg, MD) without any exogenous growth factors. All manipulations of purified CD34$^+$ cells are performed using plasticware coated with 0.5% human serum albumin (HSA) or human AB serum to prevent nonspecific cell loss. For culture on the device,

10-mL or 100-mL media, respectively, containing any desired combination of cytokines and growth factors can be used.

$CD34^+$ cells are isolated from PBPC leukapheresis products using a modification of the above procedure. The PBPC products are centrifuged at 800 *g* for 10 minutes to deplete platelets from the cell preparations. The cell pellet is suspended in Dulbecco's phosphate-buffered saline containing 1 mM EDTA (DPBSE) and 5% human AB serum or autologous plasma at a concentration of 4–5 × 10^9 cells/100 mL. Four to 5 × 10^9 cells are loaded into a 624 cm^2 RPR CELLector CD34 and incubated for 1 hour with a slow (19 rpm) circular rocking motion. After incubation, the nonadherent $CD34^-$ cells are removed and the device is washed four to five times using 50 mL DPBSE. The $CD34^+$ cells are released as described earlier. If further purification of the peripheral blood $CD34^+$ cells is desired, the $CD34^+$ cell fraction is loaded into a RPR CELLector SBA at 2 × 10^7 total cells per 25 cm^2 device and incubated for 30–60 minutes at room temperature. The nonadherent $CD34^+SBA^-$ fraction is collected and used as the final purified $CD34^+$ cell population.

B. Human $SBA^-CD34^+CD38^-$ Primitive Hematopoietic Stem Cells

For the isolation of $SBA^-CD34^+CD38^-$ cells, SBA^-CD34^+ cells are isolated as above and stained with 1 μg/10^6 cells of the unconjugated CD38 monoclonal antibody (clone HB-7, Becton Dickinson, Milpitas, CA). After two washes to remove free CD38 antibody, the stained cells are depleted of $CD38^+$ cells using serial capture on two 25 cm^2 RPR CELLector Goat antimouse Ig devices containing a goat polyclonal antibody to mouse immunoglobulin. The final, nonadherent, $CD38^-$ cell–depleted fraction is collected as the $SBA^-CD34^+CD38^-$ population. The $CD38^+$-adherent populations can be released by the 2-hour incubation method as detailed in the above section.

C. Human $CD8^+$, $CD4^+$, $CD5/8^+$, $CD3^+$, T-Cell Receptor and Vβ-Specific T Cells

T-cell populations are isolated from peripheral blood, bone marrow, ascites, or tumor-infiltrating lymphocyte (TIL) preparations using polystyrene devices containing either the CD8, CD4, CD5 + CD8, or CD3 mAbs. Except in cases where leukapheresis products are used, mononuclear cells are isolated from the cell suspensions before cell separation. For leukapheresis products, only a low-speed centrifugation step to remove platelets is required before cell separation. Depending on the particular application, different cell loads are used for the cell separation devices. If maximal depletion of the target populations is desired, white blood cell loads of 2×10^7, 1.2×10^8, 5×10^8, and 2×10^9 are used for the 25, 150, 624, and 3000 cm^2 MacroCELLector devices, respectively. If even further depletions are required, the nonadherent populations are depleted over a second serial cell separation device. For certain cell culture applications where maximal saturation of surface-cell binding is desired, white blood cell loads of 7×10^7, 6×10^8, 2×10^9 are used for the 25, 150, 150, and 624 cm^2 devices, respectively. These cell loads for maximal saturation are used when the target population is 20–30% of the starting cell population. When the target populations is < 20% or > 30%, these loads should be adjusted based on the percentage of Ab-binding cells in the initial input population. After a 1-hour room temperature incubation, the nonadherent T-cell–depleted fraction is collected and the adherent T-cell population is manipulated as appropriate. To remove the adherent populations, the T cells are cultured 24–72 hours in AIMV medium (Life Technologies) in the presence of 100–1200 IU/mL recombinant human IL-2 (rHuIL-2) with or without phytohemagglutinin (0.1 μg/mL). The specifics of the T-cell culture protocols are modified depending on the particular Ag used for capture and the intended use of the purified cells. For instance, $CD8^+$ TIL cells can be recovered and expanded on the device by culture for 48–72 hours in the presence of 1200 IU/mL rHuIL-2. Because of the activation signal imparted by the immobilized CD3 Ab, purified $CD3^+$ cells are activated, released, and expanded in the device with AIMV medium containing 100 IU/mL IL-2. In some cases, binding of the target Ag with the immobilized Ab causes the target Ag to cap and modulate. As a result, detection of the target Ag on the purified cells is difficult immediately after cell release, but is restored after 2–3 days of culture outside the cell-separation device.

IV. CELL-SEPARATION RESULTS USING THE RPR CELLector

Human peripheral blood or bone marrow mononuclear cells from normal healthy volunteers were loaded into polystyrene cell separation devices containing a variety of covalently immobilized mAbs. Table 1 shows that these devices specifically and quantitatively bind cells bearing Ags recognized by the mAb. In more than 400 experiments, the average depletion of the targeted T-cell populations ranged from 80 to 89%. Depletion of $CD34^+$ cells was somewhat lower at 50% due to the low CD34 Ag density on these cells. With all of the devices, target cell depletion was reproducible as evidenced by the low standard deviations for depletion. This depletion was also adjustable by varying the cell load and by adding sequential devices. After adherent cell release, the average purity of the captured cells was 97, 92, 95, 97, and 89% for the CD3, CD4, CD5/8, CD8, and CD34 devices, respectively (Table 1). The purity of the adherent population was measured using fluorescein- or phycoerythrin (PE)-conjugated Abs specific for the targeted Ag. With most of the T-cell Ags, binding by the immobilized Ab stimulated modulation and capping, and, as a result, detection of the target Ag on the surface of the newly released cells was difficult. Upon culture of the cells for 48–72 hours in standard culture devices, the target Ag reappeared and was easily measured by many standard immunodetection assays (Table 1).

Table 1 Performance of Cell-Separation Devices

Antibody on cell separation device	Number of evaluations	Absolute depletion of cell bearing target antigen (%)	Purity of captured cells (%)
CD3	177	80.4 (2.2)	97 (3)
CD4	167	87.5 (7.3)	92 (5)
CD5/8	114	CD5 = 88.8 (1.4)	95 (5)
		CD8 = 85.7 (5.0)	
CD8	71	85.0 (4.3)	97 (1)
CD34	78	50.7 (11.3)	89 (7)

In these experiments, peripheral blood or bone marrow mononuclear cells were isolated from normal healthy volunteers and loaded into 25, 150, or 624 cm^2 CELLector devices containing covalently immobilized antibodies targeting the CD3, CD4, CD5/8, CD8, or CD34 antigens. The absolute depletion of the targeted cells from the starting population is calculated as:

$$100\% \times \frac{N_S P_S - N_N P_N}{N_S P_S}$$

where: N_S = # cells in the starting cell fraction; N_N = # cells in the nonadherent depletion fraction; P_S = purity (%) of the target cell population in the starting fraction; P_N =purity (%) of the target cell population in the nonadherent depleted fraction.

The purity of the captured fraction is also noted in the table and was measured 1 ($CD34^+$) or 7 (CD3, CD4, CD5/8, and CD8) days after culture. Please note that human SBA^- bone marrow was used to evaluate the CD34 cell separation devices. The number of evaluations of the purity of the bound cells from the CD3, CD4, and CD5/8 devices is 6, 10, and 3, respectively. The purity of cells bound to the CD5/8 devices was defined as those cells that were $CD5^+$.

V. CELLS RELEASED FROM THE IMMUNOAFFINITY DEVICE: NO RESIDUAL MOUSE IMMUNOGLOBULIN

Cells captured and released from the immunoaffinity devices do not have detectable mouse immunoglobulin on their surface. To determine whether residual mouse immunoglobulin present on the devices remained on the purified cells, isolated CD8$^+$ and CD34$^+$ cells were incubated with fluorochrome-conjugated Abs that recognize either mouse or rabbit immunoglobulins. As measured by flow cytometry, fluorescence using either polyclonal or monoclonal antibodies to mouse immunoglobulin always remained within the background staining of the antirabbit immunoglobulin Abs at 0–7% dull positive cells ($n = 248$, data not shown).

VI. SEQUENTIAL CELL SEPARATIONS: ISOLATION OF SBA$^-$CD34$^+$CD38$^-$ HEMATOPOIETIC PROGENITOR CELLS

Since the purified cells do not retain mouse immunoglobulin, they can be directly subfractionated, even using two-step Ab-staining procedures. Such purification is demonstrated in Figure 1 with the isolation of primitive SBA$^-$CD34$^+$CD38$^-$ hematopoietic progenitor cells. Bone marrow CD34$^+$ cells were isolated using RPR CELLector SBA and CD34 devices. After recovery of the CD34$^+$ cells from the cell-separation device, the population was incubated with an anti-CD38 mAb and separated into SBA$^-$CD34$^+$38$^+$ and SBA$^-$CD34$^+$38$^-$ populations using RPR CELLectors containing an immobilized goat polyclonal Ab–recognizing mouse immunoglobulin. Figure 1 shows two-color flow cytometric analysis of the starting bone marrow, SBA$^-$, CD34$^+$, and the final SBA$^-$CD34$^+$38$^+$ and SBA$^-$CD34$^+$38$^-$ fractions. After isolation, the starting CD34$^+$ cell population was 85.9% CD34$^+$ and 89.5% CD38$^+$, and after separation into the CD38$^+$ and CD38$^-$ subfractions, the final SBA$^-$CD34$^+$38$^+$ and SBA$^-$CD34$^+$38$^-$ fractions were 95.4 and 15.7% CD38$^+$, respectively. The final purity of the SBA$^-$CD34$^+$38$^-$ fraction was 61.0%, compared with its starting purity of 0.04% in the original bone marrow mononuclear cell fraction.

VII. CULTURE AND ACTIVATION OF PURIFIED CELLS ON THE IMMUNOAFFINITY DEVICE

Cells purified using the RPR CELLector can be directly cultivated in those devices. Figure 2 shows four growth curves of CD8$^+$ T cells isolated from bulk

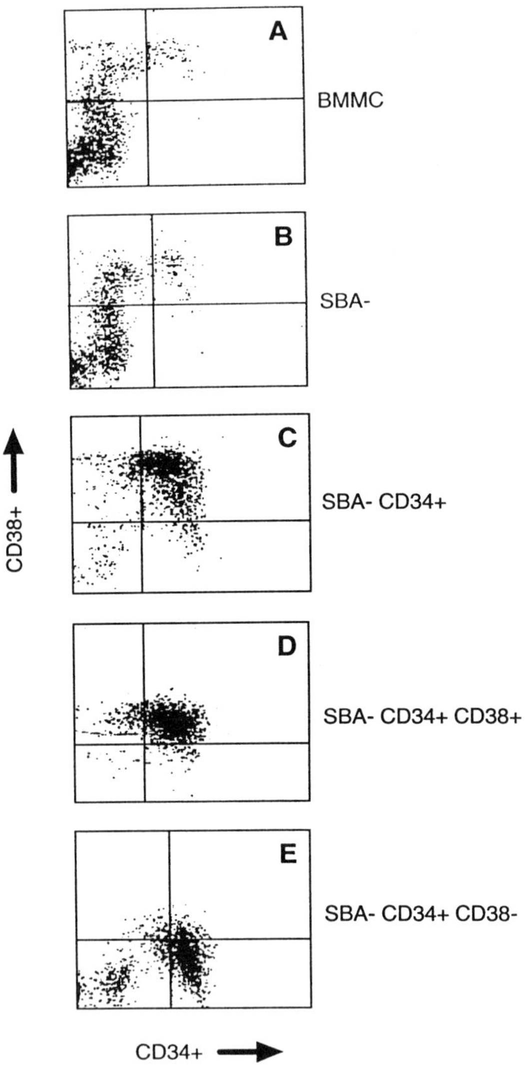

Figure 1 Isolation of SBA$^-$CD34$^+$CD38$^-$ primitive stem cells. Human bone marrow CD34$^+$ cells were subfractionated into CD38$^+$ and CD38$^-$ populations. The figure shows fluorescence and histograms of the input bone marrow mononuclear cells (BMMC), soybean agglutinin depleted (SBA$^-$), SBA$^-$CD34$^+$, SBA$^-$CD34$^+$CD38$^+$, and SBA$^-$CD34$^+$CD38$^-$ populations as analyzed for the presence of the CD34 (x-axis) and CD38 (y-axis) antigens. Please note that fluorescence of the purified CD38$^+$ cells was somewhat dimmer than that seen in the starting CD34$^+$ fraction, since the same CD38 monoclonal antibody was used for the isolation and detection of the CD38$^+$ cells.

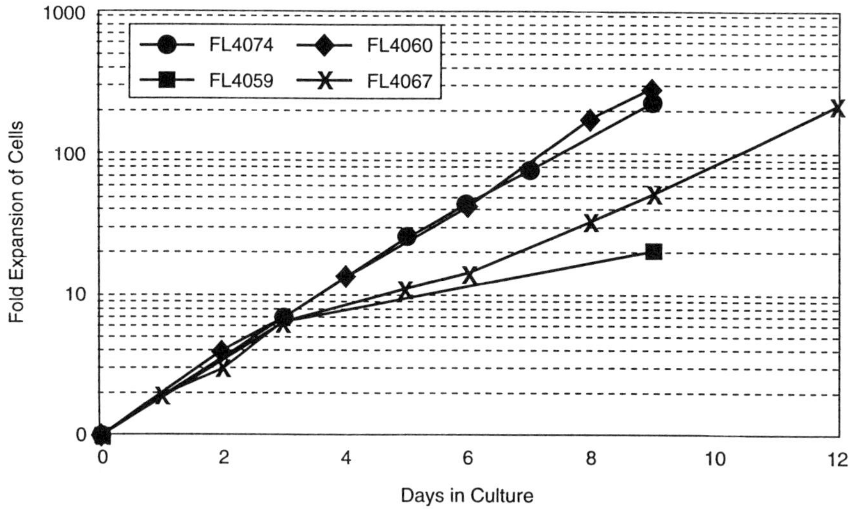

Figure 2 Culture of $CD8^+$ TIL cells using the CELLector CD8. Bulk TIL cells cultured were established from renal cell carcinoma samples using a modification of the methods of Steger et al. (21). Serum-free AIMV media was used for all culture steps. $CD8^+$ TIL cells were isolated and cultured as described in the text. The cell number was determined at various time points during the culture period. The $CD8^+$ TIL cells were retained in the CELLector CD8 during the first 3 days of culture. After that point, the cells were transferred to standard culture devices for further expansion.

cultures of TIL from primary renal cell carcinoma tumors. In these experiments, $CD8^+$ T cells were captured from 14- to-21-day cultures of bulk TIL cells and subsequently incubated in serum-free AIMV medium containing 1200 IU/mL rHuIL-2. After 3 days of culture, the $CD8^+$ cells were removed from the cell separation device and cultured in gas-permeable bags. Cell counts were made periodically throughout the culture. As seen in Figure 2, cell numbers in four experiments increased 5- to 8-fold during the 3 days of culture in the immunoaffinity device. Subsequent culture of the $CD8^+$ cells in gas-permeable bags produced a 20- to 168-fold expansion of cells by 8 days. The purity of the $CD8^+$ cells at the end of these cultures was 96–99%.

Similar results were observed when purified $CD34^+$ bone marrow cells were cultivated in the CD34 cell-capture devices. In two representative experiments (Figure 3), $CD34^+$ cells were isolated on the RPR CELLectors CD34. After isolation, serum-free medium containing rHuIL-3, rHuIL-1, and rHu stem cell factor (SCF) was added to the cells that were subsequently

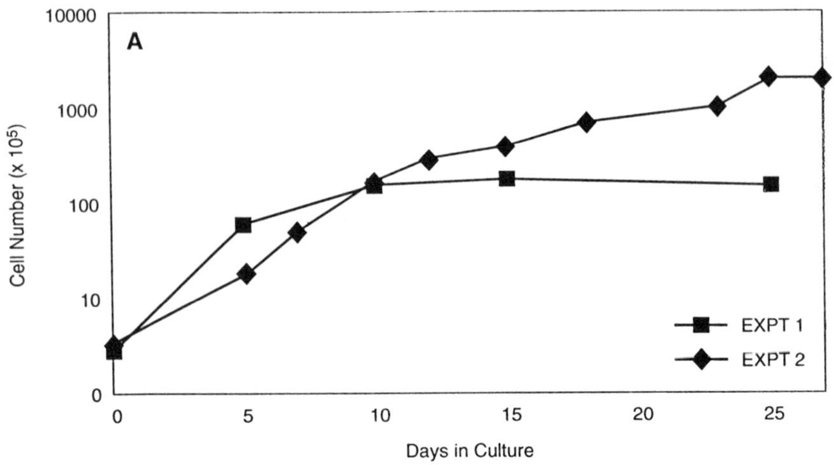

Figure 3 Culture of human bone marrow CD34$^+$ cells in the CELLector CD34. CD34$^+$ cells isolated from human bone marrow were cultured directly in the CELLector CD34 for 25 days in Stempro-34 media (Life Technologies) containing 10 ng/mL IL-3, 10 ng/mL IL-1, and 100 ng/mL SCF. Fresh medium was added to the cultures twice weekly. By 24 hours of incubation, the vast majority of the purified CD34$^+$ cells detached from the cell separation device. At the indicated time points, an aliquot of the cell culture was retrieved and used to determine cell counts, cell phenotype, and progenitor function. (A) Cell numbers; (B) CFU-GM numbers; (C) BFU-E numbers. Note that experiment 1 is plotted on the Y1 axis; experiment 2 is plotted on the Y2 axis.

cultivated for 25 days. Cell samples were collected periodically. As shown in the growth curves in Figure 3A, cell numbers increased 100- to 300-fold by 14 days of culture. Under these growth conditions, the purified CD34$^+$ cells proliferate and differentiate, producing committed progenitors and, at later time points, maturing myeloid cells. At the start of the culture, > 65% of the cells were CD34$^+$. After 7 days of culture, only 20% of the cells were CD34$^+$ despite a 10- to 20-fold increase in cell number and a 2- to 3-fold increase in CD34$^+$ cell number (data not shown). By 12–14 days of culture, the percentages of CD34$^+$ cells decreased to below the limits of sensitivity of the flow cytometer, making quantitation of CD34$^+$ cell numbers very difficult. Increases in progenitor cell number paralleled those of CD34$^+$ cell number. Both GM-CFC (Fig. 3B) and erythroid blast-forming cell (E-BFC) (Fig. 3C) number peaked at approximately day 7 to day 10 of CD34$^+$ cell culture and averaged 6.2- and 36.6-fold in 18 separate experiments. At later time points, GM-CFC and E-BFC numbers decreased as maturing myelocytes, promyelocytes, and

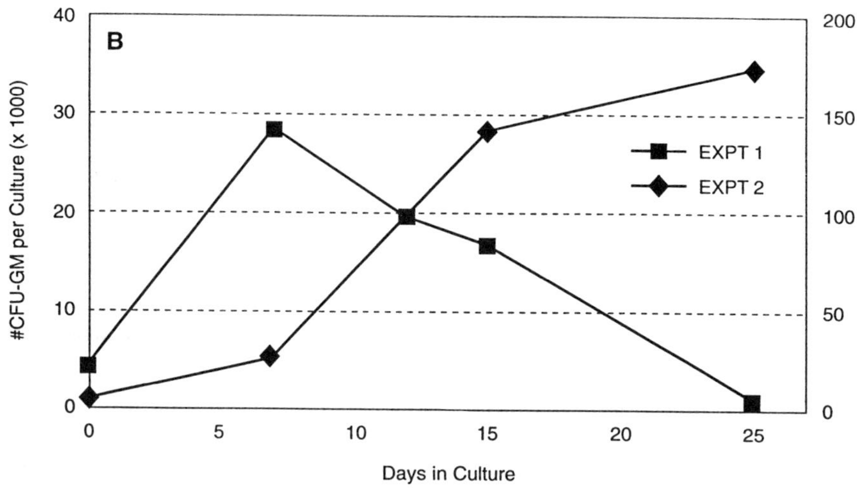

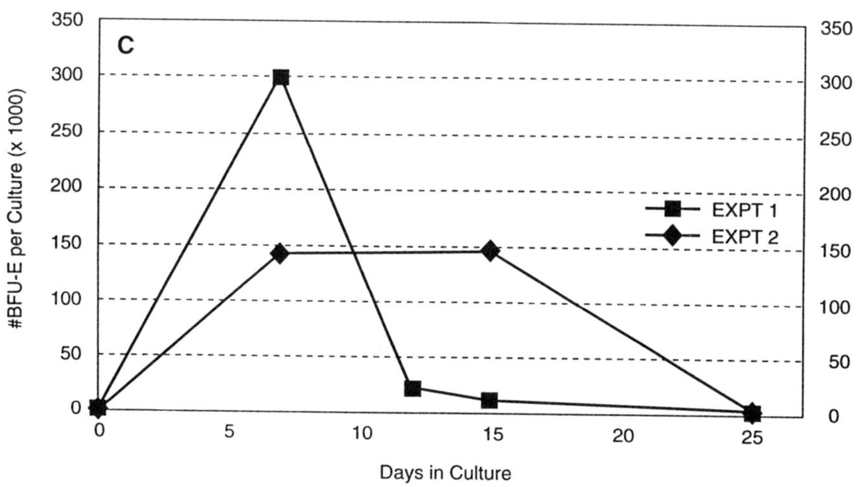

neutrophils began to dominate the cultures. The magnitude of expansion of the cells and progenitors was highly dependent on the growth-factor combinations used for expansion (data not shown).

The immobilized Ab devices can also trigger signal transduction in cells. For instance, polystyrene devices coated with an anti-CD3 mAb were used to simultaneously purify and activate $CD3^+$ T cells. $CD3^+$ T cells captured from normal peripheral blood mononuclear cells expanded 10- to 30-fold over a 7-day period when cultured on a CD3 CELLector together with 100 IU/mL rHuIL-2. Similarly, T-cell subsets bearing specific T-cell receptor (TCR) β-chain variable region (Vβ) rearrangements were also purified, expanded, and activated using the polystyrene cell-separation devices containing mAbs targeting particular TCR β-chain variable region segments. Specifically, $CD8^+$/Vβ3 and $CD8^+$/Vβ13 T cells were directly captured from the peripheral blood of patients with ovarian cancer. The immobilized TCR Vβ3 and 13 mAbs activated the captured T cells and stimulated cell proliferation in the presence of low-dose rHuIL-2. After 7–10 days of expansion, the cultured TCR Vβ-specific T cells were phenotyped for a variety of activation and adhesion molecules. Table 2 summarizes such data from one patient. In the starting peripheral blood mononuclear cell sample, TCR $V\beta3^+$ and $V\beta13^+$ cells comprised 6.2 and 7.8% of the population, respectively. After isolation and expansion, the purity of the TCR $V\beta3^+$ and $V\beta13^+$ populations were 97 and 99%, respectively. These proliferating TCR Vβ-specific T cells expressed several activation molecules such as CD25, HLA-DR, CD69, and CD45RO, and the adhesion molecules CD54 (ICAM-1) and CD58 (LFA-3). Some differences in the activation marker phenotype of TCR Vβ3 and 13^+ T cells were observed. CD45 RA^+ cells predominated the TCR $V\beta13^+$ population (99%), whereas this marker was present on only 15% of the $V\beta3^+$ population. In contrast, the CD45RO marker was present on all TCR $V\beta3^+$ T cells and on only half of the TCR Vβ13 cells. Both $V\beta3^+$ and $V\beta13^+$ T cells cultivated from the peripheral blood of HLA-$A2^+$ ovarian cancer patients showed restricted cytotoxicity against HLA-$A2^+$ ovarian tumor cell lines.

VIII. CELL SEPARATION AND COCULTURE ON THE RPR CELLector: ACTIVATION OF ANTIGEN-SPECIFIC T CELLS

Antigen-specific activation of defined T-cell subsets can be triggered on the cell separation devices. As a specific example, $CD8^+$ T cells were isolated from the peripheral blood of human immunodefiency virus (HIV)–seropositive individuals. The $CD8^-$ population was collected and infected for 1 hour with the recombinant canarypox vector, vCP300, which expresses the *gag/pol*, *env*, and *nef* genes of the MN strain of HIV. After infection, the $CD8^-$ Ag-presenting cells were added at a ratio of 1 to 10 to the adherent $CD8^+$ cells. Recombinant HuIL-7 also was added, and the cocultures were incubated at 37°C. After $CD8^+$ cell release and expansion, the cultures were transferred to conventional

Table 2 Phenotypic Analysis of VB3+ and VB13+ T Cells (% of Cells with Designated Antigen)

T-Cell antigens	VB3+ cells	VB13+ cells
CD3	100	100
CD4	0	1
CD8	100	80
Activation markers		
CD25	48	55
CD122	41	13
HLA-DR	100	100
CD69	58	84
CD71	39	59
CD28	40	88
CD45RA	15	99
CD45RO	99	51
Adhesion molecules		
CD54	82	33
CD58	100	100
Other markers		
CD16	0	4
CD80	100	76
CD56	16	1

In this experiment, TCR Vβ3+ and TCR Vβ13+ CD8+ cells were isolated from the peripheral blood of ovarian cancer patients using the following procedure. CD8+ cells were captured and released from a CD8 RPR CELLector and subsequently loaded onto cell separation devices containing monoclonal antibodies targeting either the TCR Vβ3 or Vβ13 segment of the TCR β chain. The cells were subsequently cultured in this device for 7 days in the presence of 100 IU/mL IL-2. The cells were phenotyped with a variety of monoclonal antibodies targeting several T cells, activation molecules, or adhesion molecules.

culture devices and restimulated weekly with cryopreserved aliquots of the infected CD8− Ag-presenting cells.

After 2–6 weeks of stimulation and culture, the cells were collected and tested for cytolytic activity against chromium-labeled autologous CD4+ cells that were infected with recombinant vaccinia virus constructs expressing specific HIV genes. The cytotoxicity results from one such coculture are shown in Figure 4. When CD8+ cells were cultured with phytohemagglutinin and rHuIL-2 in the absence of the Ag-presenting cells, little cytotoxicity was observed against autologous CD4+ cells whether or not they expressed HIV genes. Likewise, when the purified CD8+ cells were cocultured with non-

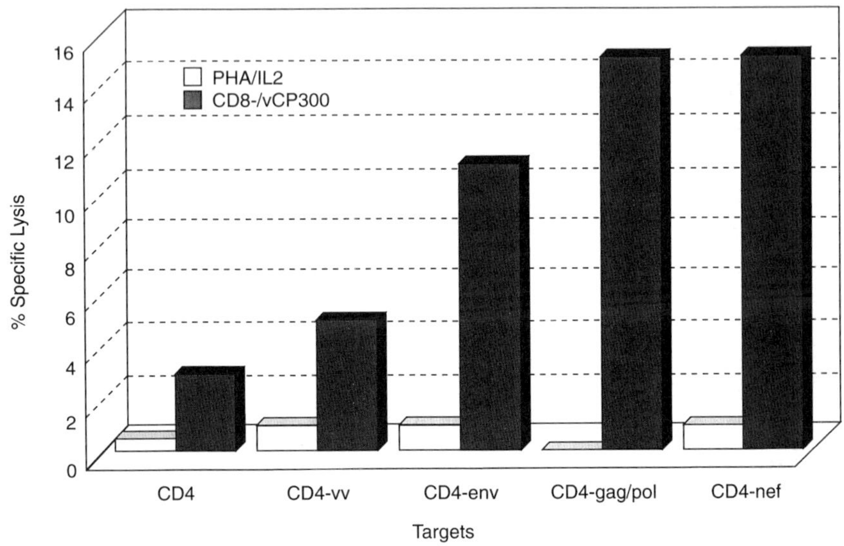

Figure 4 Culture of HIV specific $CD8^+$ cells. $CD8^+$ and $CD8^-$ cells were separated from a peripheral bone marrow mononuclear cell preparation from a HIV-seropositive individual using an RPR CELLector CD8. The $CD8^-$ cells were collected, infected with vCP300 at an MOI of 10. These $CD8^-$ stimulators were added to the $CD8^+$ responders at a responder to stimulator ratio of 1:10. The cultures were maintained in AIMV media containing 20 ng/mL IL-7 and were stimulated weekly for 2 weeks with the cryopreserved infected $CD4^+$ cells. As a control culture, the $CD8^+$ T cells were cultured with 0.1 μg/mL phytohemagglutinin and 600 IU IL-2/mL for 14 days. Periodic media changes were made to each culture. At the end of the culture, the final cell product was tested in a chromium release assay against purified autologous $CD4^+$ cells which had been infected with various recombinant vaccinia constructs containing no gene inserts (CD4-vv), the HIV *env* gene (CD4-*env*), the HIV-*gag/pol* gene (CD4-*gag/pol*), or the HIV *nef* gene (CD4-*nef*). Uninfected CD4 cells are represented by the CD4 bars. The % specific lysis of those indicated targets by the PHA + IL-2 cultured (light bars) and $CD8^-$ cell cocultured $CD8^+$ cells (dark bars) is indicated. The effector-to-target ratio in this assay was 40:1.

infected $CD8^-$ cells, little cytotoxicity towards the HIV gene expressing $CD4^+$ cells was observed (data not shown). However, when the $CD8^+$ cells were cocultured with vCP300-infected $CD8^-$ cells, the final cell product showed enhanced specific cytotoxicity towards the autologous $CD4^+$ cell targets that expressed the HIV gene products. Analogous results were observed when a CD3 RPR CELLector and a Mage1-expressing canarypox vector were used to culture and expand $CD3^+$ Mage1-specific T cells from a breast cancer TIL preparation (22). After Ag-presenting cell stimulation, the final $CD3^+$ cell product showed enhanced killing of Mage1-expressing autologous B-cell lines and restricted oligoclonal T-cell receptor expression.

IX. GENE TRANSFER INTO CAPTURED CELLS

The RPR CELLector cell separation technology also can be used to facilitate genetic modification of target cells and to isolate those cells successfully transduced or transfected. Purified $CD8^+$ cells were genetically modified directly after isolation using DDAB:DOPE liposomes together with the adeno-associated virus (AAV)–based plasmids pMP6cat (23,24). pMP6cat carries the chloramphenicol acetyltransferase gene driven by the CMV immediate early gene promoter. The gene cassette also carries a hybrid intron comprised of splice donors and acceptors from the adenovirus major late promotor leader sequences and the Ig heavy chain intron. The expression cassette is inserted between two copies of the natural inverted terminal repeats of AAV. When $CD8^+$ cells were lipofected directly after capture in the device without previous growth-factor stimulation, CAT activity was readily detectable in the T cells 72 hours after transfection (Fig. 5, lanes 1 and 2). Stimulation of the purified $CD8^+$ cells for 3–11 days before transfection using phytohemagglutinin and 100 IU/mL rHuIL-2 failed to increase the amount of transgene CAT expression observed (Fig. 5, lanes 3–8).

Similar results were observed when purified bone marrow $CD34^+$ cells were transduced with rAAV vectors carrying the CAT gene. As above, $CD34^+$ cells from human bone marrow were isolated and directly transduced with the rAAV virus, A1CMVIXcat (23), without previous cytokine stimulation of the cells. The expression cassette of A1CMVIXcat also used the CMV immediate early gene promotor to drive CAT gene transcription. After overnight exposure to the recombinant virus, the $CD34^+$ cells were detached from the device, washed, and resuspended in DMEM containing IL-3, IL-6, and SCF. Three days after infection, the transduced cells were assayed for CAT enzyme activity (Fig. 6). Those cells exposed to the A1CMVIXcat rAAV stocks

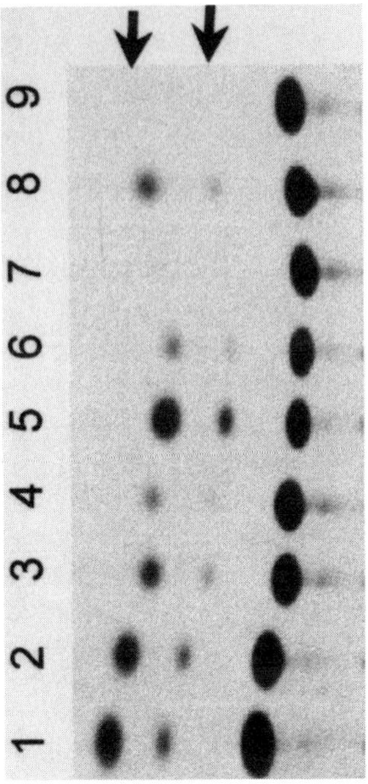

Figure 5 Lipofection of purified $CD8^+$ cells. $CD8^+$ T cells from two separate donors were purified from peripheral blood mononuclear cells and lipofected. In this experiment, the $CD8^+$ cells were either not cultured prior to lipofection (lanes 1,2) or were prestimulated with phytohemagglutinin and 100 IU/mL IL-2 for 3 (lanes 3,4), 8 (lanes 5,6), or 11 (lanes 7,8) days before lipofection. Lane 9, no transfection negative control. For transfection, 1 mL of serum-free RPMI-1640 medium was added to the adherent cells in the cell-separation device. The DNA-liposome complex was prepared by adding 30 nmol of DDAB as DDAB:DOPE 1:1 liposomes to 30 μg of the plasmid pMP6cat (24). After incubation at room temperature for 10–20 minutes, 1 mL of serum-free RPMI-1640 was added to the DNA-liposome complex. The complex was then added to the adherent $CD8^+$ cells. After incubation at room temperature for 10–20 minutes, 2 mL of RPMI-1640 containing 10% fetal calf serum and 100 IU/mL IL-2 was added to the cells. The cells were cultured in a humidified, 37°C 5% CO_2 incubator. Three days after transfection, the cells were harvested and assayed for CAT enzyme activity using standard thin-layer chromatography assays. The arrows indicate the positions of the two monoacetylated forms of chloramphenicol, the products of the CAT enzyme.

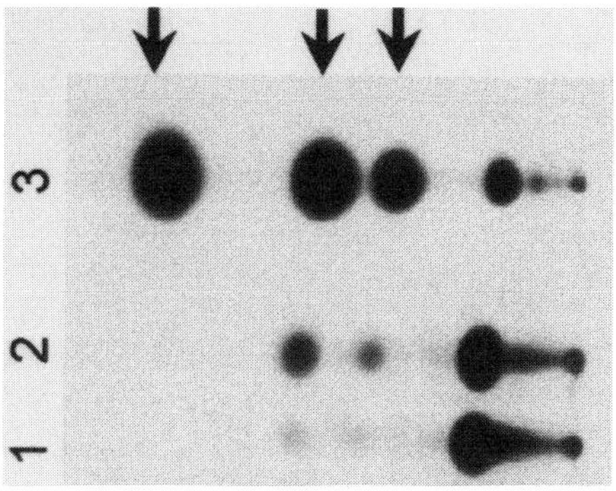

Figure 6 Recombinant adeno-associated virus vector (rAAV) infection of purified $CD34^+$ cells. Human bone marrow $CD34^+$ were isolated but were not released from the 25 cm^2 RPR CELLector CD34. Two mL of DMEM containing 10% fetal calf serum was added to the bound $CD34^+$ cells. An additional 2 mL of the A1CMVIXcat rAAV virus stock prepared in DMEM containing 10% fetal calf serum was added to the device. The recombinant AAV A1CMVIXcat was prepared with the plasmids pA1CMVIXcat (23) and pΔBal (25) using 293 cells and the procedures detailed in Lebkowski et al. (25). After overnight incubation at 37°C, the detached $CD34^+$ cells and media were collected, pelleted, and resuspended in DMEM containing 10% FCS, 100 ng/mL SCF, 10 ng/mL IL-3, and 10 ng/mL IL-1 at 10^5 cells/mL. After culture at 37°C for 3 days, CAT enzyme activity was assayed using standard thin-layer chromatography procedures. The three arrows indicate the position of the monoacetylated and diacetylated chloramphenicol CAT enzyme products. (Lane 1) Mock infected $CD34^+$ cells; (lane 2) A1CMVIXcat infected $CD34^+$ cells; (lane 3) positive control CAT enzyme preparation.

showed detectable CAT activity (Fig. 6, lane 2), whereas those $CD34^+$ cells infected with "mock" unpackaged A1CMVIXcat plasmid (Fig. 6, lane 1) failed to show transgene activity. These collective results indicate that cells can be genetically modified while still immobilized on the Ab surface, providing a convenient platform to quickly manipulate cells.

X. ISOLATION OF GENE-MODIFIED CELLS

Successfully transfected or transduced cells can be purified using this cell-separation technology. As a specific example, purified $CD4^+$ cells were infected with the recombinant retrovirus MFG-NGFR that carries a deleted version of the gene for the cell-surface protein, nerve growth factor receptor (NGFR) (26–28). This nonfunctional truncated version encodes the extracellular and transmembrane domains of the receptor and, when expressed, can be readily detected on the surface of cells using Abs to NGFR. After MFG-NGFR infection of $CD4^+$ cells, approximately 22% of the cells expressed the NGFR transgene (Fig. 7). After cell separation using a polystyrene device containing an immobilized NGFR mAb, the adherent and nonadherent cell fractions were examined for expression of NGFR. Only 5% of the nonadherent $NGFR^-$ depleted fraction even weakly expressed NGFR, whereas the adherent $NGFR^+$ fraction was highly enriched for successfully transfected cells, being 78% positive for the cell surface molecule. Similar results were observed when purified $CD34^+$ cells were infected with the MFG-NGFR retrovirus vector (data not shown).

XI. CELL SEPARATION ON POLYSTYRENE FIBERS

To facilitate the above described manipulations of purified cells from a wide range of sources, the MacroCELLector technology was devised. The core of the MacroCELLector is a woven pad of 80-μm fibers that yields 100 cm^2 of surface area for each cm^3 of volume. The first MacroCELLector devices have been specifically designed to deplete $CD8^+$ and $CD5/8^+$ cells from peripheral blood or bone marrow for allogeneic cell infusion. In experiments where 3×10^{10} PBPCs were manipulated using a CD5/8 MacroCELLector, $94 \pm 5\%$ of $CD5^+$ and $96 \pm 3\%$ of $CD8^+$ cells were depleted duplicating the performance of the smaller cell separation devices. Figure 8 shows a scanning electronmicrograph of $CD8^+$ cells bound to fibers with the immobilized CD8 mAb. As is readily apparent, the cells are much smaller than the fibers, and despite the high surface area of the pad there is sufficient area around the fibers to provide unimpeded flow and ample nutrients to the cells. We are now further developing this fiber technology to fully automate the capture, activation, modification, culture, and release of target cells, making complex cell-processing schemes reproducible and manageable for a broad array of cell therapies.

XII. CLINICAL APPLICATIONS OF THE CELL SEPARATION TECHNOLOGY

The discovery of simple and controllable methods to covalently attach proteins to polystyrene has led to the development of a series of cell separation devices

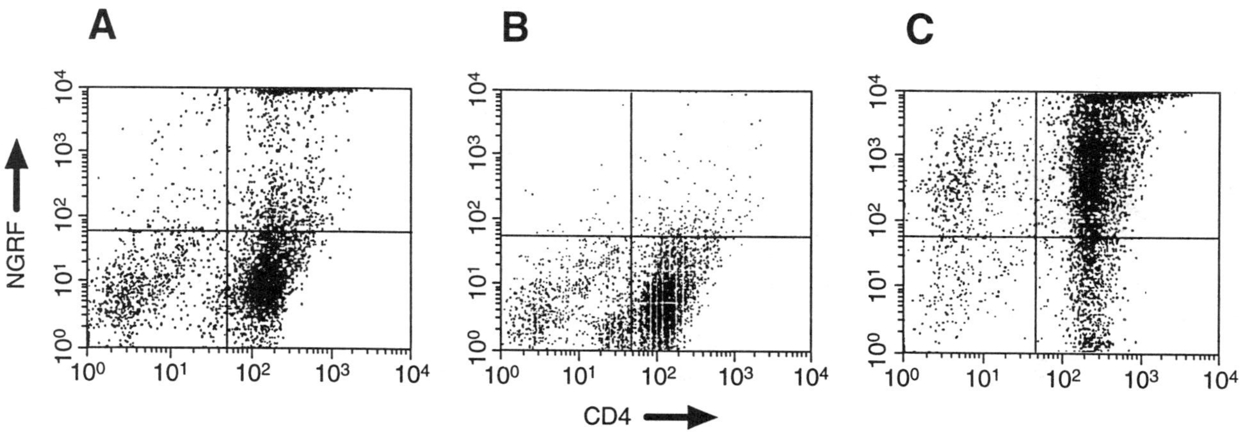

Figure 7 Separation of NGFR-transfected cells. Purified peripheral blood $CD4^+$ cells were isolated and prestimulated for recombinant retroviral infection using the methods of Rudoll et al. (29). For $CD4^+$ cell infection, the cells were prestimulated with AIMV medium containing 50 ng/mL anti-CD3, 50 ng/mL anti-CD28, and 100 U/mL IL-2 for 48 hours at 37°C. The $CD4^+$ cells were infected at a concentration of 2×10^6 cells/mL in 1 part viral supernatant and 3 parts fresh media. Protamine sulfate was added to a final concentration of 4 μg/mL. The cells were centrifuged at room temperature at 2500 rpm for 90 minutes in a Sorvall T6000 centrifuge. After centrifugation, the cells were incubated overnight at 37°C. The process was repeated two additional times. Three days after infection, the $CD4^+$ population was loaded into 25 cm^2 RPR CELLectors containing anti NGFR monoclonal antibody (ATCC HB8734). After separation of the nonadherent, $NGFR^-$, and adherent $NGFR^+$ populations, the cells were recultured in the same media and tested by flow cytometry for NGFR expression using the fluorescein-onjugated NGFR antibody. (A) MFG-NGFR infected cells prior to NGFR cell separation; (B) the nonadherent, $NGFR^-$, cells after cell separation; (C) the adherent, $NGFR^+$, population after cell separation. CD4 and NGFR fluorescence are measured on the x and y axes, respectively.

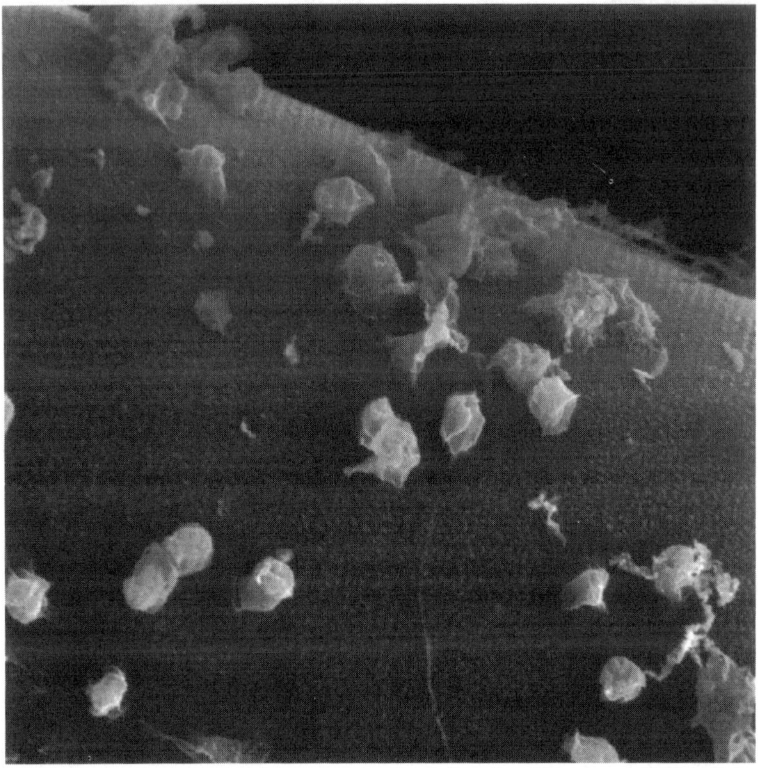

Figure 8 Scanning electron microscopy of $CD8^+$ cells bound to polystyrene fibers. $CD8^+$ cells were bound to polystyrene fibers containing an immobilized CD8 monoclonal antibody.

that can be used to process cells for a variety of clinical applications. Based on the particular Ab or ligand immobilized, these devices have been shown to both specifically and quantitatively capture cells bearing the target Ags. In addition, residual immobilized protein/Ab is not found on the surface of the harvested cells. Moreover, the protein surfaces can be used to activate, culture, and gene-modify specific cell types. The solid phase attachment of cells during such manipulations can greatly add to the ease, reproducibility, and speed of cell-processing protocols.

These cell separation devices have been used in numerous clinical protocols for several applications. Devices containing both the CD5 and CD8 mAbs have been used for the depletion of T cells in allogeneic bone marrow transplantation using matched-related, mismatched-related, and matched-unrelated donors (30,31). In these trials, approximately 1.0–2.0 logs of T cells were depleted from bone marrow allografts, depending on the exact protocol chosen. Furthermore, depletion of cells was found to lower the incidence of graft-versus-host disease (GVHD), as compared with historical controls, without negatively impacting engraftment (30,31). We are now in the follow-up phase of a multicenter, randomized, blinded study directly comparing CD5/8$^+$ T-cell depletion to no depletion for the incidence of GVHD and engraftment. In this trial, chronic myelogenous leukemia (CML), acute lymphocytic leukemia (ALL), and acute myelogenous leukemia (AML) patients with allografts from matched-unrelated and mismatched-related donors are being studied. As an extension of these studies, the cell separation devices are also being used to deplete CD8$^+$ cells from donor lymphocyte infusions to reduce the incidence of GVHD associated with this cell therapy for relapsing CML (32).

For other adoptive immunotherapy strategies, the CD8$^+$ cell separation devices have been used to positively select CD8$^+$ cells from the peripheral blood of AIDS patients. After ex vivo polyclonal T-cell expansion using pytohemmaglutinin antigen (PHA) and IL-2, the purified CD8$^+$ cells were administered in five monthly infusions along with 2×10^6 IU/m^2/day rHuIL-2, for the treatment of opportunistic infections, Kaposi's sarcoma, and HIV viral burden (33–35). Phase 1 and 2 trials showed the reproducibility of CD8$^+$ cell production, the safety of the administration of even 1×10^{10} expanded CD8$^+$ cells, and efficacy of such infusions for the treatment of opportunistic infections (33–35).

In another adoptive immunotherapy application, CD8$^+$ TIL cells were isolated from patients with renal cell carcinoma for the treatment of metastatic disease (21). In initial studies, 1×10^8 to 3×10^{10} CD8$^+$ TIL were infused along with multiple 5-day cycles of rHuIL-2. The partial and complete response rate was 46% (10/22). Results are now pending from a multicenter randomized study in metastatic renal cell carcinoma where CD8$^+$ TIL with rHuIL-2 infusions were directly compared with the administration of rHuIL-2 alone.

These cell separation devices also have been used in a series of gene-therapy protocols. Hoogerbrugge et al. (36) used the cell separation devices for the isolation of CD34$^+$ cells for subsequent adenosine deaminase (ADA) gene transduction as a treatment for ADA deficiency. In addition, Nabel et al. negatively depleted CD8$^+$ cells for the enrichment and subsequent gene transduction of CD4$^+$ cells using a dominant negative *rev* construct for the inhibition of HIV replication (37). Our future clinical study plans include the use of these devices for the enrichment of primary tumor cells for immediate IL-2 gene lipofection in the generation of therapeutic tumor vaccines (24).

Likewise, we are using these devices to isolate dendritic cells for subsequent manipulation and use as active immunotherapies. Finally, these devices also are serving as the platform to isolate and expand Ag-specific and targeted T-cell populations for the treatment of infectious diseases and cancer.

The flexibility of this covalent attachment methodology allows the immobilization of any protein or peptide to the polystyrene surface. One can easily envision the attachment of extracellular matrix proteins, growth factors, stromal factors, and receptors for the creation of novel substrates for the activation and growth of cells. The versatility of the polystyrene base also allows for the molding, extrusion, weaving, and fabrication of device formats that can be tailored to specific applications and tolerances. It is this technology that will serve as the fundamental platform for the automation of cell-processing schemes.

ACKNOWLEDGMENTS

We would like to give special thanks to Elisa Brunette, Bill Cox, Lynley Donovan, Charisse Eary-Reeves, Larry Fama, Melinda Hall, Mark Harvey, Sonia Jain, Don Lamons, Maureen Mc Nally, Ashok Muniappan, Gerard O'Donoghue, Farshid Oshidari, Kevin Page, Enzo Paoletti, Peter Pedrotti, Rukmini Pennathur-Das, Mohan Philip, Martin Sorich, Clay Smith, Janis Stagnaro, Marisa Tantoco, Jim Tartaglia, Beth Vertin, Kent Weinhold, Jeff Wolf, and Josephine Yau for providing data and/or key reagents for the work described in this chapter. We would also like to thank Elaine Andrews for preparing the manuscript.

REFERENCES

1. DeVita VT Jr, Hellman S, Rosenberg SA, eds. *Biologic Therapy of Cancer.* Philadelphia: L.B. Lippincott Company, 1995.
2. Schlom J. Monoclonal antibodies in cancer therapy: basic principles. In: DeVita VT Jr, Hellman S, Rosenberg SA, eds. *Biologic Therapy of Cancer.* Philadelphia: LB Lippincott Company, 1995:507–521.
3. Pai LH, Pastan I. Immunotoxin and recombinant toxin. In: DeVita VT Jr, Hellman S, Rosenberg SA, eds. *Biologic Therapy of Cancer.* Philadelphia: LB Lippincott Company, 1995:521–533.
4. Larson SM, Cheung N-K, Sgouros G. Radioisotope Conjugates. In: DeVita VT

Jr, Hellman S, Rosenberg SA, eds. *Biologic Therapy of Cancer.* Philadelphia: LB Lippincott Company, 1995:534–552.

5. Kwak LW, Urba WJ, Grossbard ML. Clinical applications of monoclonal antibodies in cancer: B-cell lymphomas. In: DeVita VT Jr, Hellman S, Rosenberg SA, eds. *Biologic Therapy of Cancer.* Philadelphia: LB Lippincott Company, 1995:553–565.
6. Waldman TA. T-cell leukemia/lymphoma. In: DeVita VT Jr, Hellman S, Rosenberg SA, eds. *Biologic Therapy of Cancer.* Philadelphia: LB Lippincott Company, 1995:566–575.
7. Houghton AN, Chapman PB. Melanoma. In: DeVita VT Jr, Hellman S, Rosenberg SA, eds. *Biologic Therapy of Cancer.* Philadelphia: LB Lippincott Company, 1995:576–590.
8. Bordin EC, Anderson T. Adenocarcinomas. In: DeVita VT Jr, Hellman S, Rosenberg SA, eds. *Biologic Therapy of Cancer.* Philadelphia: LB Lippincott Company, 1995:590–595.
9. Foon KA. Biologic therapy with interferon-α and β: clinical applications. Hairy cell leukemia, chronic myelogenous leukemia, and myeloproliferative disorders. In: DeVita VT Jr, Hellman S, Rosenberg SA, eds. *Biologic Therapy of Cancer.* Philadelphia: LB Lippincott Company, 1995:365–373.
10. Savage PD, Muss HB. Renal cell cancer. In: DeVita VT Jr, Hellman S, Rosenberg SA, eds. *Biologic Therapy of Cancer.* Philadelphia: LB Lippincott Company, 1995:373–387.
11. Kirkwood JM. Melanoma. In: DeVita VT Jr, Hellman S, Rosenberg SA, eds. *Biologic Therapy of Cancer.* Philadelphia: LB Lippincott Company, 1995:388–411.
12. John WJ, Foon KA. Clinical applications of interferon in other tumors. In: DeVita VT Jr, Hellman S, Rosenberg SA, eds. *Biologic Therapy of Cancer.* Philadelphia: LB Lippincott Company, 1995:427–433.
13. Quesada JR. Biologic therapy with interferon-γ. In: DeVita VT Jr, Hellman S, Rosenberg SA, eds. *Biologic Therapy of Cancer.* Philadelphia: LB Lippincott Company, 1995:435–442.
14. Lotze MT. Biologic therapy with interleukin-2: preclinical studies. In: DeVita VT Jr, Hellman S, Rosenberg SA, eds. *Biologic Therapy of Cancer.* Philadelphia: LB Lippincott Company, 1995:207–233.
15. Schwartzentruber DJ. Biologic therapy with interleukin-2: clinical applications. Principles of administration and management of side effects. In: DeVita VT Jr, Hellman S, Rosenberg SA, eds. *Biologic Therapy of Cancer.* Philadelphia: LB Lippincott Company, 1995:235–249.
16. Marincola FM, Rosenberg SA. Melanoma. In: DeVita VT Jr, Hellman S, Rosenberg SA, eds. *Biologic Therapy of Cancer.* Philadelphia: LB Lippincott Company, 1995:250–262.
17. Yang JC. Renal cell cancer. In: DeVita VT Jr, Hellman S, Rosenberg SA,

eds. *Biologic Therapy of Cancer.* Philadelphia: LB Lippincott Company, 1995:262–269.

18. Sznol M. Other cancers. In: DeVita VT Jr, Hellman S, Rosenberg SA, eds. *Biologic Therapy of Cancer.* Philadelphia: LB Lippincott Company, 1995:269–278.
19. Lebkowski JS, Schain LR, Okrongly D, Levinsky R, Harvey MJ, Okarma TB. Rapid isolation of human CD34 hematopoietic stem cells-purging of human tumor cells. *Transplantation* 1992; 53:1011–1019.
20. Schain LR, Okrongly D, Okarma TB, Lebkowski JS. Separation of lectin-binding cells using polystyrene culture devices with covalently immobilized soybean agglutinin. *J Hematother* 1994; 3:37–46.
21. Steger GG, Pierce WC, Figlin R, et al. Patterns of cytokine release of unselected and $CD8^+$ selected renal cell carcinoma tumor-infiltrating lymphocytes (TIL). Evidence for enhanced specific killing of tumor necrosis factor-secreting/IL-6 nonsecreting TIL in vitro and correlation with complete response in vivo. *Clin Immunol Immunopathol* 1994; 72:237–247.
22. Toso JF, Oei C, Oshidari F, et al. MAGE-1 specific precursor cytotoxic T-lymphocytes present among tumor-infiltrating lymphocytes from a patient with breast cancer: characterization and antigen-specific activation. *Cancer Res* 1996; 56:26–20.
23. Philip R, Brunette E, Kilinski L, et al. Efficient and sustained gene expression in primary T lymphocytes and primary and cultured tumor cells mediated by adeno-associated virus plasmid DNA complexed to cationic liposomes. *Mol Cell Biol* 1994; 14:2411–2418.
24. Philip R, Clary B, Brunette E, et al. Gene modification of primary tumor cells for active immunotherapy of human breast and ovarian cancer. *Clin Cancer Res* 1996; 2:59–68.
25. Lebkowski JS, McNally MM, Okarma TB, Lerch LB. Adeno-associated virus: a vector system for efficient introduction and integration of DNA into a variety of mammalian cell types. *Mol Cell Biol* 1988; 8:3988–3996.
26. Riviere I, Brose K, Mulligan RC. Effects of retroviral vector design on expression of human adenosine deaminase in murine bone marrow transplant recipients engrafted with genetically modified cells. *Proc Natl Acad Sci* 1995; 92:6733–6737.
27. Krall WJ, Skelton DC, Yu X-J, et al. Increased levels of spliced RNA account for augmented expression from the MFG retroviral vector in hematopoietic cells. *Gene Ther* 1996; 3:37–48.
28. Mavilio FG, Gerrarik S, Rossini S, et al. Peripheral blood lymphocytes as target cells of retroviral vector-mediated gene transfer. *Blood* 1994; 83:1988–1997.
29. Rudoll T, Phillips K, Lee S-W, et al. High efficiency retroviral vector mediated gene transfer into human peripheral blood $CD4^+$ T-lymphocytes. *Gene Ther* (in press).

30. Gajewski J, Przepiorka D, Giralt S, et al. Effective partial depletion of T-cells by panning using anti-CD5/8 antibody coated flasks in HLA-mismatched bone marrow transplants (BMT). *Blood* 1994; 84:2154a (abstract).
31. Carabasi MH, Collins NH, Small TN, et al. Long term follow up of recipients of HLA-matched related bone marrow transplants depleted of T cells by soybean lectin agglutination combined with AIS CELLectors. *Blood* 1995; 86:367a (abstract).
32. Giralt S, Hester J, Huh Y, et al. CD8-depleted donor lymphocyte infusion as treatment for relapsed chronic myelogenous leukemia after allogeneic bone marrow transplantation. *Blood* 1995; 86:4337–4343.
33. Ho M, Armstrong J, McMahon D, et al. A phase 1 study of adoptive transfer of autologous $CD8^+$ T lymphocytes in patients with acquired immunodeficiency syndrome (AIDS)-related complex or AIDS. *Blood* 1993; 81:2093–2101.
34. Whiteside TL, Elder EM, Moody D, et al. Generation and characterization of ex vivo propagated autologous $CD8^+$ cells used for adoptive immunotherapy of patients infected with human immunodeficiency virus. *Blood* 1993; 81:2085–2092.
35. Klimas N, Patarca R, Walling J, et al. Clinical and immunological changes in AIDS patients following adoptive therapy with activated autologous CD8 T cells and interleukin-2 infusion. *AIDS* 1994; 8:1073–1081.
36. Hoogerbrugge PM, Beusechem V, Valerio D, et al. Gene therapy in 3 children with adenosine deaminase deficiency. *Blood* 1993; 82:1245a (abstract).
37. Woffendin C, Ranga U, Yand Z, Xu L, Nabel GJ. Expression of a protective gene prolongs survival of T cells in human immunodeficiency virus-infected patients. *Proc Natl Acad Sci USA* 1996; 93:2889–2894.

5

The CEPRATE® SC System: Technology, Clinical Development, and Future Directions

Shelly Heimfeld, Karen Auditore-Hargreaves, Mark Benyuenes, Michael Emde, Cindy Jacobs, Mark Jones, Nicole Provost, Grant Risdon, and Joe Tarnowski
CellPro, Incorporated, Bothell, Washington

I. INTRODUCTION

An emerging clinical strategy is therapy based on the use of specific cell types to provide selective therapeutic benefit. The development of cellular therapy is one of the newest and most exciting concepts currently being investigated for the treatment of cancer, autoimmune diseases, and the correction of certain genetic abnormalities. Cell therapy now is used in many different clinical settings, such as the infusion of peripheral blood progenitor cells (PBPC) to enhance hematopoietic recovery after high-dose chemotherapy, to transfer genetically modified cells to correct certain inherited and acquired diseases, and for use in adoptive immunotherapy to specifically boost immune responsiveness. While there are still important scientific, clinical, and regulatory hurdles to be overcome, it seems clear that over the next few years there will be significant progress in bringing the promise of cell therapy into clinical practice.

Key to the advancement of cell-based strategies are procedures and technologies for the enrichment of specific cell populations capable of the appropriate therapeutic response. Areas where cell selection could play an important clinical role include stem cell enrichment, tumor cell purging, T-cell depletion for allogeneic transplantation, ex vivo expansion, gene therapy, and immunotherapy. Using avidin-biotin immunoadsorption and the unique continuous-flow methodology originally developed by Berenson et al. (1–4), CellPro (Bothell, WA) has refined and developed several products for the rapid purification of a variety of different cell types from bone marrow, mobilized PBPC, or cord blood. These include the CEPRATE® SC Stem Cell Concentration System (CellPro, Bothell, WA) for clinical stem cell separations and the CEPRATE® LC Cell Separation System, available for purifying stem cells and other cell populations for nonclinical research. In this chapter, we will give an overview of the unique aspects of the CEPRATE® system, the results obtained using this system in a wide range of clinical settings, and future applications of this technology for the advancement of cell therapy.

II. CEPRATE® CELL-SELECTION TECHNOLOGY

Avidin-biotin immunoadsorption and the proprietary continuous-flow technology of CellPro, Inc. form the basis for enrichment of the various target cells. This section will give a short description of the CEPRATE® SC instrumentation hardware and disposable set and some of the operational steps used to standardize and automate this positive cell selection procedure for the purification of cells from bone marrow or mobilized peripheral blood apheresis products.

A. Continuous-Flow Avidin-Biotin Immunoadsorption

To enrich for particular target cells, CellPro's technology takes advantage of the specificity of monoclonal antibodies (mAbs) combined with a unique continuous-flow, avidin-biotin–based immunoadsorption system (Fig. 1). The starting cell population is incubated for 20–25 minutes with a directly biotinylated mAb that recognizes an antigen (Ag) on the target cells of interest. Importantly, no preprocessing steps such as density gradients or adherence depletion steps are required to preenrich for the target cells. After washing to remove unbound Ab, the incubated cell mixture is perfused through a disposable column containing large polyacrylamide beads covalently linked with avidin. Because of the very high affinity between avidin and biotin (approximately 1 million times stronger than most Ag/Ab interactions), the biotin-labeled cells specifically adhere to the avidin beads, while the negative cells flow through the column and can be collected separately for other uses. After

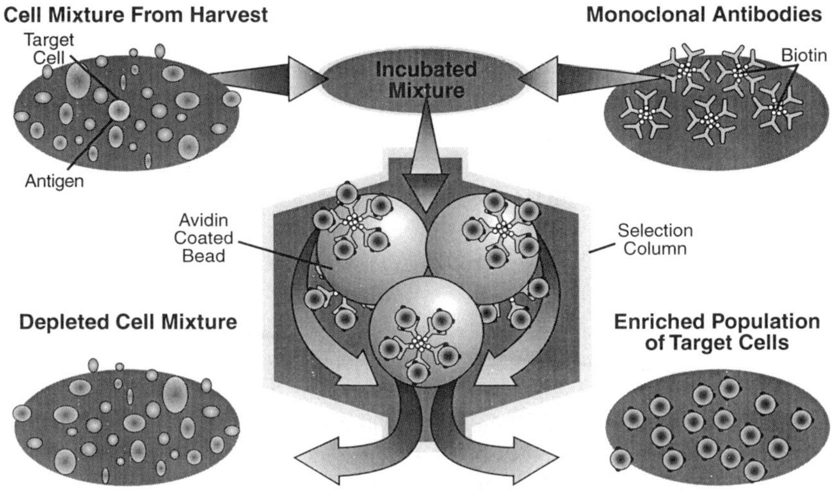

Figure 1 Avidin-biotin–based continuous flow immunoadsorption.

a brief wash to further remove residual negative cells, the biotin-labeled target cells then are eluted from the beads by very gentle agitation and flow out of the column.

Because of the much stronger biotin-avidin bond, release occurs at the Ag/Ab site, resulting in the majority of biotinylated mAbs remaining attached to the beads. Thus, unlike some other cell-selection technologies, at the end of the CEPRATE® procedure, one obtains an enriched population of target cells free of any remaining beads. These target cells can then be further processed for cryopreservation, secondary tumor purging, gene therapy, or ex vivo expansion.

Since many different Abs that recognize a variety of Ags on the cell surface can be directly or indirectly biotinylated, this basic CEPRATE® technology can be used to purify many distinct cell populations. The unique continuous flow aspects of the system also provide other advantages. First, there is minimal nonspecific binding, which leads to higher final purity of the target cells. Second, since negative cells are continuously removed from the avidin beads, there is a significant reduction in the size of the column required for clinical-scale processing. With the current CEPRATE® system, as many as 120×10^9 cells have been processed over a single column. Finally, continuous flow allows for fast throughput and very rapid enrichment of the large cell numbers generally required for therapeutic applications.

B. The CEPRATE® SC Stem Cell Concentration System

Instrument and Disposable Components

The CEPRATE® system consists of a single-use, sterile, prepackaged disposable kit and a computer-controlled automated instrument (Fig. 2). Included in the kit are one vial of anti-CD34 biotinylated mouse mAb, a column containing the avidin beads, and an integrated tubing set with collection bags attached.

Bone marrow or PBPC samples are collected using standard medical techniques. If bone marrow is harvested, the nucleated cells first must be concentrated and the red cells and plasma removed. This is usually done by buffy coat preparation on a cell washer. No such procedure is required for PBPC preparations. Samples are washed once and resuspended for Ab incubation. At the beginning of a procedure, the processing laboratory will take the vial of Ab and begin the 25-minute incubation and washing of the clinical sample. During this time the column and tubing set can be installed on the CEPRATE® instrument and the automated priming program run.

When the bone marrow or PBPC sample is ready, it is attached to the tubing set on the machine as shown in Figure 2. The optical sensor detects when the sample bag has been drained, and the software switches the instrument to the next phase of the program. The pump controls the rate of fluid

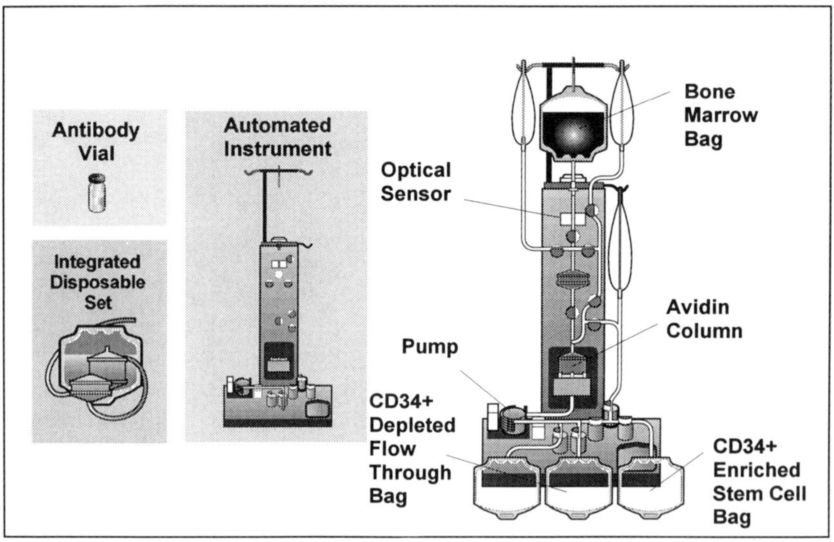

Figure 2 CEPRATE® SC stem cell concentration system.

flow, and computer-controlled valves direct the fluid path so that $CD34^-$ cells are collected in one bag and $CD34^+$ cells are collected in another, enriched stem cell bag. Not shown in Figure 2 are the main computer and software program that controls all aspects of the instrument operation.

Principles of Operation

Figure 3 depicts the initial loading of the cells into the CEPRATE® system. Cells begin to flow from the sample bag through the tubing set and into the avidin column. The target $CD34^+$ cells bind to the beads, while the negative cells pass through and are collected in the flow-through collection bag. This part of the procedure is quick, typically taking less than 10 minutes to complete. After the optical sensor detects that the sample bag has been drained, the instrument switches automatically to wash the avidin column and flush out residual negative cells. Figure 4 shows scanning electron micrographs of avidin beads taken from a column. On the left are beads before any loading of cells, while on the right are $CD34^+$ cells attached by the biotin-avidin interaction.

After this washing phase, the $CD34^+$ cells are released from the beads by mild agitation (Fig. 5). The avidin column contains a built-in magnetic

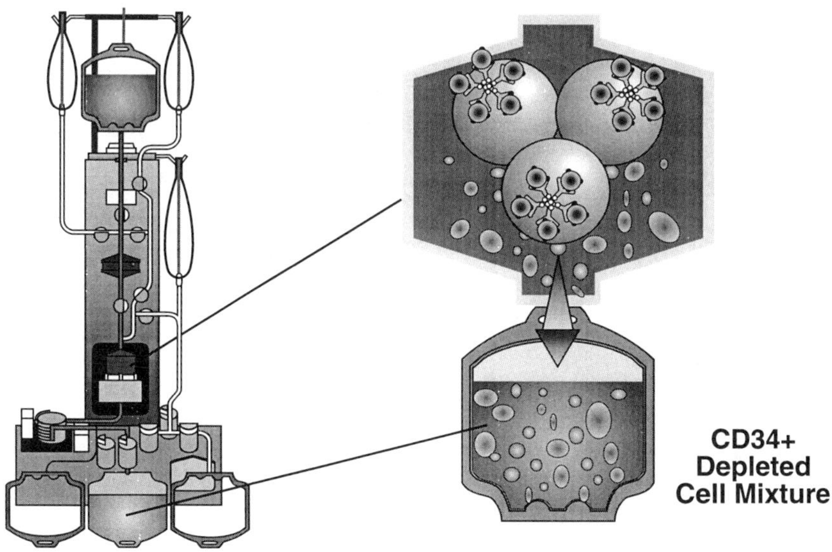

Figure 3 CEPRATE® SC automated operation: initial loading of cells.

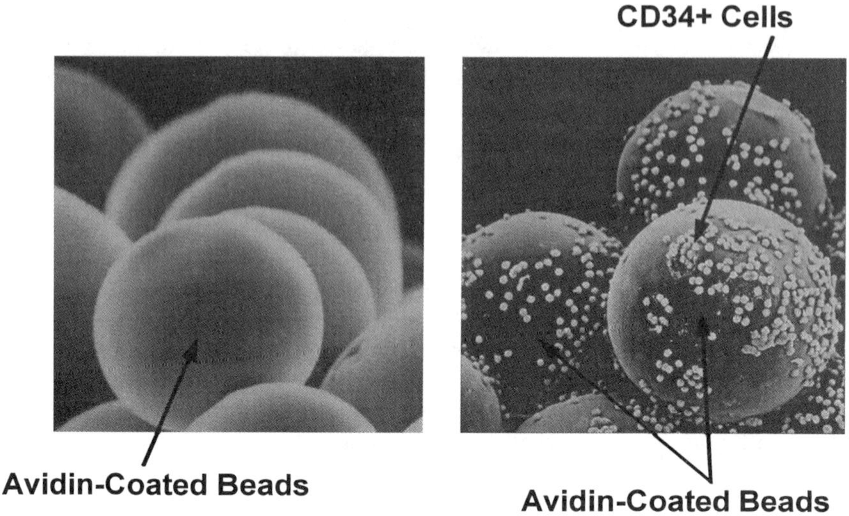

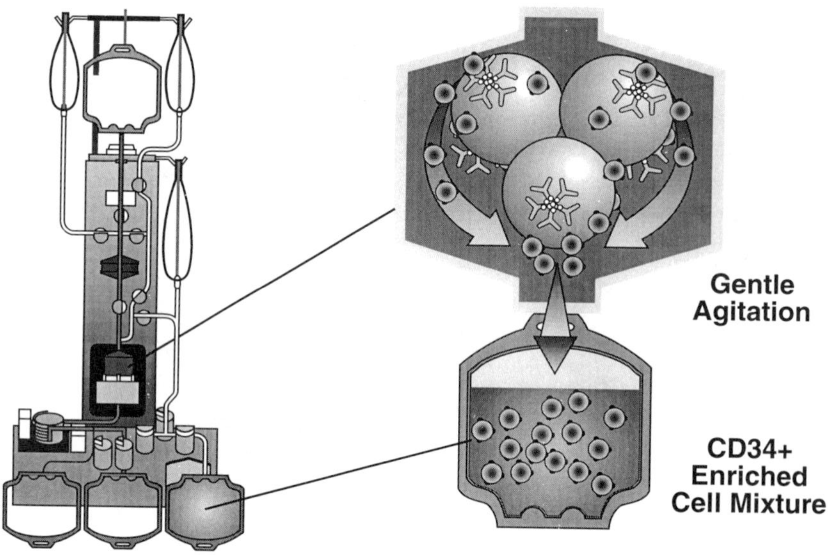

Figure 5 CEPRATE® SC automated operation: elution and collection.

impeller, and agitation consists of rotating the impeller at 20 rpm for 1 minute. The released cells are flushed out of the column and collected in the stem cell bag. If required as part of the clinical protocol, these cells then are ready for further volume reduction and cryopreservation. The entire time on the instrument, from the beginning of loading the sample to the enriched cells in the bag, is 25 minutes. The entire CD34$^+$-enriched fraction can be frozen in a small 5-mL vial, a dramatic reduction from the 100 to 250 mL or greater volumes used for standard bone marrow or PBPC products (Fig. 6).

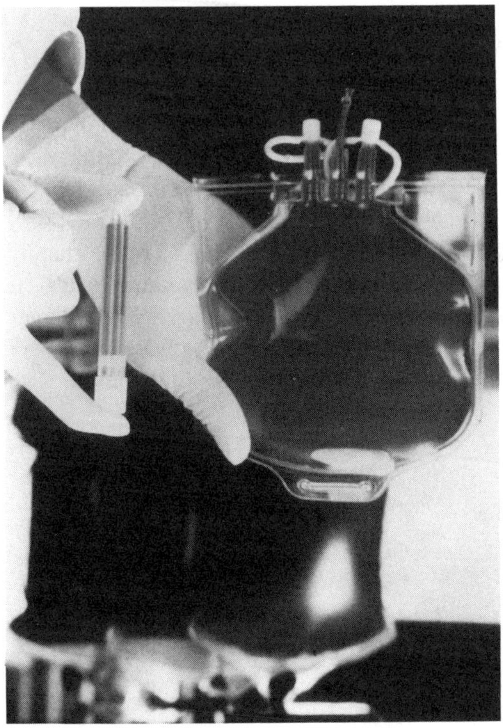

Figure 6 A standard bone marrow harvest of 1.5–2 L is shown in background. This material usually is concentrated by buffy-coat centrifugation to yield 100–200 mL of product, which is diluted 1:1 with cryoprotective media containing 10% dimethyl sulfoxide (DMSO) and then frozen. After CEPRATE processing, the CD34-enriched cells can be cryopreserved in a 5-mL vial, which represents a significant volume reduction and greatly facilitates any further manipulation such as ex vivo expansion or gene therapy.

III. CLINICAL EXPERIENCE

The CEPRATE® SC Stem Cell Concentration System has been used to treat more than 5000 patients in over 200 transplant centers worldwide in a variety of clinical settings. These include autologous and allogeneic bone marrow and PBPC transplantations after high-dose chemotherapy, ex vivo expansion of purified progenitor cells, and gene therapy. Some of our experiences using this system for the enrichment of CD34$^+$ stem and progenitor cells for these different applications are summarized in the following sections.

A. Autologous Transplantation

The safety of the CEPRATE® SC Stem Cell Concentration System for selection of CD34$^+$ cells was initially evaluated in a single-center, open-label, phase 1-2 clinical trial in women with stage II, III, or IV breast cancer who were receiving autologous bone marrow transplantation (5,6). Engraftment data for various cohorts treated in that initial study are listed in Table 1. Patients receiving CD34-selected cells from bone marrow alone took approximately 3 weeks to recover their neutrophil and platelet counts to safe levels. If recombinant human granulocyte colony-stimulating factor (rHuG-CSF) was given posttransplant, neutrophil recovery was more rapid but platelet recovery was unaffected. Using CD34-selected PBPC, either in combination with

Table 1 Engraftment Results of Transplantation of Patients with Breast Cancer with CD34$^+$-enriched Cells

Cell type	N	Days to ANC $>0.5\times10^9$/L (range)	Days to PLT $>20\times10^9$/L (range)
CD34$^+$ bone marrow alone	30	22 (15–34)	22 (16–32)
CD34+ bone marrow with rHuG-CSF post-transplantation	11	11 (10–13)	20 (11–31)
CD34+ bone marrow and CD34+ PBPC with rHuG-CSF posttransplantation	12	11 (10–12)	14 (10–25)
CD34+ PBPC with rHuG-CSF posttransplantation	101	10 (9–12)	15 (10–25)

ANC = Absolute neutrophil count; PLT = platelet count.

selected bone marrow or alone, yielded rapid recovery of both neutrophil and platelet counts.

Based on the positive results of that study, a multicenter, randomized, open-label, phase 3 clinical trial in women with stage II, III, or IV breast cancer receiving autologous bone marrow transplantation was completed (7). Shown in Table 2 are the processing results for the bone marrow samples combining the phase 1-2 and phase 3 data. Those data indicate an approximate 60-fold increase in the CD34 and granulocyte-macrophage colony-forming unit (CFU-GM) content of the CEPRATE®-selected cells, with an overall recovery of 33–40% of the starting values.

The primary objectives of this phase 3 study were to demonstrate that patients who received CD34-selected cells experienced less toxicity associated with the infusion of the cryopreserved cells but had comparable hematological recovery when compared with patients who received marrow processed by standard buffy-coat preparation. Both primary objectives were met; patients in the CEPRATE® group had significantly less toxicity associated with the infusion of the cryopreserved cells, and neutrophil engraftment (ANC $> 0.5 \times 10^9$/L) was equivalent. For CEPRATE®-selected cells, 91% of the patients engrafted by day 20, compared with 88% of the patients who received buffy coat marrow. Based on these and other data, the U.S. FDA has approved use of the CEPRATE® system for CD34 selection from bone marrow.

Additional trials have documented equivalent or even improved enrichment results using mobilized PBPC for CD34 selection. Furthermore, these trials have also shown that selection for $CD34^+$ cells results in a significant reduction (2 to > 5 logs) of contaminating tumor cells, as documented using

Table 2 Results Obtained with the CEPRATE® System Using Bone Marrow from 70 Patients with Stage II to IV Breast Cancer[a]

	Start	Enriched
CD34 purity (%range)	1.3 (0.1–3.5)	75.3 (16–92)
CD34 cells ($\times 10^6$/kg) (range)	4.2 (0.7–10.6)	1.4 (0.4–4.5)
CFU-GM/10^5 cells (range)	28 (4–112)	1,680 (177–10,000)
CFU-GM ($\times 10^4$/kg) (range)	9.9 (0.9–60)	4.0 (0.4–21)

[a]Flow cytometry data were obtained using a central reference laboratory, granulocyte-macrophage colony-forming units (CFU-GM) are from individual clinical processing laboratories. Data are given as median values (and ranges). Selection resulted in approximately a 60-fold enrichment for both $CD34^+$ cells and CFU-GM activity, with 33% overall recovery of the $CD34^+$ cells and 40% recovery of CFU-GM.

Table 3 Tumor Cell Depletion Results of Transplantation of Patients with Breast Cancer with CD34$^+$-Enriched Cells

	Bone marrow	Peripheral blood progenitor cells
Preselection	2,810,000	61,500
Postselection	27,600	43
Log reduction	2 (1 to > 4)	2 (1 to > 3)

sensitive immunocytochemistry staining techniques in the setting of breast cancer (Table 3) (6) or quantitative PCR analyses for the detection of myeloma cells (7). Hematopoietic and immune reconstitution after transplantation with CD34$^+$-enriched cells appears similar to unmanipulated bone marrow or PBPC. In summary, it has been shown that the CEPRATE® system is a safe and effective technology for the enrichment of CD34$^+$ cells for autologous transplantation.

B. Allogeneic Transplants

In the allogeneic setting, use of the CEPRATE® system as a method for T-cell depletion has consistently shown a 3-log reduction in overall T-cell numbers (8). More rapid hematological recovery has been found after CD34$^+$ PBPC infusion compared with bone marrow, and in the matched-related setting, a low incidence of graft-versus-host disease (GVHD) has been observed (9). However, in situations where the donor is unrelated or is a mismatched family member, there may be a need for additional T-cell depletion that can be achieved by CD34$^+$ selection alone.

IV. FUTURE DIRECTIONS

A. Secondary T-Cell Depletion

Using the same avidin-biotin immunoadsorption technology employed for CD34$^+$ selection, CellPro has developed a T-cell–depletion system that incorporates a CD2-depletion column after the CD34$^+$-selection column. The CD34$^+$-adsorbed fraction from the CEPRATE® SC is incubated for 30 minutes at room temperature with a biotinylated, murine mAb to the pan-T–cell Ag CD2 (the E rosette receptor). The resultant mixture of cells and Ab is applied directly to a second, smaller column of avidin-conjugated polyacrylamide beads. CD2$^+$ T cells are retained on the column, while CD34$^+$CD2$^-$ cells flow through and are collected. Using this system, we have achieved a median of

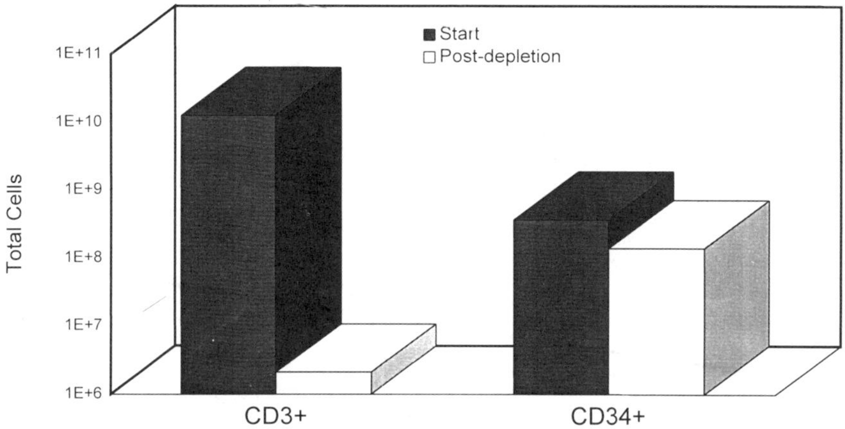

Figure 7 Depletion of T cells and recovery of CD34$^+$ cells after CEPRATE® SC and TCD processing.

3.5 log reduction in T cells and recovered almost half of the CD34$^+$ cells, relative to the starting PBPC product (Fig. 7). Table 4 compares the performance of three methods of T-cell depletion on rHuG-CSF–mobilized PBPC collections. Performance of the CEPRATE® TCD system is equivalent to or better than CD34$^+$E$^-$ and SBA-E$^-$ methods of depletion and is considerably less resource- and time-intensive, requiring only 3 hours (including CD34$^+$ selection).

Table 4 Comparison of T-Cell Depletion Methods on Mobilized PBPC

Method	CD34$^+$ cells in final product (× 10^7) median	CD34$^+$ recovery (%) median	CFU-GM in final product (× 10^6) median	T cells in final product (× 10^6) median	TCD ($\log_{10}$) median	n
CD34$^+$ E-[a]	11.1 ± 7.1	24	37.4 ± 16.4	1.3 ± 1.2[b]	4 ± 5	4
SBA-E-[a]	8.3 ± 5	not done	46.1 ± 38.3	32 ± 2[b]	2.6 ± 0.3	4
CEPRATE® TCD	5.9 ± 7.4	41 ± 9.2	not done	2.4 ± 0.4[c]	3.5 ± 0.1	4

[a]From Ref. 10.
[b]Determined by limiting dilution analysis.
[c]CD3$^+$ determined by flow cytometry.
TCD, T-cell depletion.

B. Gene Therapy

The CEPRATE system has also been used in more than 20 different clinical gene therapy protocols, with approximately 150 patients treated at leading academic institutions such as the National Institutes of Health, MD Anderson Cancer Center, Fred Hutchinson Cancer Research Center, University of Pittsburgh, Columbia University, and Children's Hospital of Los Angeles. Investigations have included both marking studies that have documented that infused tumor cells can contribute to relapse (11), as well as therapeutic studies to insert the adenosine deaminase gene into severe chronic immunodeficiency patients (12), the glucocerebrosidase gene to treat Gaucher's disease, and the multidrug-resistant gene to enhance chemotherapy resistance in breast and ovarian cancer patients. In a collaborative study with Cell Genesys, the CEPRATE® system has been used to enrich for CD8 cells for transduction with a "universal receptor" to target and lyse human immunodeficiency virus (HIV)–infected cells. In conclusion, many gene-transfer protocols are underway using CEPRATE® technology. These studies have shown that retroviral vectors can be used safely and long-term persistence can be demonstrated. However, the clinical effectiveness of gene therapy still remains to be demonstrated.

C. Ex Vivo Expansion Using Enriched Human CD34$^+$ Progenitor Cells

The ability to grow human progenitor cells ex vivo has broad potential for human cellular therapy. Potential therapeutic benefits include faster hematopoietic reconstitution after myeloablative therapy, hematopoietic support for multiple repeated cycles of high-dose chemotherapy, a reduced harvest requirement for bone marrow or PBPCs, generation of sufficient cells from cord blood to support adult transplantation, and enhanced tumor cell or T-cell purging. Our initial studies focused on defining optimal conditions for clinical scale ex vivo expansion after selection of CD34$^+$ progenitor cells. The CEPRATE® system was used to isolate large numbers of CD34$^+$ progenitor cells from bone marrow or PBPC. Expansion was performed in static liquid culture systems. The cultures were analyzed for ^{3}H-thymidine incorporation, for overall expansion (defined as fold-increase over the inoculated value) in total cell numbers, total CD34$^+$ cells and expression of other "lineage-specific" markers, and various types of colony-forming activity. Our results indicate that CD34 enrichment enhances expansion 10-fold compared with unseparated marrow and that combinations of several cytokines are required for optimal growth. In particular, stem-cell factor (SCF) + interleukin (IL)-3 + IL-6 + G-CSF results in a 20- to 50-fold increase in progenitor cells by 7–14 days of culture (13).

Based on these types of in vitro findings, a clinical trial was done to investigate the ability of PBPCs cultured ex vivo to restore hematopoiesis in patients with cancer who had undegone high-dose chemotherapy (14). Approximately one tenth of a normal apheresis collection was used for the expansion. The results showed that ex vivo expanded cells could safely be infused into patients and that those cells promoted rapid hematopoietic recovery identical to that obtained with non-expanded PBPC. Because the chemotherapy given was not completely myeloablative, no information on the long-term reconstitution ability of ex vivo expanded cells was obtained.

D. Immunotherapy

T cells play a central role in normal immune surveillance and in regulating many aspects critical to transplantation and immunotherapy. Specific T-cell populations control tumor cell eradication, GVHD, rejection, and tolerance. The isolation of specific T-cell subsets has become an increasingly important requirement for experimental immunology, and the CEPRATE® system is being developed to enrich for these types of cells. The protocol is essentially the same as that for CD34 enrichment, with an automated, easy-to-use process and a sterile, single-use disposable column and tubing set. The only modifications are that the specificity of the biotin-conjugated Ab is changed and the total process time is even more rapid (< 1 hour) because there is no need to wash out residual unbound Ab before applying to the avidin column. Figure 8 shows the results using the CEPRATE® system for CD4, CD8, CD19, and CD56 enrichment from peripheral blood. In general, the CEPRATE® system gives excellent purities of > 90%, with high yields of > 70%. With the clinical-scale system, very large numbers of $> 5 \times 10^9$ purified cells can be obtained (Table 5). These enriched cells do not appear to be activated by the selection process and retain their full functionality properties.

V. SUMMARY

Avidin-biotin immunoadsorption technology and the CEPRATE® SC system permit the rapid purification of specific cell populations in clinically relevant quantities. In the case of $CD34^+$ cell enrichment, this system has been used to treat more than 5000 patients in a wide variety of clinical settings. Engraftment

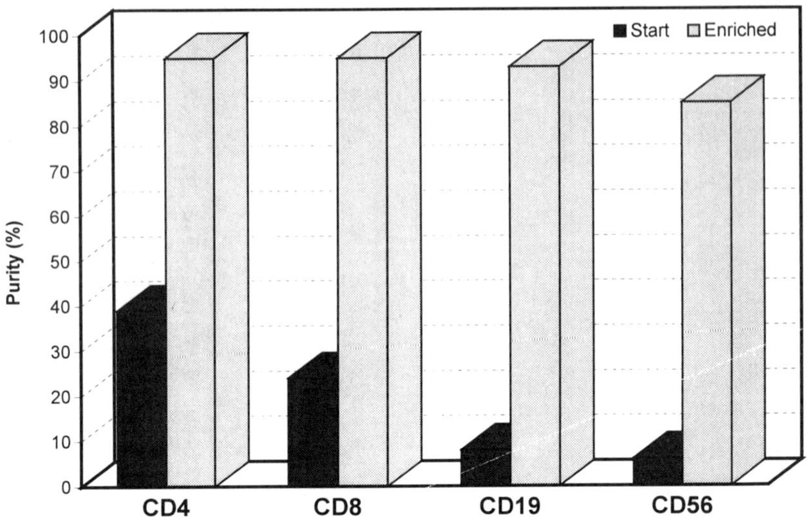

Figure 8 Selection of various lymphocyte subsets with the CEPRATE® system.

is essentially equivalent to that seen with unselected transplants. Demonstrated clinical benefits include, in the case of autologous transplantation, 2–5 logs of tumor cell purging with the additional benefit of much less toxicity, as has been documented with the infusion of cryopreserved unselected bone marrow or PBPC. A further advantage of $CD34^+$ enrichment relates to the large reduction in volume and total numbers of cells. Volume reduction also results in less freezer storage space to maintain the cryopreserved samples. Furthermore, the

Table 5 CD4 Selection with the CEPRATE® SC System

Experiment 1	Start	Enriched
TNC ($\times 10^9$)	91.4	7.4
CD4 purity (%)	10.9	89.3
CD4 cells ($\times 10^9$)	9.9	6.6
Experiment 2	Start	Enriched
TNC ($\times 10^9$)	79.5	8.5
CD4 purity (%)	10.4	87.9
CD4 cells ($\times 10^9$)	8.3	7.5

TNC, total nucleated cells.

reduction in total cell numbers means that further processing manipulations such as additional tumor purging or other ex vivo culture procedures, such as expansion or gene therapy, now become much more feasible. In the allogeneic setting, CD34 enrichment yields approximately a 3-log depletion of T cells, and this depletion can be further enhanced by use of a secondary T-cell depletion column. The CEPRATE® SC system also is being developed to enrich specific lymphocyte subsets such as CD4 and CD8 cells, which opens up the exciting area of immunotherapy as a new modality for cancer therapy. The ability to purify unique cell fractions, at clinically relevant numbers, has widened the field of cancer and gene therapy to the concept of graft engineering, where specific cells are used for a specific therapeutic benefit.

REFERENCES

1. Berenson RJ, Bensinger WI, Kalamasz D. Positive selection of viable populations using avidin-biotin immunoadsorption. *J Immunol Methods* 1986; 91:11–19.
2. Berenson RJ, Andrews RG, Bensinger WI, et al. Antigen $CD34^+$ marrow cells engraft lethally irradiated baboons. *J Clin Invest* 1988; 81:951–955.
3. Berenson RJ, Bensinger WI, Hill RS, et al. Engraftment after infusion of $CD34^+$ marrow cells in patients with breast cancer or neuroblastoma. *Blood* 1991; 77:1717–1722.
4. Berenson RJ, Heimfeld S, Krieger MS, Jacobs CA, Auditore-Hargreaves K, Shpall EJ. Selection and transplantation of autologous hematopoietic progenitor cells. In: Mertelsmann R, Levitt D, eds. *Hematopoietic Stem Cells: Biology and Therapeutic Applications.* New York: Marcel Dekker, 1995:519–533.
5. Shpall EJ, Jones RB, Bearman SI, et al. Transplantation of enriched CD34-positive autologous marrow into breast cancer patients following high-dose chemotherapy: influence of CD34-positive peripheral-blood progenitors and growth factors on engraftment. *J Clin Oncol* 1994; 12:28–36.
6. Shpall EJ, LeMaistre CF, Holland K, Ball E, Jones RB, Saral R, Jacobs C, Heimfeld S, Berenson R, Champlin R. A prospective randomized trial of buff coat versus CD34-selected autologous bone marrow support in high-risk breast cancer patients receiving high-dose chemotherapy (submitted).
7. Vescio RA, Hong CH, Cao J, Kim A, Schiller GJ, Lichtenstein AK, Berenson RJ, Berenson JR. The hematopoietic stem cell antigen, CD34, is not expressed on the malignant cells in multiple myeloma. *Blood* 1994; 84:3283–3290.
8. Noga SJ, Davis JM, Berenson RJ, et al. Graft engineering III: $CD34^+$ augmentation/elutriation (A/E) allows reduction in GVHD prophylaxis without affecting engraftment. *Blood* 1994; 84:89a (abstract 346).
9. Link H, Arseniev L, Bähre O, et al. Combined transplantation of allogeneic bone marrow and $CD34^+$ blood cells. *Blood* 1995; 86:2500–2508.

10. Fernandez JM, Yan Y, Bleau S, Collins NH, O'Reilly RJ. *Blood* 1995; 86:2491.
11. Deisseroth AB, Zu Z, Claxton D, et al. Genetic marking shows that Ph^+ cells present in autologous transplants of chronic myelogenous leukemia (CML) contribute to relapse after autologous bone marrow in CML. *Blood* 1994; 83:3068–3076.
12. Kohn DB, Weinberg KI, Nolta JA, Heiss LN, Lenarsky C, Crooks GM, Hanley ME, Annett G, Brooks JS, el-Khoureiy A, et al. Engraftment of gene-modified umbilical cord blood cells in neonates with adenosine deaminase deficiency. *Nat Med* 1995; 1:1017–1023.
13. Heimfeld S, Fei R, Tsui Z, Fogarty B, Thompson P, Berenson RJ. Ex-vivo expansion of $CD34^+$ progenitor cells. In: Lasky L, Warkentin P, eds. *Marrow and Stem Cell Processing for Transplantation.* Bethesda, MD: American Association of Blood Banks, 1995:201–208.
14. Brugger W, Heimfeld S, Berenson RJ, Mertelsmann R, Kanz L. Reconstitution of hematopoiesis after high-dose chemotherapy by autologous progenitor cells generated ex vivo. *N Engl J Med* 1995; 333:283–287.

6

High-Density Particles: A Novel, Highly Efficient Cell Separation Technology

Norma S. Kenyon and Camillo Ricordi
Diabetes Research Institute, University of Miami School of Medicine, Miami, Florida

John G. Gribben and Lee M. Nadler
Dana Farber Cancer Institute, Boston, Massachusetts

Robert K. Zwerner
Coulter Corporation, Miami, Florida

Thomas R. Russell
Coulter Cellular Therapies, Inc., Roxbury, Massachusetts

I. INTRODUCTION

With the advent of technologies that exploit monoclonal antibodies (mAbs) for the purposes of positive or negative selection of various cell subpopulations, it has become possible to manipulate, or "engineer," cellular grafts of autologous or allogeneic origin for use in a variety of cell-based therapies. Such therapies include autologous or allogeneic stem cell transplantation for treatment of solid tumor or hematological malignancies (1,2); T-cell depletion for prevention of graft-versus-host disease (GVHD) after allogeneic bone marrow transplantation (3–5); infusion of cytokine-activated or antigen (Ag)-specific killer cells (6) or tumor-infiltrating lymphocytes (TIL) (7); vaccination with antigen-pulsed dendritic cells (8); donor lymphocyte infusion for treatment of relapsed leukemia (9–13); and purging of malignant cells from autologous bone marrow or peripheral blood stem cell harvests (14). Many groups also have begun to explore the potential for selective differentiation of CD34$^+$ cells into desired cell lineages, such as dendritic cells for use in vaccination protocols, or neutrophil or platelet precursors for use in protocols designed to decrease the period of neutropenia (15) and thrombocytopenia after bone marrow transplantation in cytoablated recipients.

One emerging application of cell-based therapy is the use of donor bone marrow cell infusion into noncytoablated recipients of solid organ and/or cellular allografts, in protocols designed to induce a state of donor-specific tolerance, thus obviating the need for life-long, generalized immunosuppression of the recipient (16–18). As the number of applications for allogeneic stem cell transplantation increases, it has become evident that larger numbers of cells, on the order of 5- to 10-fold more than have been traditionally used for bone marrow transplantation in cytoablated recipients, will need to be processed to achieve engraftment in noncytoablated patients or in recipients of allogeneic T cell–depleted stem cell transplants (19–21). In addition, the ability to select stem cells and stem cell subpopulations has facilitated the implementation of gene therapies (22) and will clearly have implications for clinical protocols involving stem cell expansion (23–25).

Obstacles to the successful implementation of cell-based therapies include the requirement for time-consuming, costly, and labor-intensive processing to obtain critical cells, and, especially in cases of rare event or infrequent populations, poor cell recoveries. Ideally, cell-separation procedures should be rapid and simple to perform, with optimal recovery and viability of desired subpopulations and elimination of undesired cells. A novel, dense particle technology has been developed that meets these requirements (26,27).

II. HIGH-DENSITY PARTICLES

A. Description

Figure 1 is a scanning electronmicrograph of the high-density particles (HDP). These nickel particles are approximately 10 μm in diameter, with a density of 9 g/cm^3. In contrast to other particles used in cell-separation procedures, the dense particles have a highly irregular surface, which effectively increases the surface area available for binding of mAbs. Due to the high density of the particles, gravity settling will occur within 4 minutes after suspension of the HDP in a test tube containing media, as illustrated in Figure 2. Before coating with Ab, the particles are heated for 3–5 hours at 250°C, thus effectively sterilizing the preparation. In addition, the amount of nickel that leaches from the particles is significantly decreased to nontoxic levels after heating. Monoclonal or polyclonal Ab-coated (e.g., goat antimouse) HDP are produced by

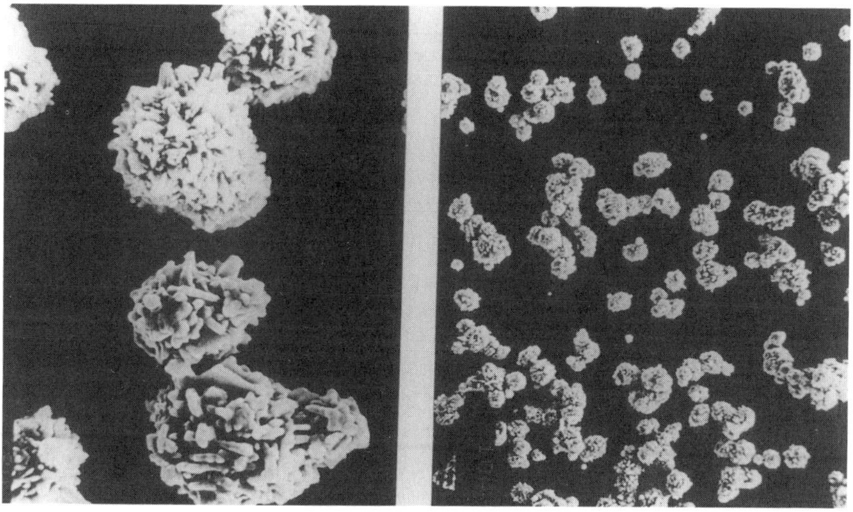

Figure 1 Scanning electronmicrograph (SEM) of high-density particles (HDP). The nickel particles are approximately 10 μm in diameter, with a density of 9 g/cm^3. This SEM reveals the highly irregular surface of the HDP.

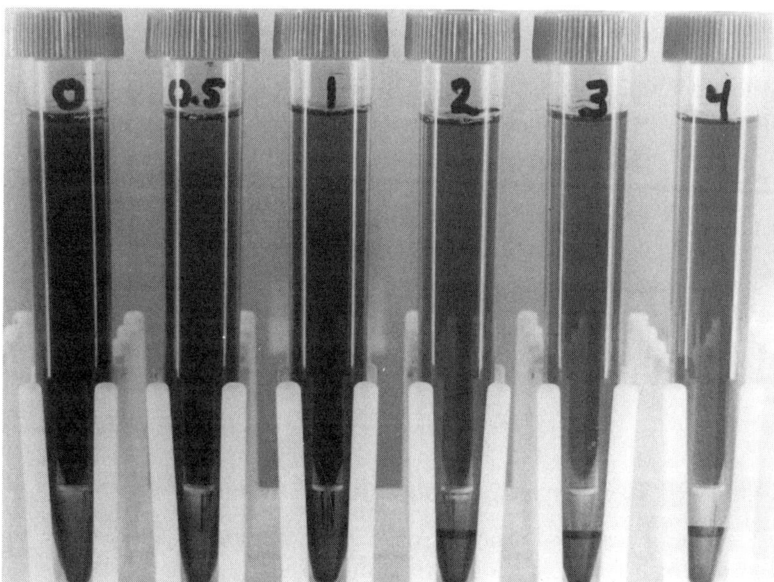

Figure 2 Settling time of HDP. Dense particles were mixed end-over-end for 4 minutes, followed by gravity settling. As can be seen in the figure, the particles (first tube equals time 0) have formed a pellet within 4 minutes.

passive adsorption, and the characteristic biochemical properties of the individual mAb determine whether the HDP is coated directly with the reagent or is coated onto a goat antimouse HDP. The particles are stored at 4°C in a Tris NaCl buffer containing 0.02% bovine serum albumin (BSA). Preliminary data suggest that the particles are stable for as long as 3 years, and the studies to verify stability for various Ab-coated particles are currently underway.

B. Cell Processing with Dense Particles

Dense-particle technology can be applied directly to research or clinical-scale samples of whole blood, apheresis blood products (obtained from patients mobilized with cytokines), bone marrow aspirates, or vertebral body marrow. Depending on the application (see Sec. II.C), the need for preprocessing of the sample by density-gradient centrifugation, elutriation, or red blood cell lysis or sedimentation is eliminated. The volume of dense particles required for the particular application is determined and HDP are placed into the processing tube. After allowing the particles to settle for 4 minutes, the supernatant is removed and fresh buffer is added, followed by resuspension of the HDP. This process is repeated, and after removal of the wash buffer the sample is added directly to the pelleted dense particles, followed by end-over-end mixing for 4 minutes at room temperature. The tube is placed upright and the particles settle over 4–10 minutes, followed by removal of the processed cells from the HDP pellet. Multiple rounds are extremely easy to perform in that no centrifugation or washing is required between rounds; the sample is simply removed from one pellet and added directly to the next round of prewashed, pelleted HDP. Due to the short period of mixing and the rapid settling of the particles, billions of cells can be processed rapidly and simply, with a minimum of nonspecific cell loss. Processing of whole blood samples with this technology has characteristically resulted in elimination of targeted cell populations to undetectable levels, as judged by flow cytometric and polymerase chain reaction analysis (see Sec. III.D). Nonspecific cell loss has been approximately 2% per round, thus resulting in excellent recovery. Viability has been maintained or, as seen in some experiments, enhanced.

C. Results of Toxicity and Mutagenicity Studies

Because of the inherent applications of HDP in therapeutic approaches, the potential for toxic and mutagenic effects due to the leaching of nickel ions from the particles has been extensively studied in vitro, and preliminary studies in small and large animals are ongoing. Initially, the solubility of nickel in blood, under the standard conditions of processing with HDP, was investigated. Blood was processed with BSA-coated HDP, using a protocol defined for depletion of specific cell populations, which involved incubation of the blood with HDP, removal of the particles by gravity sedimentation, and density-gradient centrifugation of the treated blood over Ficoll-paque. The Ficoll-interface cells were removed and washed three times. Analysis of nickel ion levels in blood, Ficoll-interface cells, or supernatants from the cell washes revealed that minimal concentrations of nickel remained associated with the white blood cell fraction (Table 1). In fact, the amount of nickel in the final cell preparation was comparable to the levels detected in the wash buffer itself. Additional experiments have demonstrated that heating of the particles for 3 hours at 250°C, used for purposes of sterilization, results in the formation of a metal oxide that effectively reduces the release of nickel ion by a factor of 10.

The potential toxicity of HDP on cell growth has been tested on the MOLT-4 cell line, through the addition of 0–500 mg/mL of nickel particles to cell cultures; 25 mg of particles are used per mL of sample in the standard cell-separation procedure. Cell concentration and time of exposure to the HDP also were varied. While incubation of the cells with unheated particles, for 30 minutes, clearly inhibited cell growth at 250 or 500 mg/mL, heating of the HDP eliminated this effect, even with exposure times as long as 2 hours.

In studies performed at Coulter, the Diabetes Research Institute (University of Miami), and the Dana Farber Cancer Institute (Harvard University), processing of various stem cell–containing samples (normal or mobilized peripheral blood, vertebral body marrow, or bone marrow aspirates) with lineage-specific HDP was followed by culturing of processed and depleted material to assess the colony-forming capacity of the stem cells. The data from short-term methylcellulose cultures (enumerated at day 14) and long-term culture-initiating cell studies (LTC-IC, analyzed at 6 weeks) clearly demon-

Table 1 Nickel Ion Levels in Samples Treated with Dense Particles

Sample	Nickel ion in sample (μg/L)
Whole blood (nontreated)	25
Blood treated with nickel particles	521
Supernatant after first cell wash	277
Supernatant from second cell wash	10
Final cell preparation (after third wash)	20

strate that processing with HDP does not affect the ability of the resulting cell populations to grow in short- or long-term culture.

A murine, syngeneic bone marrow transplantation model was used to assess the effect of dense particles on the in vivo repopulating capacity of marrow processed with HDP. Bone marrow was harvested from normal Balb/c mice and mixed end-over-end with unheated, mouse serum albumin-coated HDP for 20 minutes. After allowing for gravity settling of the particles, a magnet was applied to insure that all HDP were removed from the marrow before transplantation, and the bone marrow cells were washed twice in RPMI. Cells were infused into the tail vein of lethally irradiated (925 rads) Balb/c mice. Control animals consisted of irradiated animals that received no marrow. The various experimental and control groups, as well as the survival data, are detailed in Table 2. There was no significant difference in survival between animals that received control, unprocessed marrow, and mice transplanted with HDP-processed bone marrow cells (HDP-BMC). To further address the potential effect of nickel particles on the repopulating capacity of bone marrow, a group of mice transplanted with HDP-BMC were sacrificed to serve as marrow donors for another set of lethally irradiated animals. The results of this study, shown in Table 3, clearly revealed that processing with HDP had no adverse effect on the in vivo repopulating capacity of murine hematopoietic stem cells.

Using accepted in vitro methodologies and analysis criteria (28), Evert Nieboer of McMaster University has extensively analyzed HDP for potential mutagenicity. The mutagenicity studies (24-hour exposure) conducted indicate that the nickel particles are not mutagenic by accepted criteria (personal communication).

Table 2 Survival of Lethally Irradiated Mice Transplanted with HDP-Treated Syngeneic Bone Marrow

Gene	*n*	Cell number	Cell treatment	Survival[a]
Radiation	1	None	NA	0/1
Control 1	4	10×10^6	No particles	4/4
Control 2	4	20×10^6	No particles	3/4
Experimental 1	10	10×10^6	Particles	7/10
Experimental 2	10	20×10^6	Particles	9/10

[a]214 days (5/1/95).

Table 3 Syngeneic Bone Marrow Derived from Mice Previously Transplanted with HDP–Bone Marrow Cell Reconstitutes Lethally Irradiated Mice

Group	*n*	Cell number	Cell source	Survival[a]
Radiation	1	None	NA	0/1
Control	9	20 × 106	Marrow from control animals	7/9
Experimental	16	20 × 106	Marrow from HDP-marrow–transplanted mice	13/16

[a]179 days (5/1/95).

III. APPLICATIONS OF DENSE-PARTICLE TECHNOLOGY

A. Processing of Peripheral Blood

Enrichment for Specific Cell Populations

Dense particles can be applied directly to unprocessed whole blood, thus eliminating the need for time-consuming and labor-intensive preprocessing to obtain a cell preparation that can be used in the majority of negative-selection procedures. As examples, HDP have been used to produce lymphocyte- or granulocyte-enriched preparations from whole blood. Shown in Figure 3 is an example of a lymphocyte-enriched population, obtained after depletion of platelets, granulocytes, and monocytes by processing with CD41, CD15, and CD14 plus 33 specific HDP, respectively. The blood was processed, as described earlier, by adding HDP directly to the blood, and samples were taken before and after depletion for evaluation on a Coulter STKS (Coulter) hematology analyzer. In a similar manner, a granulocyte-enriched population can be obtained (Fig. 4) by processing whole blood with platelet (CD41)-, monocyte (CD14 + CD33)-, T-cell (CD2 + CD5 + CD7 + CD8)–, and B-cell (CD19 + CD20)–specific HDP. Red blood cells can be eliminated from the sample using standard cell lysis techniques or 1 g Ficoll sedimentation. Alternatively, HDP coated with a mAb specific for glycophorin A can be used to remove the red blood cells.

T-Cell Depletion and Recovery After Multiple Rounds

The results for both of the aforementioned examples were obtained after one round of processing with the various combinations of particles described. While these initial experiments were designed to assess the efficacy of

WHOLE BLOOD

	$(\times 10^3/\mu l)$
WBC	6.2
PLT	197
NEU	3.2
LYM	2.3
MON	0.5
EOS	0.3
BAS	0.0

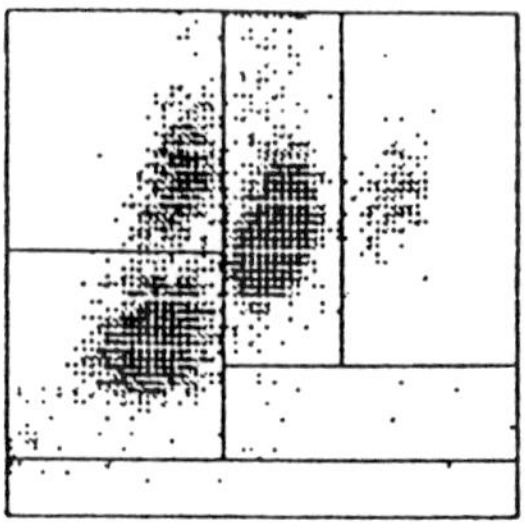

DEPLETED BLOOD

	$(\times 10^3/\mu l)$	% WHOLE BLOOD
WBC	2.1	33.9
PLT	6	3.0
NEU	0.0	0.0
LYM	2.0	87.0
MON	0.0	0.0
EOS	0.0	0.0
BAS	0.0	0.0

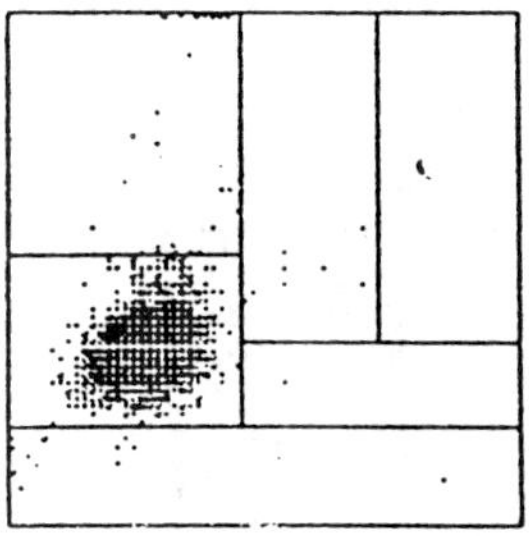

Figure 3 Enrichment of lymphocytes by negative depletion. Enriched lymphocyte populations were obtained by depleting platelets (PLT), granulocytes (NEU and Lym), and monocytes (MON), from whole blood by processing with CD41, CD15, and CD14, plus 33 specific HDP, respectively. Samples were analyzed on a Coulter STKS hematology analyzer. (Top) Unprocessed whole blood; (bottom) lymphocyte-enriched blood. (EOS-eosinophils, BAS-basophils.)

depletion, subsequent studies addressed the issue of cell recovery after one or multiple rounds of depletion. Shown in Figure 5 are the results obtained for recovery of $CD4^+$ cells and efficacy of depletion of $CD8^+$ cells after six rounds of processing of whole blood. The particles used in each round of depletion, as well as the absolute counts for $CD4^+$ and $CD8^+$ cells and other white blood cells subsets, are detailed in Table 4; 94% of the $CD4^+$ cells were recovered after six rounds of depletion. On average, experience with whole blood processing at Coulter indicates that 2–3% nonspecific cell loss occurs with each round of processing.

WHOLE BLOOD

	$(x10^3/\mu l)$
WBC	10.2
PLT	287
NEU	7.3
LYM	2.3
MON	0.4
EOS	0.1
BAS	0.1

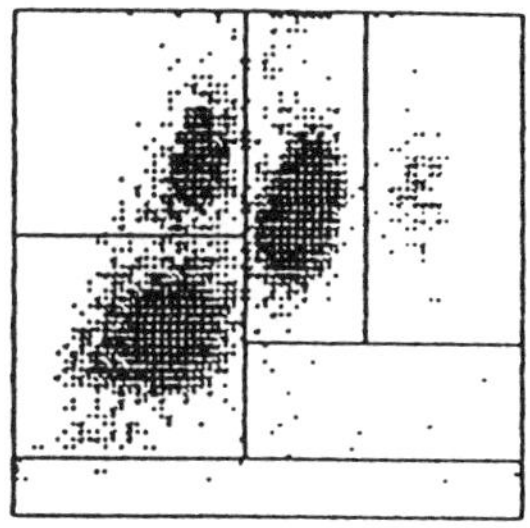

DEPLETED BLOOD

	$(x10^3/\mu l)$	% WHOLE BLOOD
WBC	6.7	65.6
PLT	28	1.0
NEU	6.3	86.3
LYM	0.2	8.7
MON	0.1	25.0
EOS	0.1	100
BAS	0.0	0.0

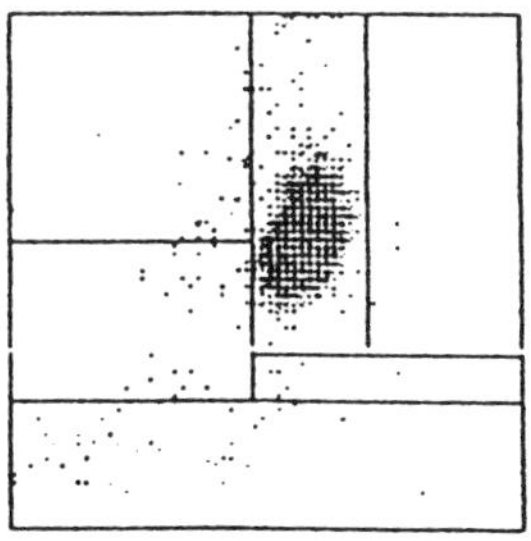

Figure 4 Enrichment of granulocytes by negative selection. Enriched granulocyte populations were obtained by processing whole blood with HDP specific for platelets (CD41), monocytes (CD14 + 33), T cells (CD2 + CD5 + CD7 + CD8), and B cells (CD19 + CD20). Samples were analyzed on a Coulter STKS hematology analyzer. (Top) Unprocessed whole blood; (bottom) granulocyte-enriched blood.

B. Platelet Removal from Peripheral Blood and Bone Marrow

Extensive cell clumping, which can interfere with recovery after freeze/thaw and during positive or negative cell selection, is frequently caused by platelets. Dense particles are extremely powerful as a means to remove platelets from samples of peripheral blood obtained by leukapheresis and from bone marrow aspirates, as demonstrated in Figure 6. In this experiment, whole blood was processed using density-gradient centrifugation over Ficoll-Paque, and the

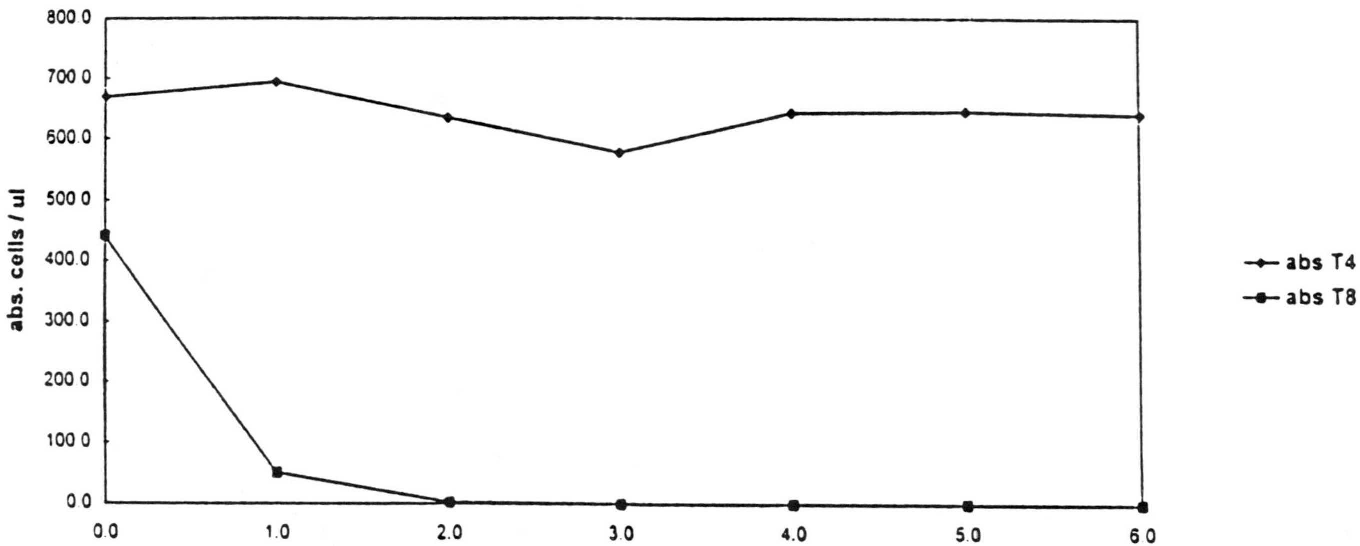

Figure 5 Recovery of CD4$^+$ cells from whole blood after multiple rounds of processing with HDP. Using the particles described in Table 4, whole blood was processed through six rounds of depletion. As depicted in the graph, 94% recovery of the CD4$^+$ cells was obtained, with elimination of the CD8$^+$ cells to nondetectable levels. (abs=absolute.)

Table 4 Recovery of CD4$^+$ Cells After Multiple Rounds of Processing with High-Density Particles

Cycles of depletion, particles used	Absolute CD4	Absolute CD8
Undepleted	666	441
1x = CD8,CD14,CD15,CD20, and CD41	694	51
2x = CD8 and CD14	635	3
3x = CD8 and CD14	579	0
4x = CD15	644	0
5x = CD15	646	0
6x = CD8,CD14, and CD41	642	0

94% of CD4$^+$ cells were recovered after the six cycles of depletion.

Ficoll-interface cells were removed. The sample was not washed to retain platelet contamination and was subsequently processed with CD41-coated HDP. Shown in Figure 6 are photomicrographs of the sample before and after platelet depletion. The platelet-specific HDP have been effectively used to process peripheral blood (normal and mobilized) and bone marrow aspirates for platelet removal.

C. Enrichment of Human Hematopoietic Stem Cells

Vertebral Body Marrow

Results from both experimental and clinical studies indicate that hematopoietic chimerism may be essential for the establishment and maintenance of donor-specific tolerance to solid organ and/or cellular allografts. Application of bone marrow transplantation as a means to induce tolerance, however, has not been used clinically because of the previously held notion that cytoablation of the recipient is required to obtain engraftment. This paradigm has recently been challenged by the demonstration that engraftment can be achieved in noncytoablated mice by escalating the dose of donor bone marrow cells (17,29). The finding of donor-derived hematopoietic cells, or microchimerism, throughout the body of long-term liver transplantation recipients (27–29 years after transplantation) also lends support to the concept that cytoablation is not required to achieve stem cell engraftment (17). With regards to the induction of donor-specific tolerance, the nature of key tolerogenic cells in the bone marrow inoculum remains to be elucidated, but veto cells, natural suppressor cells, and facilitator cells have all been postulated to play a role in

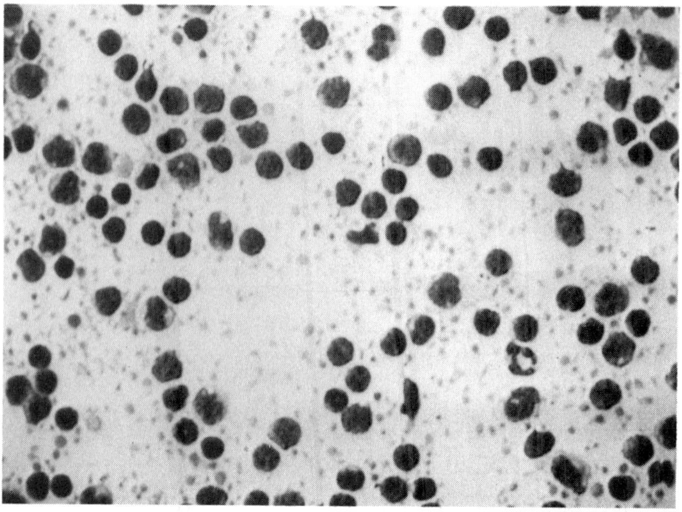

(a)

Figure 6 Platelet removal from peripheral blood leukocyte samples. A platelet-containing preparation of peripheral blood leukocytes was processed with CD41-coated HDP. (a) Photomicrograph of the unprocessed, platelet-rich sample, (b) Removal of the platelets is clearly evident in the CD41-treated sample.

this effect. One common property of these cell types is that they are class II dim or negative.

Ongoing clinical trials at the University of Miami involve the infusion of 10^9 nucleated donor bone marrow cells/kg recipient body weight, which translates to $1.5–2 \times 10^9$ total cells/kg. The donor bone marrow cells are retrieved from the vertebral bodies of the cadaver organ donor; unlike bone marrow aspirates, the cells obtained from vertebral bodies are uncontaminated by peripheral blood and therefore do not contain platelets or large numbers of mature lineage-specific cells. Currently, whole vertebral body marrow (VBM) is being used in the clinical protocols. To debulk the transplant, thus decreasing the risk of dimethyl sulfoxide (DMSO)–related toxicity, and to remove erythroid- and myeloid-lineage cells that can interfere with recovery after freeze/thaw, we began to assess various available cell-separation technologies. It rapidly became apparent that density-gradient centrifugation was not readily applicable to the large numbers of cells (70–100 billion nucleated cells per recipient) required in our clinical trials; likewise, the specialized centrifuges that have evolved for the processing of bone marrow aspirates and for

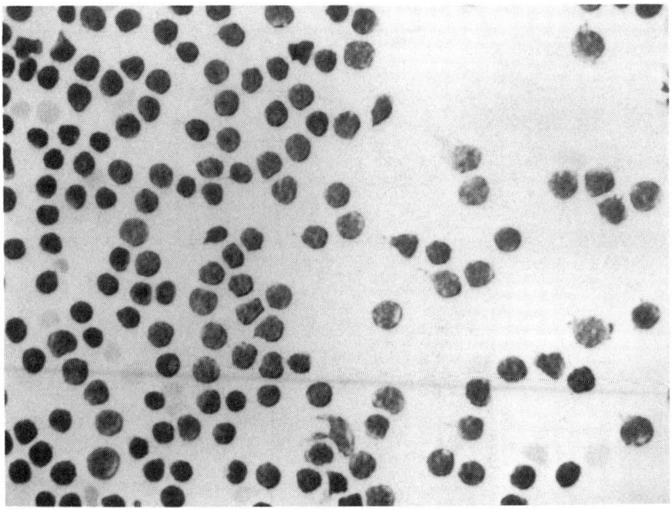

(b)

leukapheresis procedures did not lend themselves to the manipulation of VBM. We did not wish to pursue positive selection for stem cells, as another key cell type is thought to be required for the induction of tolerance. Additionally, we wished to eliminate the bright class II positive, directly immunogenic cells from the bone marrow inoculum (30), and technologies available for negative selection of these cells could not be applied to VBM without time-consuming, labor-intensive preprocessing of the marrow.

With the knowledge that dense particles could be applied directly to whole bone marrow, we initiated studies to assess the potential utility of HDP technology for the fractionation of human VBM and to determine if the procedure could be scaled up to accommodate large cell numbers. We used HDP coated with a CD15-specific mAb, KC48, for removal of myeloid lineage cells; $KC48^+$ myeloid cells are routinely $CD34^-$. The CD15-HDP were highly effective, in that myeloid cells were clearly depleted, as determined by flow cytometry and microscopic examination of cytospin preparations (Figs. 7 and 8). Morphological analysis of the CD15-depleted cells revealed that 70–80% of the remaining cells were of the erythroid lineage. High-density particles coated with a mAb specific for glycophorin A (KC16), present on mature and nucleated red blood cells, were subsequently used to further process the marrow. Processing with these two particles routinely resulted in preparations composed of 15.3 ± 3.7% $CD34^+$ cells ($n = 10$); furthermore, a mean debulk-

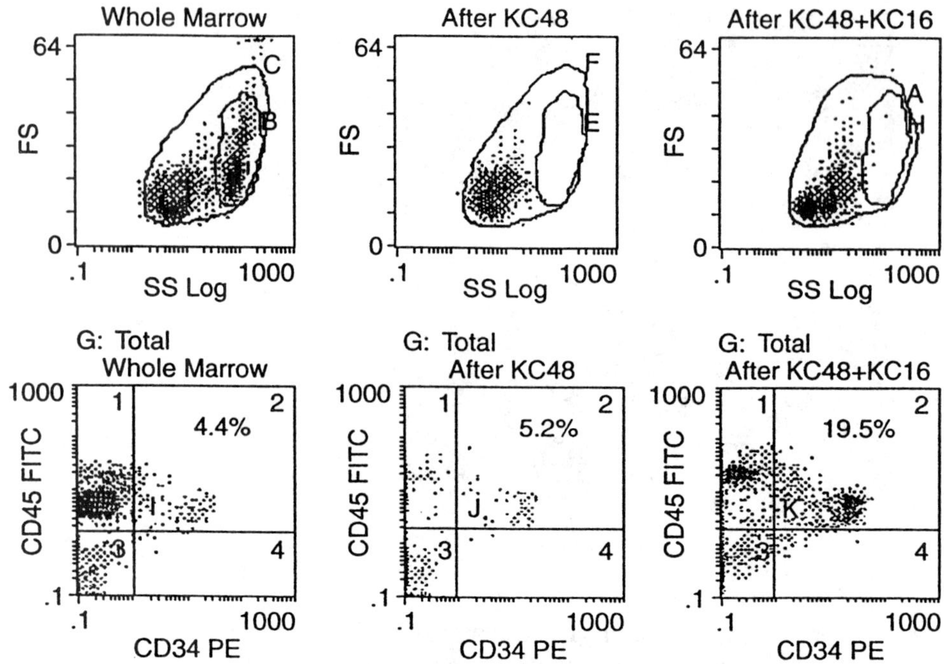

Figure 7 Dual-parameter flow cytometric histograms of whole vertebral body marrow (VBM), KC48-depleted VBM, and KC48/KC16-depleted marrow. (Top) Light scatter histograms; (bottom) dual-color histograms of CD45-FITC (fluorescein isothiocynate) versus CD34-PE (phycoerythrin). Flow cytometric analysis of unlysed, whole VBM (left) revealed "lymphoid" and myeloid populations of cells, with 4.4% CD45/34$^+$ stem cells present. After elimination of KC48$^+$ cells (middle), the myeloid population was eliminated (see light scatter histogram) and 5.2% CD45/34$^+$ stem cells were present. Further processing to remove mature and nucleated red blood cells with KC16-HDP (right) resulted in a preparation consisting of 19.5% CD45/34$^+$ cells. Note that the "lymphoid" population appears denser in the light scatter histogram, reflecting elimination of erythroid lineage cells; also, a "monocytoid" population is now clearly visible.

ing of 88.5 ± 4.4% ($n = 10$) was achieved (31,32). These types of results can be obtained in less than 1 hour, and the procedure has been scaled up for processing of the 70 billion cells used in our clinical trials. By adjusting cell concentration and particle volume, it was determined that 100 million cells/mL could be effectively processed with 50 μL/mL HDP. This decreased the HDP required (initial experiments were done at 5 million cells/mL) from 700 to 35 mL. Results from the analysis of colony-forming capacity revealed that the number of colonies obtained after elimination of CD15$^+$ (KC48) and glycophorin A$^+$ (KC16) cells was not as high as predicted based on the number of input CD34$^+$ cells. It is well known that various cytokines produced by T and B cells, which constitute approximately 70% of the KC48/KC16 negative preparation, can have an inhibitory effect on in vitro colony-forming capacity. Using T- and B-cell–specific HDP, we eliminated the lymphoid populations and analyzed the resulting preparations for colony-forming capacity. A clear correlation between the input number of CD34$^+$ cells and colony-forming capacity was observed, thus verifying that T and B cells, as opposed to an inhibitory effect of the particles themselves, were responsible for the results obtained with the KC48$^-$/KC16$^-$ cells.

We have subsequently performed experiments usng additional lineage-specific HDP and have rapidly and simply obtained VBM preparations composed of approximately 60% CD34$^+$ cells. Either the preenriched, KC48/KC16-negative preparation (approximately 15% CD34$^+$) or the stem cell–rich (approximately 60% CD34$^+$) sample can easily be processed to recover stem cell preparations of 99% purity, as demonstrated in Figure 9. In this particular example, cell sorting was used to achieve purity, but the cells could easily be processed using currently marketed technologies for positive selection of CD34$^+$ hematopoietic stem cells.

Blood

High-density particles have been effectively used to rapidly enrich for rare event hematopoietic stem cells from both normal and mobilized peripheral blood, as well as from cord blood. In these experiments, blood was first processed with a combination of CD15 (KC48) and CD41 HDP to remove myeloid populations and platelets, respectively. Red blood cells were subsequently removed by density-gradient centrifugation over Ficoll-paque, followed by processing with glycophorin A–specific HDP to remove residual red cells. The final step involved processing with a mixture of HDP specific for (i.e., coated with a mAb specific for) CD2, 4, 5, 8, 14, 16, 19, and/or 20. For leukapheresed peripheral blood (normal or stimulated), however, red cells can easily be removed without the need for gradient centrifugation. With starting percentages of 0.5% and 0.7%, means of 16.5% ($n = 13$) and 45.9% ($n = 4$) CD34$^+$ cells were obtained for normal peripheral blood and cord blood, respectively. Shown in Figure 10 is a representative example obtained using a similar protocol to process rHuG-CSF–stimulated, leukapheresed whole blood. In this particular example, a starting population of 3.7% CD34$^+$ cells was increased to 35.9%, with a 94% recovery of hematopoietic stem cells. As

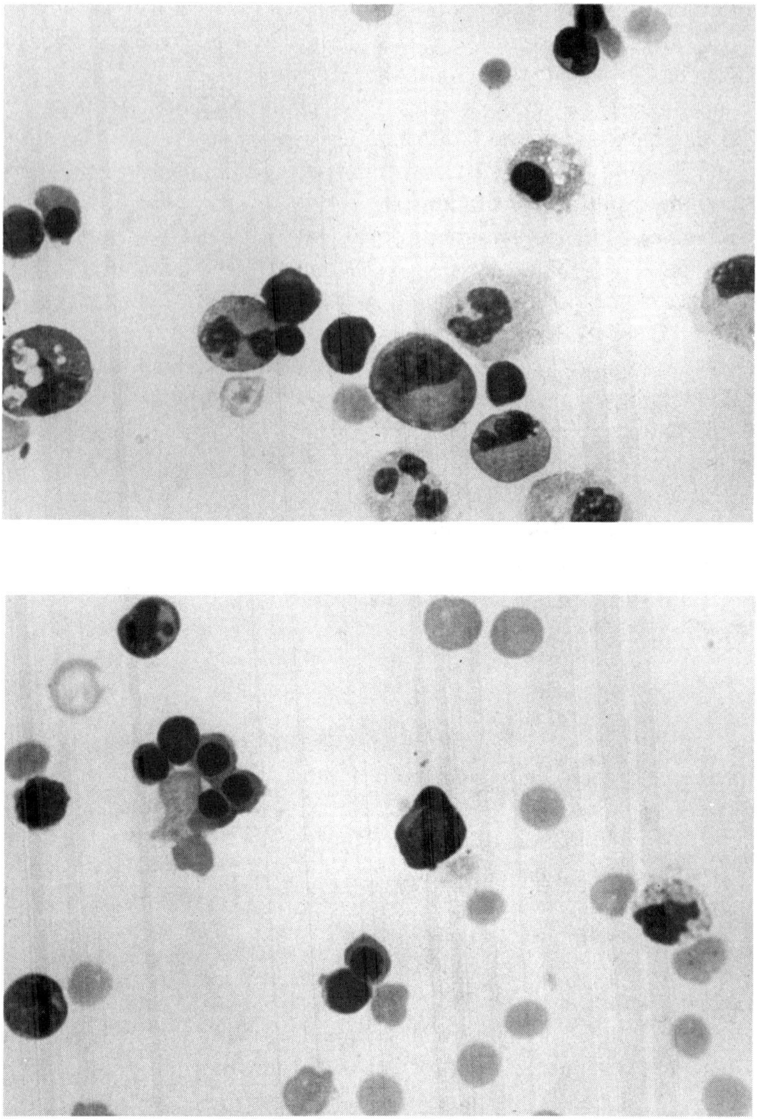

Figure 8 Photomicrographs of whole vertebral body marrow (VBM), KC48-depleted VBM, and KC48/KC16-processed VBM. Shown at the top is a photomicrograph of whole VBM. The heterogeneous nature of the marrow is evident; note that VBM is

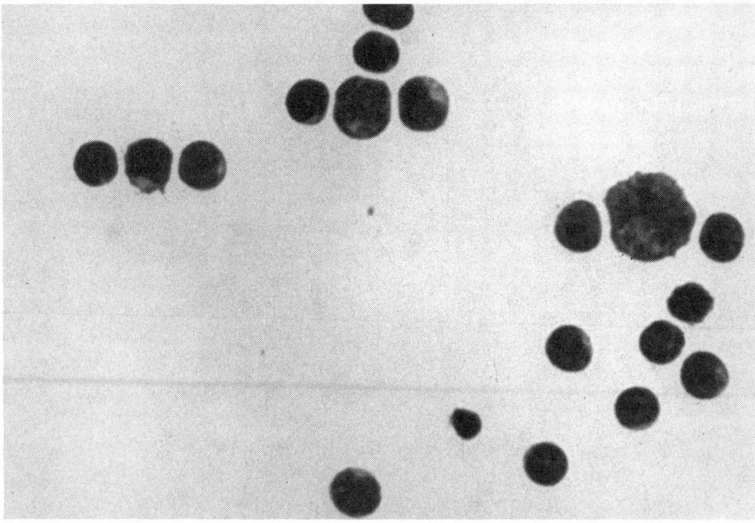

devoid of platelets. The majority of cells that remain after KC48 depletion (bottom of p. 118) are comprised of erythroid-lineage cells; removal of these populations with KC16 HDP results in a relatively homogeneous population of lymphoidlike cells (above).

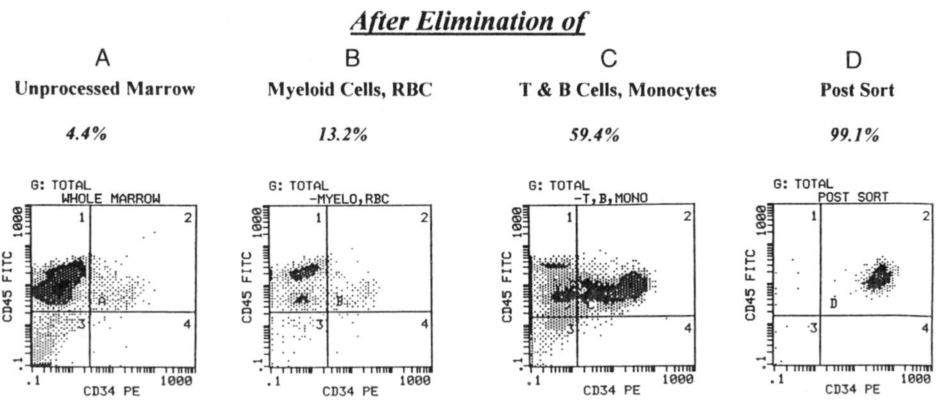

Figure 9 Flow cytometric analysis of vertebral body marrow (VBM) after enrichment for hematopoietic stem cells. (A) Whole marrow was processed to remove myeloid and erythroid cells; (B) preparation consisting of 12.9% CD45/34$^+$ stem cells. After removal of T and B cells plus monocytes with lineage-specific HDP, a preparation consisting of 59.2% (C) stem cells was obtained, which was rapidly sorted to yield 99.1% CD45/34$^+$ stem cells (D).

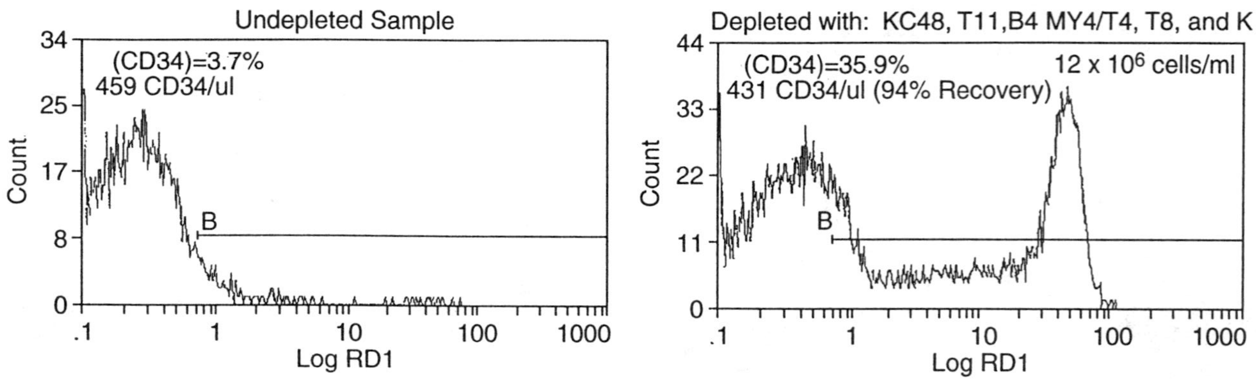

Figure 10 Enrichment for $CD34^+$ stem cells from leukapheresed, rHuG-CSF–stimulated peripheral blood. A sample of mobilized, leukapheresed peripheral blood was processed with HDP coated with mAb specific for CD15, CD2, CD19, CD4 + 14, CD8, and glycophorin A to remove granulocytes, T cells, B cells, monocytes, $CD8^+$ T cells, and red blood cells, respectively. $CD34^+$ stem cells were enriched from 3.7% (left) in the starting population to 35.9% (right) in the processed sample, with 94% recovery of stem cells.

mentioned for VBM, preenrichment of $CD34^+$ cells can be followed by a positive selection step; Figure 11 shows the data obtained when normal or CD34 preenriched peripheral blood was sorted to collect the stem cells. The unprocessed (i.e., density gradient, no HDP) peripheral blood sample contained 0.4% $CD34^+$ cells, and more than 1 hour of cell sorting was required to collect a small sample containing 17.4% hematopoietic stem cells. Using a preenriched preparation of 20% $CD34^+$ cells, however, a sample containing 91.1% $CD34^+$ cells was obtained in 15 minutes.

D. Purging of Tumor Cells

Autologous stem cell support from either bone marrow or peripheral blood stem cells has permitted the use of chemotherapy dose escalation for a large number of patients with a number of hematological and solid tumors (33–39). The major obstacle to the use of autologous stem cells is that contaminating tumor cells will also be infused back to the patient and contribute to subsequent relapse. To minimize this risk, most centers obtain autologous marrow either when the patient is in complete remission or when there is no history of marrow infiltration. Others believe that peripheral blood stem cells may provide a source that is less likely to be contaminated with malignant cells and use this as a rationale to move from autologous marrow to peripheral blood transplantation. However, it is now clear that a number of patients with no morphological evidence of disease in the marrow or peripheral blood have contaminating malignant cells when assessed by more sensitive techniques. Therefore, increasing effort has been made to determine the relative contribution of infusion of minimal numbers of malignant cells to subsequent relapse of disease. An alternative approach to decrease tumor contamination of stem cells is to "purge" malignant cells. The aim of purging is to eliminate any contaminating malignant cells yet leave intact hematopoietic stem cells necessary for engraftment. The development of purging techniques has led subsequently to a number of studies of autologous bone marrow transplantation in patients with either a previous history of marrow infiltration or even overt marrow infiltration at the time of marrow harvest (35,36,39,40). These clinical studies have demonstrated that purging can deplete malignant cells in vitro without significantly impairing hematological engraftment.

Whereas the rationale for removing any contaminating tumor cells from the autologous marrow appears to be compelling, the issue of purging remains highly controversial. To date, there have been no clinical trials testing the efficacy of purging by comparison of infusion of purged versus unpurged autologous marrow, due primarily to the large number of patients required for

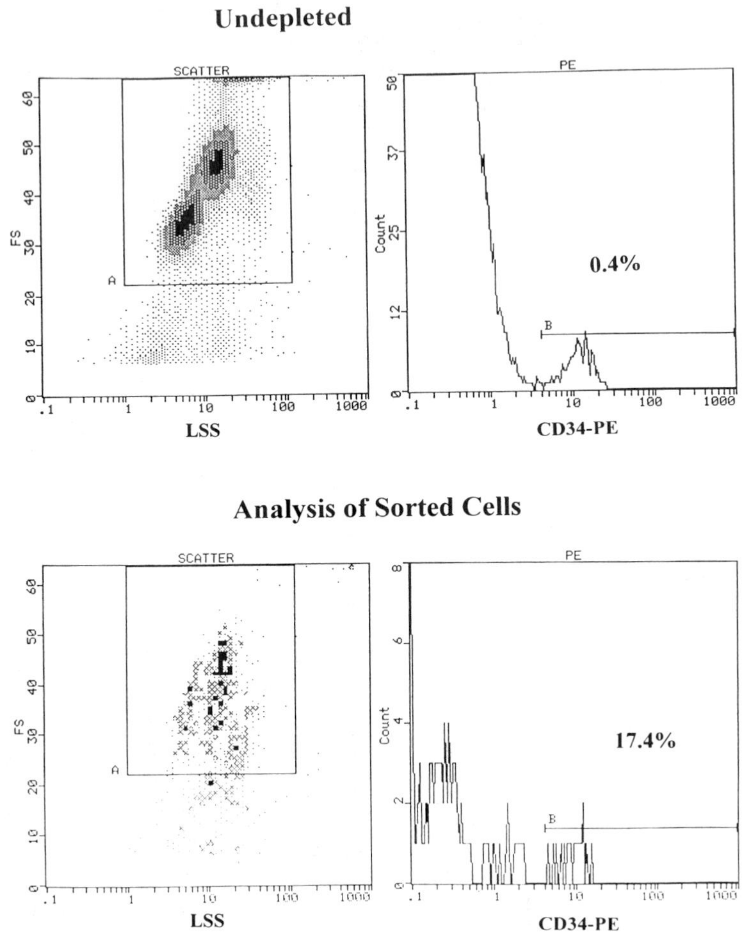

(a)

Figure 11 Preprocessing with HDP can decrease sort time and result in enhanced purity of CD34$^+$ stem cells. Peripheral blood was processed over Ficoll-paque to obtain a mononuclear cell preparation, followed by sorting of CD34-phycoerythrin (PE)-labeled cells on an EPICS ELITE (right); alternatively the cells were preprocessed with HDP and then sorted (left). Shown on top are the light-scatter histograms and single-parameter histograms (CD34-PE) of unsorted cells. Results from analysis of sorted cells

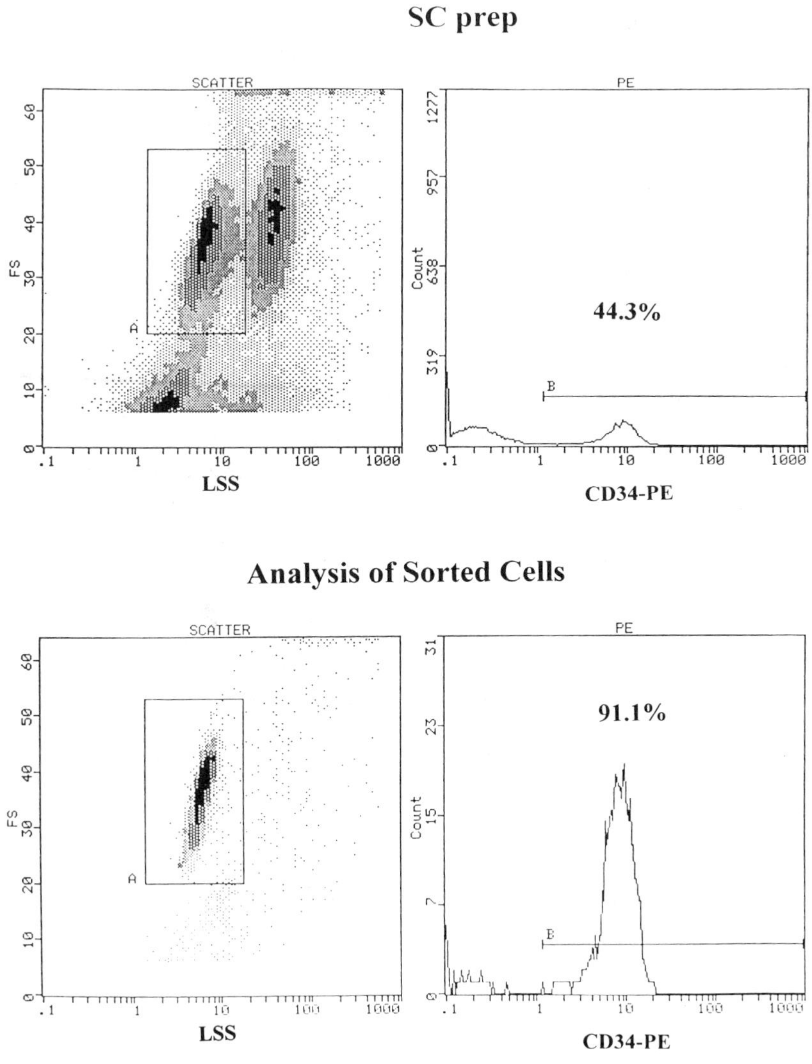

(b)

are shown on the bottom. A starting population of 0.4% CD34⁺ cells was sorted for more than 1 hour to obtain a small sample for analysis; a purity of only 17.4% CD34⁺ stem cells was obtained. Preprocessing with lineage-specific HDP (T cell, B cell, granulocyte, monocyte) resulted in an enrichment to 20% CD34⁺ HSC, thus allowing for collection of 91.1% pure CD34⁺ cells after approximately 15 minutes of sorting.

such studies. Intense argument therefore persists as to whether attempts to remove residual tumor cells from the harvested marrow have contributed to improving disease-free survival in these patients.

However, three independent lines of evidence have suggested that the infusion of tumor cells in autologous marow may indeed contribute to relapse. Firstly, gene-marking studies done at St. Jude Children's Hospital have demonstrated that at the time of relapse, "marked" autologous marrow cells are detected, suggesting that the infused tumor cells contribute to relapse (41). Second, studies at the University of Nebraska have demonstrated that patients infused with morphologically normal marrow containing clonogenic lymphoma cells have an increased incidence of relapse after autologous marrow transplantation (42). Third, patients whose marrows contain polymerase chain reaction (PCR)–detectable lymphoma cells after immunological purging had an increased incidence of relapse after autologous marrow transplantation (43).

The most sensitive assays to detect minimal numbers of malignant cells within autologous marrow or stem cells are PCR based. Using PCR analysis in patients with follicular lymphoma, we found that malignant cells could be detected in every patient analyzed who had a PCR-amplifiable translocation. Similarly, we have used PCR analysis to determine the efficacy of purging using HDP. Using a cocktail of HDP coated with B cell–specific mAbs, we could reproducibly deplete all PCR-detectable lymphoma cells after three rounds of treatment (Table 5). Moreover, these results were obtained without depleting hematopoietic progenitors (Table 6) or affecting LTC-IC frequency (Fig. 12). In addition, we could deplete all PCR-detectable lymphoma cells

Table 5 Depletion of B-Lymphoma Cells Using Dense Particles

Patient	Tumor contamination	Polymerase chain reaction: Before treatment	After rounds of treatment: 1	2	3
1	0	+	–	–	–
2	> 5%	+	+	–	–
3	5%	+	–	–	–
4	5%	+	+	–	–
5	10%	+	+	+	–
6	25%	+	+	+	–

Table 6 Effect of Dense Particles on Recovery of Hematopoietic Progenitors

	Before purging	After purging
GM-CFC	122 ± 27	139 ± 34
E-BFC	374 ± 78	423 ± 82

GM-CFC = Granulocyte-macrophage colony-forming cell; E-BFC = erythroid blast-forming cell.

when the marrow was contaminated with as much as 25% lymphoma cells before purging.

While three rounds of purging can lead to PCR negativity in the B-cell lymphoma model, PCR technology may not be sensitive enough to detect the level of cells that will cause recurrence of disease. It may be possible to further deplete malignant cells with additional rounds of purging, and due to the low nonspecific cell loss associated with dense particles, the potential

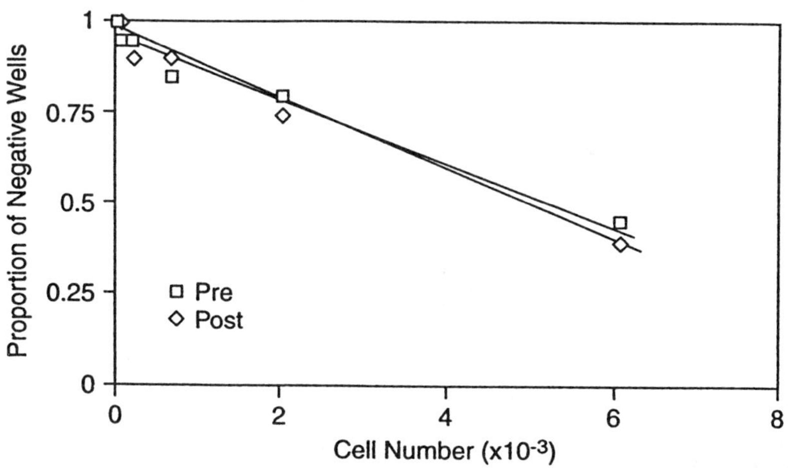

Figure 12 Analysis of long-term culture-initiating cells (LTC-IC) frequency after processing of marrow with HDP. Patient bone marrow was processed with B-cell–specific HDP to remove contaminating B-lymphoma cells, followed by analysis of LTC-IC frequency. As shown in the graph, the frequency of LTC-IC was unchanged for HDP, as compared with unmanipulated marrow.

to eliminate nondetectable tumor cells without compromising stem cell yield may be at hand.

E. Positive Selection

A number of preliminary experiments have been done to assess the potential utility of dense particles for positive selection. Representative examples of the data obtained for T-cell selection are given in Figures 13 and 14. In one case, a whole blood sample was treated with CD6-coated HDP to remove the T cells (Fig. 13). The T-cell–containing HDP pellet was collected and vortexed to remove the bound cells. As can be seen from the light scatter histogram, a pure population of lymphoid cells was obtained. Flow cytometric analysis of cells labeled with phycoerythrin (PE)-conjugated CD4 and CD3-FITC (fluorescein isothiocynate) revealed that 79.3% of the recovered cells were $CD4^+$ lymphocytes. This population was subsequently processed with CD8-specific HDP, thus resulting in a preparation consisting of 91.7% $CD4^+$ cells. Since the starting sample was whole blood, and one round of processing requires less than 10 minutes, a population consisting of > 90% CD4 cells can be obtained in less than 30 minutes. Alternatively, whole blood can be processed with CD4- or CD8-coated HDP, followed by vortexing of the HDP-cell pellet, to even more rapidly obtain enriched CD4+ (99.1% $CD3/4^+$) or CD8+ (94.1% $CD3/8^+$) cells (Fig. 14).

IV. FUTURE APPLICATIONS AND SUMMARY

The direct applicability of HDP to unprocessed blood or marrow samples, the short duration of mixing and settling, and the potential for multiple rounds of cell processing with minimal loss of nontargeted cell types are powerful attributes that facilitate the application of HDP technology to multiple research, clinical, and diagnostic strategies. Our combined experiences with HDP clearly demonstrate that the particles provide a means to rapidly and simply enrich for rare event or infrequent cell types. The resulting populations can then be further processed with HDP or with currently marketed positive- and/or negative-selection technologies. As an example, we have enriched eosinophils to more than 90% purity from whole blood. We have also used HDP to enrich for tumor cells, from mixtures containing less than 1% malignant cells, and observed that DNA analysis (ploidy and percent S phase) is subsequently enhanced. With the availability of key mAbs, it should be possible to enrich for dendritic cells and Ag-specific lymphocytes. In addition,

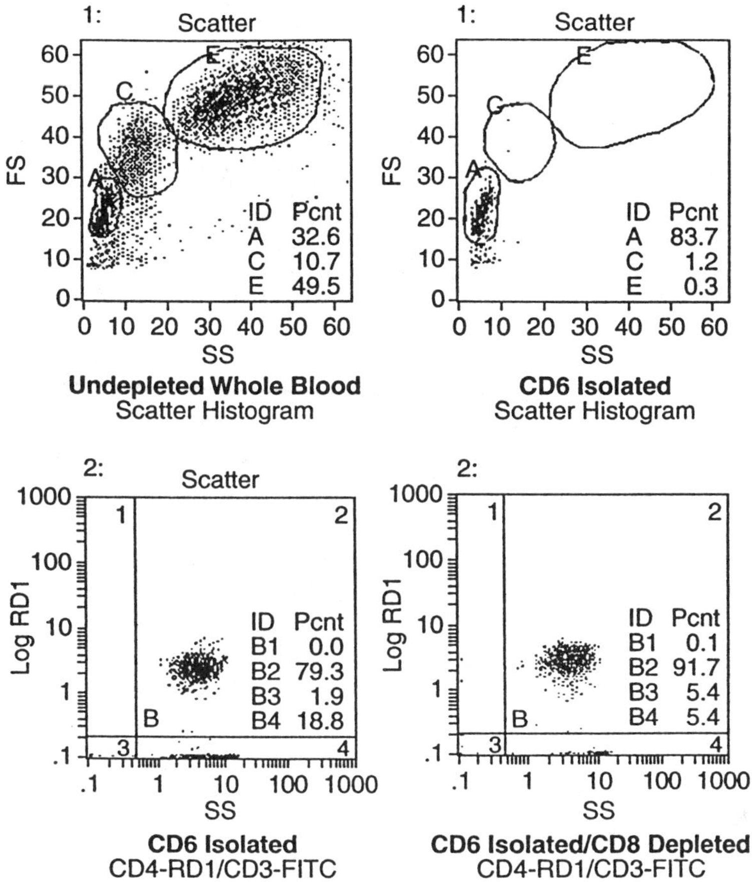

Figure 13 Positive selection of T cells and T-cell subsets. CD6-coated high-density particles (HDP) were used to process whole blood for removal of T cells. The HDP pellet was subsequently vortexed to remove the bound T cells, followed by depletion of $CD8^+$ cells with HDP. Enrichment for lymphocytes can be clearly observed by comparing the light-scatter histograms for whole blood (top left) to the histogram obtained for the cells removed from the HDP (top right). The percentage of $CD3/4^+$ T cells in the CD6 population was 79.3% (bottom left); after depletion of $CD8^+$ cells with HDP, the $CD3/4^+$ cells were enriched to 91.7% (bottom right).

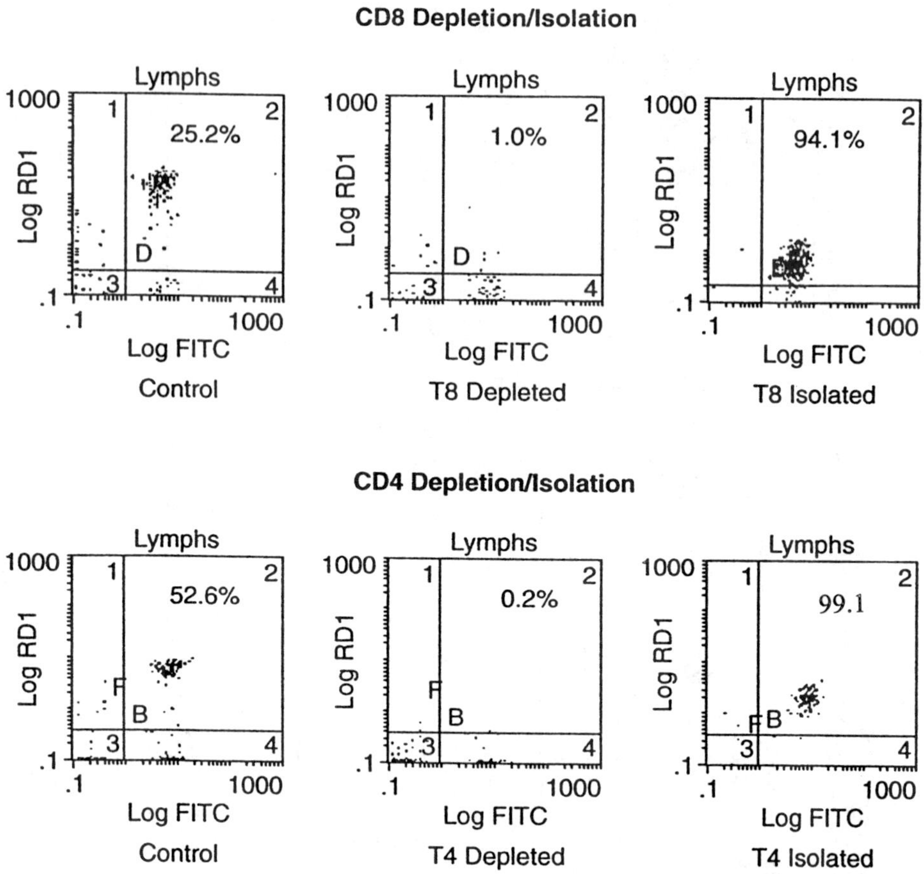

Figure 14 Positive selection of T-cell subsets. Whole blood was processed with CD4 or CD8-high-density particles (HDP), followed by vortexing to recover the bound populations. Shown are the dual parameter fluorescence histograms obtained after processing with CD8 (top) or CD4 (bottom) HDP. Purities of 94.1% and 99.1% were achieved for CD8 and CD4 cells, respectively.

the ability to rapidly and efficiently enrich for hematopoietic stem cells from a variety of clinical samples, either small or large scale, will clearly facilitate the design of protocols that call for stem cell expansion and gene therapy.

REFERENCES

1. Amos TA, Gordon MY. Sources of human hematopoietic stem cells for transplantation—a review. *Cell Transplant* 1995; 4:547.
2. Morstyn G, Sheridan W. *Cell Therapy.* Cambridge: Cambridge University Press, 1996.
3. Champlin R, Jansen J, Ho W, et al. Retention of graft-versus-leukemia using selective depletion of CD8-positive T lymphocytes for prevention of graft-versus-host disease following bone marrow transplantation for chronic myelogenous leukemia. *Transplant Proc* 1991; 231(Pt 2):1695–1696.
4. Nimer SD, Giorgi J, Gajewski JL, et al. Selective depletion of $CD8^+$ cells for prevention of graft-versus-host disease after bone marrow transplantation. *Transplantation* 1994; 57(1):82–87.
5. Champlin R. Separation of graft-versus-host disease and graft-versus-leukemia effect against chronic myelogenous leukemia. *Exp Hematol* 1995; 23(11):1148–1151
6. Malkovska V, Sondel PM. Adoptive immunotherapy with T cells. In: Morstyn G, Sheridan W, eds. *Cell Therapy.* Cambridge: Cambridge University Press, 1996:429.
7. Carson WE, Caligiuri MA. Cellular adoptive immunotherapy with natural killer cells. In: Morstyn G, Sheridan W, eds. *Cell Therapy.* Cambridge: Cambridge University Press, 1996:451.
8. Strunk D, Rappersberger K, Egger C, et al. Generation of human dendritic cells/Langerhans cells from circulating $CD34^+$ hematopoietic progenitor cells. *Blood* 1996; 87(4):1292–1302.
9. Kolb HJ, Mittermuller J, Clemm C, et al. Donor leukocyte transfusions for treatment of recurrent chronic myelogenous leukemia in marrow transplant patients. *Blood* 1990; 76(12):2462–2465
10. Slavin S, Naparstek E, Nagler A, et al. Allogeneic cell therapy with donor peripheral blood cells and recombinant human interleukin-2 to treat leukemia relapse after allogeneic bone marrow transplantation. *Blood* 1996; 87(6):2195–2204.
11. Mackinnon S, Papadopoulos EB, Carabasi MH, et al. Adoptive immunotherapy evaluating escalating doses of donor leukocytes for relapse of chronic myeloid leukemia after bone marrow transplantation: separation of graft-versus-leukemia responses from graft-versus-host disease. *Blood* 1995; 86(4):1261–1268.
12. Giralt S, Hester J, Huh Y, et al. CD8-depleted donor lymphocyte infusion as treatment for relapsed chronic myelogenous leukemia after allogeneic bone marrow transplantation. *Blood* 1995; 86(11):4337–4343.

13. Alyea E, Soiffer R, Murray C, et al. Adoptive immunotherapy following allogeneic bone marrow transplantation (BMT) with donor lymphocytes depleted of $CD8^+$ T cells. *Blood* 1995; 86:293a (abstract No. 1158).
14. Gribben J, Nadler L. Detection of minimal residual disease. In: Buckner CD, ed. *Technical and Biological Components of Marrow Transplantation*. Boston: Kluwer Academic Publishers, 1995:249.
15. Lill M, Gasson JC. Current and potential uses of defined cell populations. In: Morstyn G, Sheridan W, eds. *Cell Therapy.* Cambridge: Cambridge Univeristy Press, 1996:47.
16. McDaniel DO, Naftilan J, Huvey K, et al. Peripheral blood chimerism in renal allograft recipients transfused with donor bone marrow. *Transplantation* 1994; 57(6):852–856.
17. Starzl TE, Demetris AJ, Murase N, Ildstadt S, Ricordi C,Trucco M. Cell migration, chimerism, and graft acceptance. *Lancet* 1992; 339(8809):1579–1582.
18. Ricordi C, Karatzas T, Selvaggi G, et al. Multiple bone marrow infusions to enhance acceptance of allografts from the same donor. In: Sackstein R, Janssen W, Elfenbein G, eds. *Bone Marrow Transplantation: Foundations for the 21st Century.* New York: New York Academy of Science, 1995:345.
19. Bachar-Lustig E, Rachamim N, Li H-W, Lan F, Reisner Y. Megadose of T cell-depleted bone marrow overcomes MHC barriers in sublethally irradiated mice. *Nature Med* 1995; 1:1268.
20. Aversa F, Tabilio A, Terenzi A, et al. Successful engraftment of T-cell–depleted haploidentical “three-loci” incompatible transplants in leukemia patients by addition of recombinant human granulocyte colony-stimulating factor-mobilized peripheral blood progenitor cells to bone marrow inoculum. *Blood* 1994; 84(11):3948–3955.
21. Reisner Y, Marteli MF. Bone marrow transplantation across HLA barriers by increasing the number of transplanted cells. *Immunol Today* 1995; 16:437.
22. Dunbar CE. Clinical gene therapy. In: Morstyn G, Sheridan W, eds. *Cell Therapy.* Cambridge: Cambridge University Press, 1996:369.
23. Koller MR, Emerson SG, Palsson BO. Large-scale expansion of human stem and progenitor cells from bone marrow mononuclear cells in continuous perfusion cultures. *Blood* 1993; 82(2):378–384.
24. Zandstra PW, Eaves CJ, Piret JM. Expansion of hematopoietic progenitor cell populations in stirred suspension bioreactors of normal human bone marrow cells. *Biotechnology* 1994; 12(4):909–914.
25. Davis TA, Robinson DH, Lee KP, Kessler SW. Porcine brain microvascular endothelial cells support the in vitro expansion of human primitive hematopoietic bone marrow progenitor cells with a high replating potential: requirement for cell-to-cell interactions and colony-stimulating factors. *Blood* 1995; 85(7):1751–1761.
26. Zwerner RK, Schmittling RJ, Russell TR. Novel dense particle technology: a simple, rapid method for the removal of specific subpopulations from whole blood. *Blood* 1995; 86:230a (abstract No. 909).

27. Schmittling RJ, Fuentes S, Zwerner RK, Russell TR. Novel dense particle technology: a method which allows for multiple rounds of purging while retaining high cell recovery. *Blood* 1995; 86:991a (abstract No. 3952).
28. Fletcher GG, Rossetto FE, Turnbull JD, Nieboer E. Toxicity, uptake, and mutagenicity of particulate and soluble nickel compounds. *Environ Health Perspect* 1994; 102(suppl 3):69–79.
29. Stewart FM, Crittenden RB, Lowry PA, Pearson-White S, Quesenberry PJ. Long-Term engraftment of normal and post-5-fluorouracil murine marrow into normal nonmyeloablated mice. *Blood* 1993; 81(10):2566–2571.
30. Kenyon NS, Xu X-M, Garcia-Serra A, Ricordi C. Immunogenicity and stem cell content of class II depleted human vertebral body bone marrow. In: Sackstein R, Janssen WE, Elfenbein G, eds. *Bone Marrow Transplantation: Foundations for the 21st Century.* New York: New York Academy of Sciences, 1995:361.
31. Kenyon NS, Ricordi C, Russell TR. Novel dense particle technology: a rapid method for hematopoietic stem cell enrichment from human vertebral body marrow suspensions. *Blood* 1995; 86:985a (abstract No. 3927).
32. Kenyon NS, Ricordi C, Xu X-M, et al. Rapid enrichment of rare event hematopoietic stem cells from human vertebral body marrow. 4th International Symposium on Recent Advances in Hematopoietic Stem Cell Transplantation—Clinical Progress, New Technologies, and Gene Therapy 1996, p. 45.
33. Armitage JO. Bone marrow transplantation in the treatment of patients with lymphoma. *Blood* 1989; 73(7):1749–1758.
34. Ball ED, Mills LE, Cornwell CG, et al. Autologous bone marrow transplantation for acute myeloid leukemia using monoclonal antibody-purged bone marrow. *Blood* 1990; 75(5):1199–1206.
35. Freedman AS, Nadler LM. Developments in purging in autotransplantation. *Hematol Oncol Clin North Am* 1993; 7(3):687–715.
36. Freedman AS, Takvorian T, Anderson KC, et al. Autologous bone marrow transplantation in B-cell non-Hodgkin's lymphoma: very low treatment-related mortality in 100 patients in sensitive relapse. *J Clin Oncol* 1990; 8(5):784–791.
37. Gribben JG, Goldstone AH, Linch DC, et al. Effectiveness of high-dose combination chemotherapy and autologous bone marrow transplantation for patients with non-Hodgkin's lymphomas who are still responsive to conventional dose therapy. *J Clin Oncol* 1989; 7(11):1621–1629.
38. Peters WP, Shpall EJ, Jones RB, et al. High dose combination alkylating agents with bone marrow support as initial treatment for metastatic breast cancer. *J Clin Oncol* 1988; 6(9):1368–1376.
39. Hurd DD, LeBien TW, Lasky LC, et al. Autologous bone marrow transplantation in non-Hodgkin's lymphoma: monoclonal antibodies plus complement for ex vivo marrow treatment. *Am J Med* 1988; 85(6):829–834.
40. Takvorian T, Canellos GP, Ritz J, et al. Prolonged disease-free survival after autologous bone marrow transplantation in patients with non-Hodgkin's lymphoma with a poor prognosis. *N Engl J Med* 1987; 316:1499.

41. Brenner MK, Rill DR, Moen RC, et al. Gene-marking to trace origin of relapse after autologous bone-marrow transplantation. *Lancet* 1993; 341(8837):85–86.
42. Sharp JG, Joshi SS, Armitage JO, et al. Significance of detection of occult non-Hodgkin's lymphoma in histologically uninvolved bone marrow by culture technique. *Blood* 1992; 79(4):1074–1080.
43. Gribben JG, Freedman AS, Neuberg D, et al. Immunologic purging of marrow assessed by PCR before autologous bone marrow transplantation for B-cell lymphoma. *N Engl J Med* 1991; 325(22):1525–1533.

7

Antibody- and Complement-Mediated Cell Separation

Adrian P. Gee
University of South Carolina School of Medicine, Columbia, South Carolina, and M. D. Anderson Cancer Center, Houston, Texas

I. INTRODUCTION

The availability of monoclonal antibodies (mAbs) has added a degree of specificity to selective cell separation that was difficult to achieve with polyclonal reagents. Sensitization of the cell subpopulations with mAb is both rapid and technically simple; however, the choice of which cells to remove from a mixed population has been less straightforward. Two approaches are available—positive and negative selection (1). In the former, the target cells are identified by mAb, which is then used to mediate their enrichment. In negative selection, the enrichment is achieved by using mAb to identify and mediate the removal of all cells in the mixture, except those to be enriched. Each technique has specific advantages and disadvantages that determine its suitability for a particular application. A major consideration is the availability and cost of suitable mAbs. For high-purity enrichment of a cell subpopulation from complex cell mixtures, positive selection is usually the method of choice, since it requires a single mAb with specificity for that target cell. In contrast, negative selection requires multiple mAbs directed at a variety of epitopes to ensure complete removal of a heterogeous nontarget population (2,3). In spite of this limitation, negative selection has been used for the enrichment of rare cell populations, for example, hematopoietic cells (4). However, positive selection has now become the predominant method for this application (5) due to the widespread availability of anti-CD34 mAb, which selectively binds to primitive hematopoietic progenitor cells (6).

Monoclonal antibody–sensitized cells can be eliminated either by physical removal (7,8) or by destruction in situ (1,9). Removal techniques include panning (10,11), immunoaffinity columns (12), and various immunomagnetic techniques (13–16), which are discussed elsewhere in this volume. Destruction in situ, or purging, can be achieved by conjugation of a toxic substance (17) [e.g., ricin (18), diphtheria toxin (19), or a radioisotope (20)] to the mAb or by the addition of serum complement [C′] (21–23). This chapter describes the advantages and limitations of using mAb and C′ for cell separation and discusses points that should be considered when developing a purging procedure.

II. ADVANTAGES AND DISADVANTAGES OF C′-MEDIATED CELL SELECTION

A. Advantages

Antibody and complement treatment can be used for either positive or negative selection, as described above. It is most commonly used to destroy, or purge, an unwanted cell subpopulation in applications where that population is well characterized and no longer required for further study, e.g., elimination of tumor cells from bone marrow to be used for transplantation (21–23). The major advantage to this method is that it is technically simple, requiring a two-step incubation, first with mAb and then with C′, in the form of whole serum. Serum complement also provides a generic method for eliminating any cell that has been sensitized with a C′-fixing mAb. This facilitates the use of multiple mAbs to compensate for any antigenic heterogeneity in the target-cell population (2,3,22). In contrast, other in situ destruction methods require direct conjugation of the toxic agent to each mAb used (17–20). Serum complement is also a physiological means for eliminating cells. It exists as a family of nontoxic serum proteins requiring activation by specific mechanisms to initiate the cascade reaction that ultimately results in destruction of the target cell by colloid osmotic lysis (24,25). Since it is used in the form of whole serum, it is inexpensive, readily obtainable, and with careful handling and storage has a long shelf life.

B. Disadvantages

Target Antigen Density

There are several major disadvantages associated with using C′ for cell destruction. It is almost impossible to achieve 100% target-cell destruction when using a single IgG mAb to mediate C′ killing. This is due to the mechanism by which IgG molecules activate C′ (26). Once IgG is bound to the cell surface, a subcomponent (C1q) of the first component of C′ (C1) must bind to two adjacent IgG molecules to activate the classical complement pathway (Fig. 1) and initiate destruction of the cell (26). This configuration can be easily achieved on cell that express high levels of the target antigen (Ag), however, on cells with low Ag density, IgG doublet formation becomes increasingly rare and the chances of activating the C′ cascade decrease. As a result, a subpopulation of target cells that displays low expression of Ag can escape lysis (27–31). It has been possible to develop C′-resistant cell lines from these subpopulations (32). In many cases this phenomenon may be overcome by using multiple mAbs to target different Ags on the cell surface, with the hope that cells expressing low densities of one Ag can be effectively targeted via high expression of a second Ag (33). This does not eliminate the possibility that certain target cells may constitutively express low densities of all the target Ags, or that Ag expression may be selectively altered by exposure of the target cells to drugs, e.g., during chemotherapy (34). Under these circumstances it may prove impossible to achieve effective purging. This situation has been described when the bone marrow of patients with lymphoma was treated with a panel of mAb and C′, and residual cells were detected using a highly sensitive polymerase chain reaction (PCR) assay (35). The investigators found that after purging, PCR negativity was achieved in only 50% of the patients.

We have previously shown that cells resistant to C′, by virtue of their low Ag expression, may still be eliminated using immunomagnetic techniques (27), which are less sensitive to the effects of target Ag density (36,37). Alternatively, improved lysis may be achieved by using IgM mAbs that are more efficient at fixing C′ (Fig. 2). In class-switching experiments, use of IgM B cell-directed mAbs resulted in a 1-log increase in their ability to effect C′-mediated lysis of cell lines, in comparison to their IgG2b equivalents (37).

In some applications, for example, T-lymphocyte depletion of marrow used for allogeneic transplantation, in complete contrast to tumor purging, it is desirable to leave a small number of target cells behind after purging (8). Treatment of grafts with mAb and C′ has proved to be particularly suitable for

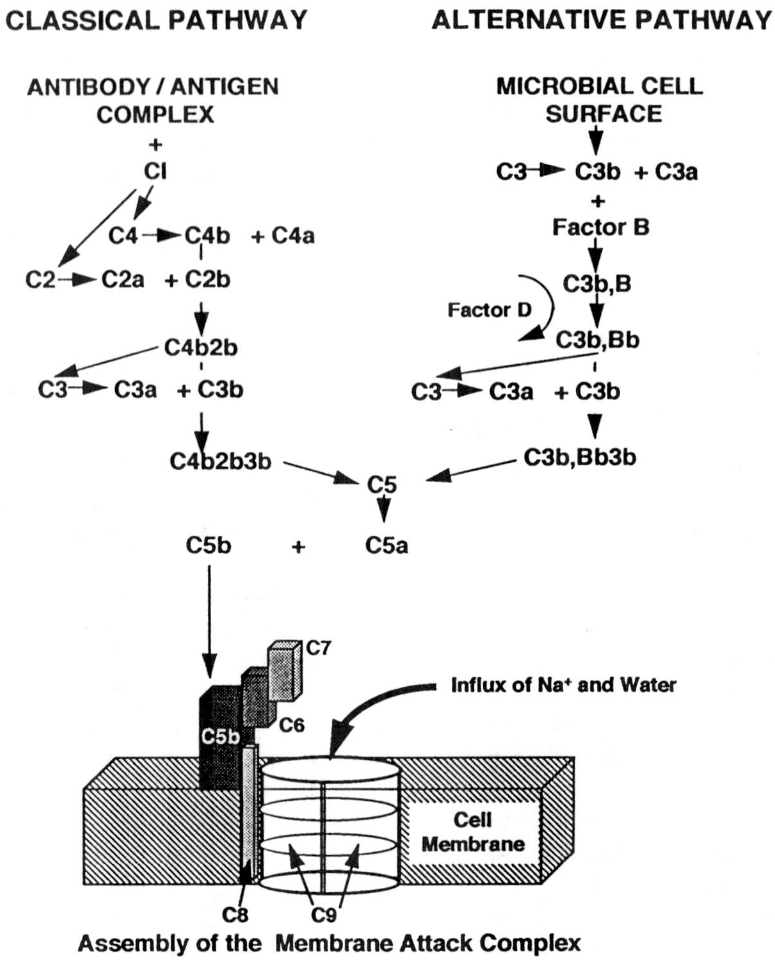

Figure 1 The classical and alternative pathways of complement-mediated cytolysis. In the majority of cell-separation applications, the classical pathway is activated using target-cell–directed mAb. These bind the C1q subcomponent of C1, which initiates the cascade, resulting in the assembly of the membrane attack complex on the cell surface and formation of a transmembrane channel.

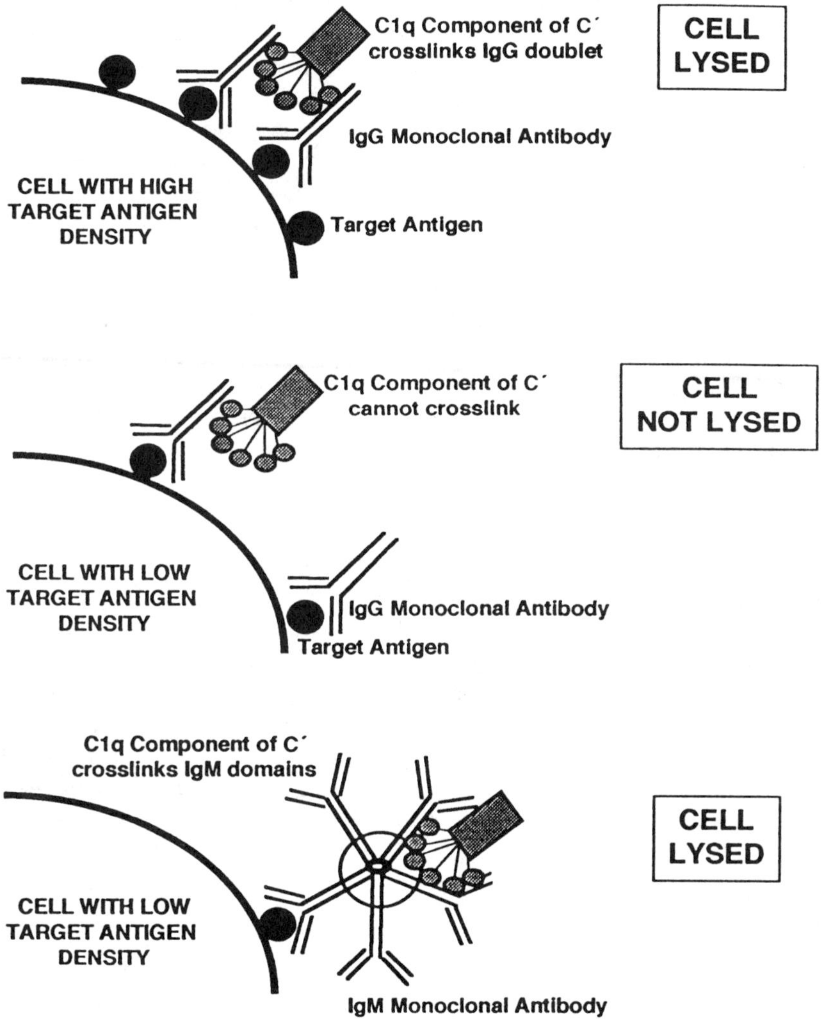

Figure 2 Effect of target antigen density on susceptibility of complement-mediated lysis. (Upper panel) At high target antigen densities, doublets of IgG are easily formed on the target cell surface. This configuration permits cross-linking of the C1q subcomponent of the C1 component of complement, initiating the cascade that results in lysis of the cell. (Middle panel) At low target-cell antigen density, it becomes progressively more difficult to form IgG doublets. As a result, the C1q is not cross-linked and lysis does not occur. (Lower panel) Under the same low-antigen conditions as shown in the middle panel, an IgM antibody molecule can internally cross-link C1q, trigger the complement cascade, and lyse the cell.

this purpose. In our laboratory, treatment of marrows with the anti-CD3 IgG mAb OKT3 and rabbit C′ results in a 2.4-log depletion of T cells that achieves the desired goal of infusing $< 1 \times 10^5$ T cells/kg body weight into the patient (38). These small numbers of T cells are insufficient to provoke a severe graft-versus-host response, but can prevent rejection of the marrow graft and may be involved in mediating a beneficial graft-versus-residual disease effect (39). The degree of sparing of target cells is usually determined by the characteristics of the particular system and may be the result of multiple effects (e.g., target Ag density and distribution and anticomplementary factors as described below). To a great extent these are predetermined and difficult to alter in a controllable manner. Unlike other methods for selective cell elimination, where experimental conditions can be easily manipulated to achieve the desired effect, in C′-mediated killing the investigator routinely uses an excess amount of both mAb and C′ and relies on the properties of the system components to obtain the desired results.

Anticomplementary Factors

In the context of bone-marrow purging, we originally described (40), and others have since confirmed (41), that a C′ decay–accelerating factor (DAF) is associated with normal marrow cells. This factor acts to hasten the decay of an intermediate complex of C′ components formed during the cascade reaction (24), before the stabilization of the complex by binding of the next C′ component. This reduces the efficiency of the lysis reaction, with the result that cells may escape destruction independent of their expression of target antigen. Decay-accelerating factor may be removed by washing the marrow cells several times before the addition of C′ (40) or by adding an anti-DAF mAb to the reaction (41). It is unlikely that these factors are only associated with human bone marrow cells (42), and their effects may explain inefficient target cell lysis in a number of systems.

"Nonspecific" Toxicity

Since C′ is routinely added to a cell mixture as a dilution of whole serum, it is not unusual to see lysis of nontarget cells due to the presence of various antibodies in the serum (2). The degree of "nonspecific" lysis is highly dependent on the species and source of serum (43). Killing of target cells by C′ obtained from the same species is usually less effective than that achieved using a xenogeneic source, and C′ from different species vary widely in their ability to be activated by and to lyse xenogeneic target cells (44,45). For lysis of human cells, rabbit C′ is particularly effective, and to avoid high levels of potentially cross-reactive immunoglobulin, the serum is usually collected from young animals (43). To reduce toxicity that may be associated with a particular animal, and to provide batches of adequate volume, it is normal to pool collections from a large number of animals. This material is then screened for both nonspecific cytotoxicity toward human cells and for its specific cytotoxic titer on mAb-sensitized human target cells (43). Although rapid screening may be performed using lymphocytes and procedures that are common in tissue-typing laboratories, the C′ source should also be tested in the specific application for which it was collected. In many cases, for simplicity, C′ is added in vast excess to the amount actually required to achieve efficient cell lysis. This can increase nonspecific activity and result in a prozone effect (46), in which lysis efficiency declines at supranormal antibody and/or C′ concentrations.

Regulatory Issues

In many countries, regulatory agencies require extensive testing of any animal product to be used for human therapeutic procedures. In the United States, the initial emphasis was on mAb (47); however, the Food and Drug Administration has now indicated that serum C′ will be required to meet the same standards (47). The level of testing required is beyond the scope and means of most academic institutions; however, at least one commercial source has indicated that its C′ will meet these requirements.

III. MATERIALS AND METHODS

As with all biological systems, the development of an efficient purging procedure using mAb and C′ requires careful optimization of experimental conditions. This should be performed for each application, and reoptimization is required if conditions are changed. The following sections describe many of the variables that should be considered when developing a purging procedure.

A. Choice of Monoclonal Antibodies

Many of the factors that should be considered when selecting a mAb for C′-mediated purging are similar to those described elsewhere in this volume for immunomagnetic cell separation. If the aim is to achieve maximum depletion of the target cell(s), the usual approach is to use multiple antibodies (Abs) directed against independently expressed Ags on the target(s) (2,3,9). As discussed previously, the efficacy of any single IgG mAb will be critically dependent on the density of expression of its particular antigen by the target cell (2). This is not true of IgM mAb since, unlike IgG Abs, a single IgM mAb bound to the cell surface is capable of binding and activating the C′ cascade (26). In experiments with chimeric Abs, in which the same Ag-combining site was linked to different human constant domains, it was found that both the target epitope density and the conditions under which the immune complex was formed (i.e., Ag or Ab excess) influenced the extent of human C′ activation and the pathway by which it was activated (46). Interestingly, IgM Abs activated the classical pathway well at high epitope densities, but showed a prozone phenomenon when the immune complexes were formed in Ab excess (46).

The choice of mAb is obviously restricted by the requirement that they are of a C′-fixing isotype (26). Mouse mAb are the most widely used for purging applications, and these must be of the IgG2a and IgM subclass. Human IgM, IgG1, and IgG3 are able to fix C′ and have been used in clinical purging applications (48). In some cases, these mAb have been infused directly into patients, or used to sensitize cells ex vivo that are then infused; reaction with C′ occurs in vivo (49). In this review, the focus is on ex vivo treatment with both mAb and C′.

In the United States, mAb that are used in clinical studies require regulatory approval as either investigational new drugs (IND) or through an investigational device exemption (IDE), if they are a component of a more extensive separation system. Antibodies approved for other indications may be used “off label” for purging at the discretion of a physician and with the approval of the institutional review board of the institution. For nonclinical applications, there are many commercially available mAb suitable for purging applications.

B. Choice of Complement Sources

The Classical and Alternative Pathways

Serum complement consists of a series of more than 20 proteins including the major C′ components and various regulators (24,25). Serum complement may be activated by the classical and/or the alternative pathway. The classical pathway is triggered by the C1q subcomponent of the first C′ component, C1, binding to either a single IgM molecule or a doublet of IgG molecules on the surface of the target cell. This initiates a cascade reaction in which subsequent components sequentially assemble on the cell surface to form enzymes that activate the next component (Fig. 1). This results, ultimately, in the assembly of a structure, known as the membrane attack complex, which is composed of components C5b, C6, C7, and C8, and multiple molecules of C9 (50). This funnel-shaped complex traverses the cell membrane resulting in colloid osmotic lysis of the cells.

Triggering of the alternative pathway is independent of Abs and is initiated by activation of the C3 component, which can occur during phagocytosis or an inflammatory reaction (25). This results in cleavage of the C3 molecule, and the C3b fragment combines with factor B to form complex that is, in turn, converted by factor D to form the membrane-bound C3 convertase, C3b,Bb. This complex fixes more C3b to form the C3b,Bb3b convertase that cleaves the C5 component (Fig. 1). It is at this point that the alternative and classical pathways merge and share common terminal steps. For purging purposes, the classical pathway is the primary route to achieve target cell lysis; however, it has been shown that activation of the alternative pathway may also determine the effectiveness of a particular C′ source in the lysis of xenogeneic tumor cells (45).

C′ Selection and Collection

As stated previously, the source of C′ can be either syngeneic or xenogeneic whole serum. Syngeneic and autologous sources may obviate the necessity of both extensive screening for adventitious agents and nonspecific cytotoxicity before use in clinical applications. In many cases, however, they are less efficient than xenogeneic sources (44,45), presumably due to a physiological mechanism designed to avoid excessive C′ activation in the host. For lysis of human cells, baby rabbit serum (43) is particularly effective. The optimal collection time appears to be 20–30 days after birth. Later collections show lower specific activity for nucleated cell lysis (43). Blood should be collected by either ear vein or cardiac puncture into sterile glass tubes on ice and allowed to clot. It is normal to pool sera from many animals to increase the lot size and dilute nonspecific toxicity associated with serum obtained from any one animal. The pooled material should be sterile filtered, aliquoted, and frozen at −80°C as soon as possible.

Rabbit complement is commercially available from a variety of vendors including C-Six Diagnostics (Mequon, Wl) and Pel-Freez (Brown Deer, Wl). Some commercial sources are tested to meet U.S. Food and Drug Administration requirements for agents used in somatic cell therapy (47). At the time of writing, no commercial source has been approved for clinical use in the United States.

Screening

Before use, each lot of C′ should be tested for specific (antibody mediated) and nonspecific (direct) cytotoxicity. Rapid, up-front screening on leukocytes may be performed using assays that are in place in most tissue-typing laboratories. Sources that pass these tests should then be tested in the specific system in which they are to be used. Target destruction and normal cell recovery and functional activity should be assessed using the tests that will ultimately be used to optimize purging efficacy. For nucleated target-cell lysis applications, the most widely used tests are limiting dilution analysis (8,38) and ^{51}Cr-release cytotoxicity assays (43).

Tests should include an examination of the effects of C′ concentration. It is normal to use C′ at high concentrations to ensure saturation of binding sites on the mAb. With some C′ sources there can be a prozone effect, such that specific lysis is reduced as the C′ concentration is increased. The use of high concentrations can also result in the appearance of nonspecific cytotoxicity. A moderate reduction in the C′ concentration can usually overcome these effects without sacrificing specific activity. In contrast, failure of lysis has also been attributed to exhaustion of C′ in the reaction, since improved cell killing was achieved by using multiple short incubations with fresh mAb and C′ (51). These results could also be due, however, to the activity of a C′ DAF (40,41), since the extent "exhaustion" was directly associated with the number of marrow mononuclear cells present in the reaction mixture.

C. Purging Procedure

Cell Preparation

The use of C′ to purge leukocyte preparations can result in problems of cell aggregation and clumping. This is usually worse in cell preparations that contain larger numbers of granulocytes. This may be either the direct result of mAb plus C′ or C′ alone on the cells (since it is usually first seen upon addition of the C′ source), or simply due to the effects of prolonged manipulation on these fragile cells. As a result, the resulting negatively selected population may be unsuitable for further studies. These effects can be avoided by using a mononuclear cell preparation (e.g., from a Ficoll Hypaque density cushion) as the starting material and by adding DNAse to the buffers during cell incubation and washing (51).

If blood or bone marrow is the initial cell source, it is important to avoid the use of Ca^{2+} chelating anticoagulants [e.g., ethylenediaminetetraacetic acid (EDTA)] during cell collection and handling, since calcium and magnesium are essential to the complement cascade reaction (24). Heparin and heparin fragments are also known to inhibit complement activity (52,53). These reagents should be removed during downstream cell processing by extensive washing with a suitable buffer, before addition of C′. Culture media and buffered salt solutions containing both Ca^{2+} and Mg^{2+} and a source of protein [e.g., heat-inactivated fetal bovine serum (FBS), plasma protein fraction, or albumin] are ideal for purging procedures. For research applications that necessitate optimal cell killing, specific veronal buffers containing defined amounts of Ca^{2+} and Mg^{2+} have been described (24).

Incubation with Antibody

Sufficient mAb must be added to the cell mixture to ensure saturation of the Ag on the target cells. The use of supersaturating concentrations is acceptable provided that it is shown that the presence of excess fluid-phase Ab does not impede C′-mediated lysis or that the cells are washed before the addition of C′. Under ideal circumstances, a titration curve for each Ab should be generated using lysis of representative target cells (e.g., cell lines with the same phenotype and Ag densities) as the assay variable. Normally, binding of labeled mAb detected by flow cytometry is used as a substitute, although the results obtained are usually not directly comparable. The concentration that is selected should be well onto the plateau portion of the curve. In our laboratory, an upper concentration of 4 μg mAb/target 1×10^6 cells is used and titrations are performed using serial threefold dilutions.

Incubation of the cells and mAb should be performed at high cell concentrations in small volumes with good mixing. This optimizes binding kinetics. The upper cell concentration will depend on the cell type. Under some circumstances blood and marrow cells will tend to aggregate at higher concentrations, especially if inadequately or roughly mixed. Incubations can be done at 4°C or ambient temperature. Ideally, mAb will be selected that do not cap or internalize at room temperature; if this is a risk, incubation on ice should be used. Incubation times depend on the application, although 20–60 minutes is a good rule of thumb. Mixing may be achieved either by slow (approximately 5 rpm) end-over-end rotation, rocking, or by occasional (every 5 to 10 minutes) manual inversion of the container. For small volumes, brief vortexing is acceptable.

In most cases it is unnecessary to wash away unbound Ab after the incubation and before addition of the C′. Prozone effects on complement activity have been reported when immune complexes are formed in IgM Ab excess (46). Each system should, therefore, be checked to determine if a wash step improves purging efficiency. In some cases, washing may improve the results primarily by removing anticomplementary factors (40,41), rather than fluid-phase mAb. Washing procedures may also be used to transition cells from the cold temperatures used for Ab sensitization to the warmer temperatures required for effective C′ action.

Incubation with Complement

Complement contains a number of heat-labile components and must, therefore, be stored at –70 to –80°C to maintain activity. Under these conditions, bovine and ovine C′ retain 80% of their activity when stored for 8 months, but lose 70–80% of activity when stored at –20°C for the same period (54). Aliquots should be thawed once and used or discarded. Aliquots required for purging should be thawed immediately before use, and any remaining should be discarded. Since C′ exerts its lytic effects through the assembly and activation of a number of enzymes, its activity is highly temperature dependent (24). Each 10°C increase in temperature, within its operating range, effectively doubles the activity. Incubations can be satisfactorily performed at ambient temperature (22–26°C) or at 37°C. The choice will again depend on the application. In our experience, incubations of bone marrow at 37°C can result in the appearance of nonspecific cytotoxicity, with accompanying increases in cell aggregation. We have not seen this in other applications. The use of lower temperatures will slow the kinetics of lysis and require incubation for a prolonged period (> 90 minutes) to reach the end-point of the reaction. At 37°C, incubation for 45–90 minutes is usually adequate for destruction of nucleated cells. Cells should be mixed during the incubation as described above. Repeated treatments with C′ have been reported to increase cell-purging efficiency (55–57), possibly by sequential dilution of anticomplementary factors and/or inhibitory concentrations of mAb. The C′ may be removed after incubation is completed by three to four washes in buffer or media.

For large-scale purging for clinical applications, cell preparation and incubation with mAb and C′ can be done in a closed system, in either transfer packs (58) or in the disposable bowl of an automated cell separator (59).

D. Analysis of Results

The method used to evaluate cell lysis depends on the particular application. Although C′ ultimately kills cells by colloid osmotic lysis, this is preceded by other damage to the cell membrane (50,60). Before the formation of the terminal membrane attack complex (MAC), the C5b-8 complex produces small lesions in the membrane that can be repaired by nucleated cells (61). Damage by C5b-8 can be detected by the release of small radiolabeled molecules, e.g., ^{86}Rb, from prelabeled target cells. C5b-8 also polymerizes C9 to form a 15nM long by 10nM-diameter transmembrane tubule (50). This perturbs the lipid bilayer sufficiently to result in a net influx of sodium ions and water, increasing the internal osmotic pressure that eventually kills the cell. Damage caused by the MAC can be detected by the release of larger tracer molecules or by the inability to exclude vital stains.

In routine applications, vital staining with trypan blue or eosin is the most practicable assay for detecting cell death. In many cases, however, it is possible to detect target cells that are faintly stained and therefore difficult to classify. These probably represent cells that have incurred damage, but have not yet lost membrane integrity. It is often possible to "develop" these cells by incubation for 1 hour at 37°C, during which lesion formation is completed and dye uptake becomes unequivocal.

This difficulty in discriminating between viable and nonviable populations is also seen when flow cytometric analysis is used. It is compounded by the appearance of cell debris in C′-treated cell preparations and by shifts in staining profiles, caused by the selective destruction of high-Ag-density target cells. These effects can be minimized by a 37°C preincubation, use of a viability stain, and careful gating. Immunophenotyping assays for residual cells are discussed in detail in the chapter on immunomagnetic separation using microspheres.

When immediate results are not required, the ultimate assay for residual cells is limiting dilution clonogenic analysis (8,38). This permits quantitation of target cell frequency before and after purging. These assays are, however, tedious, and results are not available for 12–14 days. For rapid gross evaluation of purging in model systems, the target cells can be prelabeled with a radioisotope, e.g., ^{51}Cr, and killing is quantitated as specific isotope release. Molecular methods, e.g., the PCR, have also been used (35,55); however, care must be taken to ensure that nucleic acids from lysed cells do not interfere with the interpretation of the results (62). As with all selection techniques that involve in situ destruction of a cell population, analyses are restricted to an examination the residual cells. Factors that may have promoted destruction of the target cells must, therefore, be determined by inference.

IV. APPLICATIONS AND FUTURE DIRECTIONS

Treatment of cells with Ab and C′ was one of the most widely used methods for selectively eliminating a cell subpopulation from a complex mixture. Many of the initial problems relating to specificity were resolved with the advent of mAb. Applications were limited only by the availability of suitable C′-fixing mAb and the imagination and budget of the investigator. In the basic sciences, mAb and C′ were used to dissect interactions occurring between cell subpopulations in developmental biology and to purify target and effector cell populations mediating immunological responses. In clinical medicine, they were first used to modify the cellular composition of bone marrow that was to be used for transplantation (1).

It was in this application that some of the major problems were first detected (2,21). Investigators reported lack of reproducibility, which was eventually attributed, as described above, to the effects of target Ag distribution and density and to the existence of anticomplementary factors. The difficulty and expense of obtaining C′ with low nonspecific toxicity and new regulatory requirements (47) have accelerated the decline in the use of C′-mediated purging for therapeutic purposes. Today its major clinical application is in T-cell depletion of allogeneic stem cell grafts (8,38,58), where the aim is to debulk T lymphocytes but retain a small number to ensure engraftment. For tumor purging, the move has been toward either combining C′-mediated treatment with other techniques, e.g., cytotoxic agents (63), or methods that achieve higher levels of target-cell depletion and are less affected by antigen density, e.g., immunomagnetic separation (1,7,23,36,37).

The inherent limitations of mAb and C′ purging are unlikely to make it the method of choice for high-efficiency cell elimination. It can still be of value in applications where simple gross depletion of a target-cell population is required to provide information on the possible function of that population, and as a debulking procedure before the use of more efficient and specific cell separation techniques. It is likely that it will eventually disappear from clinical applications as the regulatory pressure to avoid the therapeutic use of animal sera increases.

REFERENCES

1. Gee AP. Immunologically based methods for the elimination of tumor cells from autologous stem cell grafts. *Immunomethods* 1994; 5:232–242.

2. Gee AP, Boyle MD. Purging tumor cells from bone marrow by use of antibody and complement: a critical appraisal. *J Natl Cancer Inst* 1988; 80:154–159.
3. Malpas J, Kemshead JT, Pritchard J, Greaves M. Variability of surface antigen expression on neuroblastoma cells as revealed by monoclonal antibodies. *Proc Thirteenth Meeting Int Soc Pediatr Oncol* 1981; 90a (abstract).
4. Bertoncello I, Bradley TR, Watt SM. An improved negative immunomagnetic selection strategy for the purification of primitive hemopoietic cells from normal bone marrow. *Exp Hematol* 1991; 19:95–100.
5. Shpall EJ, Gehling U, Cagnoni P, Purdy M, Hami L, Gee AP. Isolation of CD34-positive hematopoietic progenitor cells. *Immunomethods* 1994; 5:197–203.
6. Civin CI, Strauss LC, Fackler MJ, Trischmann TM, Wiley JM, Loken MR. Positive stem cell selection—basic science. *Prog Clin Biol Res* 1990; 333:387–402.
7. Gee AP. Malignant cell purging: immunophysical techniques. In: Lasky L, Warkentin P, eds. *Marrow and Stem Cell Processing for Transplantation.* Bethesda, MD: American Association of Blood Banks, 1995:83.
8. Collins NH, Gee AP, Henslee-Downey PJ. T cell depletion of allogeneic bone marrow transplants by immunologic and physical techniques. In: Lasky L, Warkentin P, eds. *Marrow and Stem Cell Processing for Transplantation.* Bethesda, MD: American Association of Blood Banks, 1995:149.
9. Gee AP, Van Hilten J, Graham-Pole J, Boyle MD. Antibody-complement killing of neuroblastoma cells. *Prog Clin Biol Res* 1985; 175:471–483.
10. Lebkowski J, Schain L, Harvey M, Moseley A, Okarma TB. Isolation and culture of human and CD34+ hematopoietic stem cells using AIS CELLectors™. In: Wunder E, Sovalat H, Henon PR, Serke S, eds. *Hematopoietic Stem Cells. The Mulhouse Manual.* Dayton, OH: Alpha Med Press, 1994:215.
11. Tricot G. T-cell depletion in unrelated bone marrow transplantation. *Semin Hematol* 1992; 29:26–32.
12. Heimfeld S, Berenson RJ. Clinical transplantation of CD34+ hematopoletic progenitor cells: positive selection using a closed, automated avidin-bioitin immunoadsorption column. In: Wunder E, Sovalat H, Henon PR, Serke S, eds. *Hematopoietic Stem Cells. The Mulhouse Manual.* Dayton, OH: Alpha Med Press, 1994:231.
13. Thomas TE, Lansdorp PM. Selective separation of cells using magnetic colloids. *Prog Clin Biol Res* 1994; 389:65–77.
14. Trickett AE, Ford DJ, Lam-Po-Tang PR, Vowels MR. Immunomagnetic bone marrow purging of common acute lymphoblastic leukemia cells: suitability of BioMag particles. *Bone Marrow Transplant* 1991; 7:199–203.
15. Gee AP, Mansour V, Weiler M. T-cell depletion of human bone marrow. *J Immunogenet* 1989; 16:103–115.
16. Miltenyi S, Muller W, Weichel W, Radbruch A. High gradient magnetic cell separation with MACS. *Cytometry* 1990; 11:231–238.
17. Uckun FM, Myers DE. Allograft and autograft purging using immunotoxins in

clinical bone marrow transplantation for hematologic malignancies. *J Hematother* 1993; 2:155–163.

18. Leonard JE, Taetle R, To D, Rhyner K. Preclinical studies on the use of selective antibody-ricin conjugates in autologous bone marrow transplantation. *Blood* 1985; 65:1149–1157.
19. Ghetie MA, Vitetta ES. Recent developments in immunotoxin therapy. *Curr Opin Immunol* 1994; 6:707–714.
20. Jones D, Goldman A, Gordon I, Pritchard J, Gregory B, Kemshead J. Therapeutic application of a radiolabeled monoclonal antibody in nude mice xenografted with human neuroblastoma, tumoricidal effects and distribution studies. *Int J Cancer* 1986; 35:715.
21. Boyle MD, Gee AP. Low antigen density tumor cells: an obstacle to effective autologous bone marrow purging? *Cancer Invest* 1987; 5:113–118.
22. Selvaggi KJ, Wilson JW, Mills LE, et al. Improved outcome for high-risk acute myeloid leukemia patients using autologous bone marrow transplantation and monoclonal antibody-purged bone marrow. *Blood* 1994; 83:1698–1705.
23. Gee AP. Monoclonal antibodies as agents to purge tumor cells from bone marrow: laboratory and clinical experiences. In: Kemshead JT, ed. *Pediatric Tumors: Immunological and Molecular Markers.* Boca Raton, FL: CRC Press, 1989:105.
24. Gee AP. Molecular titration of components of the classical complement pathway. *Methods Enzymol* 1983; 93:339–375.
25. Minta JO, Gee AP. Purification and quantitation of the components of the alternative complement pathway. *Methods Enzymol* 1983; 93:375–408.
26. Borsos T. Immunoglobulin classes and complement-fixing activity. *Prog Immunol* 1971; 1:842.
27. Janssen WE, Lee C, Gross S, Gee AP. Low antigen density leukemia cells: selection and comparative resistance to antibody-mediated marrow purging. *Exp Hematol* 1989; 17:252–257.
28. Mansour V, Weiler M, Gee A. Factors limiting the efficiency of immunomagnetic cell separation. *Prog Clin Biol Res* 1992; 377:169–179.
29. De Fabritiis P, Bregni M, Lipton J, et al. Antigenic heterogeneity among Burkitt's lymphoma cells surviving treatment with antibody and complement. *Leuk Res* 1986; 10:35–42.
30. Taupier MA, Kearney JF, Leibson PJ, Loken MR, Schreiber H. Nonrandom escape of tumor cells from immune lysis due to intraclonal fluctuations in antigen expression. *Cancer Res* 1983; 43:4050–4056.
31. Grob JP, Campana D, Timms A, Janossy G, Prentice HG. Purging in autologous marrow transplantation: importance of antigen density on leukemic blasts for the efficiency of complement-mediated cytolysis. *Proc. 2nd Int Symp Detect Treat Resid Dis Acute Leuk* 1985; 84:abstract.
32. Gee AP, Bruce KM, van Hilten J, et al. Selective loss of expression of a

tumor-associated antigen on a human leukemia cell line induced by treatment with monoclonal antibody and complement. *J Natl Cancer Inst* 1987; 78:29–35.
33. Bast RC, Montgomery B, Haleem A, et al. Bone marrow purging with antibody and complement: an American review. *Bone Marrow Transplant* 1987; 2:58.
34. Shapiro SJ, Leibson PJ, Loken MR, Schreiber H. Changes in susceptibility to cytotoxic antibody among tumor cells surviving exposure to chemotherapeutic agents. *Cancer Res* 1982; 42:2622–2627.
35. Gribben JG, Freedman AS, Neuberg D, et al. Immunologic purging of marrow assessed by PCR before autologous marrow transplantation for B cell lymphoma. *N Engl J Med* 1991; 325:1525–1533.
36. Gribben JG, Saporito L, Barber M, et al. Bone marrows of non-Hodgkin's lymphoma patients with bcl-2 translocation can be purged of polymerase chain reaction-detectable lymphoma cells using monoclonal antibodies and immunomagnetic bead depletion. *Blood* 1992; 80:1083–1089.
37. Kiesel S, Haas R, Moldenhauer G, Kvalheim G, Pezzutto A, Dorken B. Removal of cells from a malignant B-cell line from bone marrow with immunomagnetic beads and with complement and immunoglobulin switch variant mediated cytolysis. *Leuk Res* 1987; 11:1119–1125.
38. Lee C, Henslee-Downey PJ, Brouillette M, et al. Comparison of OKT3 and T10B9 for ex vivo T cell depletion of partially mismatched bone marrow transplants (BMT). *Blood* 1995; 86:625a (abstract).
39. Champlin R. T-cell depletion for allogeneic bone marrow transplantation: impact on graft-versus-host disease, engraftment, and graft-versus-leukemia. *J Hematother* 1993; 2:27–42.
40. Gee AP, Bruce KM, Morris TD, Boyle MD. Evidence for an anticomplementary factor associated with human bone marrow cells. *J Natl Cancer Inst* 1985; 75:441–445.
41. Zhong RK, Kozii R, Ball ED. Homologous restriction of complement-mediated cell lysis can be markedly enhanced by blocking decay-accelerating factor. *Br J Hematol* 1995; 91:269–274.
42. Dioguardi D, Favoino B, Laurentaci G. Anticomplementary effects developed in vitro by two xenogeneic skin extracts on human sera. *Panminerva Med* 1983; 25:191–192.
43. Roy DC, Felix M, Cannady WG, Cannistra S, Ritz J. Comparative activities of rabbit complements of different ages using an in-vitro marrow purging model. *Leuk Res* 1990; 14:407–416.
44. Higgins DA, Langley DJ. A comparative study of complement activation. *Vet Immunol Immunopathol* 1985; 9:37–51.
45. Ish C, Ong GL, Desai N, Mattes MJ. The specificity of alternative complement pathway-mediated lysis of erythrocytes: a survey of complement and target cells from 25 species. *Scand J Immunol* 1993; 38:113–122.
46. Lucisano Valim YM, Lachmann PJ. The effect of antibody isotype and antigenic

epitope density on the complement-fixing activity of immune complexes: a systematic study using chimaeric anti-NIP antibodies with human Fc regions. *Clin Exp Immunol* 1991; 84:1–8.

47. Points to consider in the manufacture and testing of monoclonal antibody products for human use. *Fed Reg* 1994; 56:14.
48. Hale G, Waldmann H. CAMPATH-1 monoclonal antibodies in bone marrow transplantation. *J Hematother* 1994; 3:15–31.
49. Dyer MJ, Hale G, Hayhoe FG, Waldmann H. Effects of CAMPATH-1 antibodies in vivo in patients with lymphoid malignancies: influence of antibody isotype. *Blood* 1989; 73:1431–1439.
50. Preissner KP, Podack ER, Muller-Eberhard HH. SC5b-7, SC5b-8, and SC5b-9 complexes of complement: ultrastructure and localization of the S-protein (vitronectin) within the macromolecules. *Eur J Immunol* 1985; 19:69.
51. Rowley SD. Recombinant human deoxyribonuclease for hematopoietic stem cell processing. *J Hematother* 1995; 4:99–104.
52. Sharath MD, Merchant ZM, Kim YS, Rice KG, Linhardt RJ, Weiler JM. Small heparin fragments regulate the amplification pathway of complement. *Immunopharmacology* 1985; 9:73–80.
53. Weiler JM, Edens RE, Linhardt RJ, Kapelanski DP. Heparin and modified heparin inhibit complement activation in vivo. *J Immunol* 1992; 148:3210–3215.
54. Pandey VS, Verhulst A, Demey F. Effect of low temperature storage on the haemolytic complement of cattle and sheep. *Vet Immunol Immunopathol* 1993; 37:337–341.
55. Negrin RS, Kiem HP, Schmidt-Wolf IG, Blume KG, Cleary ML. Use of the polymerase chain reaction to monitor the effectiveness of ex vivo tumor cell purging. *Blood* 1991; 77:654–660.
56. Le Bien TW, Stepan DE, Bartholomew RM, Strong RC, Anderson JM. Utilization of a colony assay to assess the variables influencing elimination of leukemic cells from human bone marrow with monoclonal antibodies and complement. *Blood* 1985; 65:945–950.
57. Bast RC Jr, Ritz J, Lipton JM, et al. Elimination of leukemic cells from human bone marrow using monoclonal antibody and complement. *Cancer Res* 1983; 43:1389–1394.
58. Lee C, Brouillette M, Lamb L, et al. Use of a closed system for V alpha beta-positive T cell depletion of marrow for use in partially mismatched related donor (PMRD) transplantation. *Prog Clin Biol Res* 1994; 389:523–532.
59. Ball ED. Monoclonal antibodies and complement for autologous marrow purging. In: Gee AP, ed. *Bone Marrow Processing and Purging: A Practical Guide.* Boca Raton, FL: CRC Press, 1991:281.
60. Boyle MD, Gee AP, Borsos T. Heterogeneity in the size and stability of transmembrane channels produced by whole complement. *Clin Immunol Immunopathol* 1981; 20:287–295.

61. Gee AP, Boyle MD, Borsos T. Distinction between C8-mediated and C8/9-mediated hemolysis on the basis of independent ^{86}Rb and hemoglobin release. *J Immunol* 1980; 124:1905–1910.
62. Vervoordeldonk SF, Merle PA, Steenbergen EJ, van Leeuwen EF, von dem Borne AE, Slaper-Cortenbach IC. The effect of DNAse on the detection of residual malignant cells by polymerase chain reaction after immunologic purging of autologous marrow grafts. *Prog Clin Biol Res* 1994; 389:601–609.
63. Aihara M, Aihara Y, Schmidt-Wolf G, et al. A combined approach for purging multidrug-resistant leukemic cell lines in bone marrow using a monoclonal antibody and chemotherapy. *Blood* 1991; 77:2079–2084.

8

Magnetic Cell Sorting with Colloidal Superparamagnetic Particles

Aaron B. Kantor and Ian Gibbons
AmCell Corporation, Sunnyvale, California

Stefan Miltenyi and Jürgen Schmitz
Miltenyi Biotec GmbH, Bergisch Gladbach, Germany

I. INTRODUCTION

Magnetic cell sorting (MACS) is increasing in popularity in both the clinic and the research laboratory. In this chapter, we focus on the application of small immunomagnetic particles that can be combined with a ferromagnetic matrix and high-gradient fields to simply and efficiently isolate specific subpopulations of cells from heterogeneous mixtures such as peripheral blood or bone marrow. This powerful technique allows large numbers of viable cells to be isolated with high purity and yield. The purified cells are immediately suitable for flow cytometry, molecular biology, culture, and functional transfer into animals, including humans. We review the principles of MACS and discuss applications involving multiparameter separations, rare-cell isolation, and processing of very large numbers of cells for therapeutic purposes.

II. PRINCIPLES OF MAGNETIC CELL SORTING

A. Overview

The general separation process is outlined in Figure 1. The cell-preparation and labeling methods are similar to those used in flow cytometry (1). A monodisperse suspension of viable cells is incubated with a magnetic reagent, which typically consists of a monoclonal antibody (mAb), specific for the cells of interest, covalently coupled to magnetic microbeads 20–100 nm in diameter. The cells are washed to remove excess reagent and resuspended in buffered saline containing protein such as 1% bovine serum albumin (BSA). The suspension is then passed over a column containing a ferromagnetic matrix

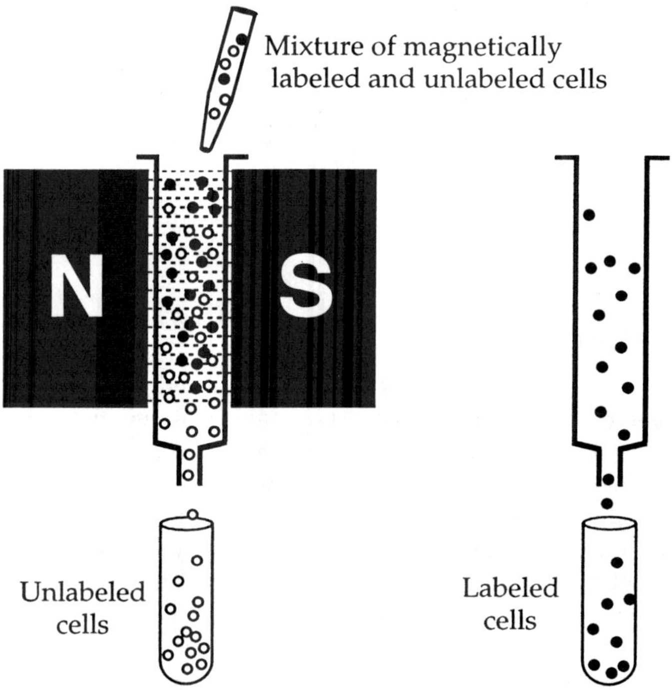

Figure 1 Isolation of magnetically labeled cells. A mixture of magnetically labeled and unlabeled cells is loaded onto a ferromagnetic matrix–separation column placed between the poles of a magnet. Magnetized cells are retained on the column while unlabeled cells pass through. Labeled cells are recovered after demagnetization of the column by removing it from the magnetic field.

placed in a strong external magnetic field. The matrix, which creates a strong magnetic field gradient, acts as a specific cell filter: magnetically labeled cells are retained on the column, while unlabeled cells flow through. After sufficient washing, the matrix and bound microbeads are demagnetized by removing the column from the magnetic field, and the retained cells are easily eluted. No harsh chemical or physical treatments are needed to remove the cells. Both the enriched (labeled) and depleted (nonlabeled) fractions are suitable for further uses.

Alternative magnetic cell separation methods use large (0.5–5 μm) magnetic spheres, with large magnetic moments, and low-strength magnets. While simple in principle, this approach has several disadvantages for the cells. The cell-bead reaction is slower with the larger particles and often requires multiple or prolonged incubations. Aggregation of cells and particles can lead to nonspecific trapping of unlabeled cells and, consequently, low purity of the selected cells. The large magnetic particles can adversely affect the viability and optical properties of the labeled cells, making this method more suitable for depletion than for enrichment. Release of the magnetic particles is often required before the cells can be used.

B. Reagents

Superparamagnetic Microbeads

There are two key properties of the particles. First, they are small enough (20–100 nm) to be colloidal, i.e., they remain dispersed due to the random molecular bombardment of Brownian motion. Second, they are superparamagnetic. The magnetically susceptible core is small enough that the particles do not retain any residual magnetism when removed from the magnetic field. These properties facilitate both the preparation of antigen (Ag)-specific reagents and separation of specific cell types.

Several types of colloidal magnetic microbeads have been used to label and separate cells, including magnetic ferritin (2), coprecipitates of protein and either iron (3) or cobalt (4) species, coprecipitates of iron oxide and methacrylate (5), and coprecipitates of iron oxide and polysaccharide (6). It is the latter class of particles that is most widely used and that has been the focus of our work at Miltenyi Biotec (Bergisch Giadbach, Germany) and AmCell (Sunnyvale, CA).

Production of iron dextran particles was described originally by Molday and MacKenzie (6) and Molday and Molday (7) and has been modified by Miltenyi et al. (8). Briefly, base is slowly added to a mixture of dextran (40 kDa), ferric and ferrous chloride in an aqueous medium to form a dispersion of particulates of mixed iron oxides (magnetite) embedded in a dextran shell. The magnetic particles are sized and purified by filtration and washing in a high-gradient magnetic field. The particle concentration of the resultant brown suspension is conveniently monitored by the optical density at 450 nm. The material, which is stable at 4°C for several months, is suitable for chemical modification. The chemical composition and physical properties of the microbeads are summarized in Table 1. The particles, which are more than 50% iron oxide, are 20–100 nm in diameter and contain embedded iron oxide particles approximately 10 nm in diameter.

Chemical Modification of Magnetic Microbeads

The microbeads can be modified by a variety of carbohydrate chemistries provided care is taken not to dissociate the iron core or aggregate the particles. Biotin (8), avidin (9), various haptens and fluorochromes, and specific mAbs have all been covalently attached to magnetic microbeads.

Molday and MacKenzie (6) originally used periodate oxidation/sodium borohydride reduction to couple protein A directly to the carbohydrate. Rela-

Table 1 Properties of Iron-Dextran Superparamagnetic Colloidal Reagents[a]

Component	% w/w
Chemical composition	
Iron oxide (as Fe_2O_3)	55–59
Dextran	35–39[b]
Antibody	2–10[c]
Physical Properties	
Size (diameter, nm)	20–100
Iron oxide core (approx. diameter, nm)	10
Density (g/mL)	2.5
Magnetization per cell (gauss μm^3)	30–500
Microbeads/cell	1000–20,000

[a]Properties are presented for a "typical" antibody (Ab) conjugate and typical staining. Various extents of substitution of the iron dextran are optimized for each Ab.

[b]The composition of the particles varies according to the degree of Ab substitution.

[c]Assuming a diameter of 30 nm for the magnetic bead, there are typically 10–200 Ab molecules/particle.

tively mild conditions were used to avoid oxidation of the iron core and aggregation of the conjugates.

A favored means to activate the microbeads is with cyanogen bromide or similar reagents and, subsequently, to introduce amino groups in the form of diaminohexane (8). Introduction of amino groups enables addition of a vast array of functional moieties using isothiocyanates, N-hydroxy-succinimide esters, and other chemistries. Heterobifunctional cross-linking agents are used to couple proteins containing sulfhydryl groups to the amino-modified microbeads. For Abs, native sulfhydryl groups in the hinge region can be exposed by mild reduction or free sulfhydryl groups can be introduced chemically. Typically, 1–20 μg of Ab or other protein is conjugated to 1 OD_{450} mL of microbeads. This corresponds to approximately 10–200 Ab molecules per microbead, assuming a bead diameter of 30 nm. The optimal amount of Ab must be determined for each Ab-Ag pair. Microbeads with low amounts of high-affinity Ab are preferred to minimize nonspecific and Fc receptor–mediated binding to irrelevant cells. The magnetic properties of the beads are retained throughout the chemical modification and are exploited to remove unreacted components at various stages by washing the particles on iron-matrix columns in a magnetic field. The final reagent is stable at 4°C for up to 1 year.

C. Ferromagnetic Columns

Paul and coworkers introduced the use of a random array steel wire (steel wool) as a magnetic filter for the isolation of red blood cells made paramagnetic by the deoxygenation of hemoglobin (10,11). When the loosely packed column is placed in between the poles of an external magnet, a high magnetic gradient is generated only in the immediate vicinity of the wire (12,13). For a permanent magnet of 0.5–1 Tesla (T) and wire with a diameter of 50–100 μm, magnetic forces of approximately 10^4 T/m can be generated compared with only 10 T/m using conventional geometries. Any microbead-labeled cell is attracted and moves toward the wire provided that the magnetic attraction exceeds the viscous force from the liquid flow in the system. Distances between the wires are only 100–200 μm, rather than the 1–20 mm for nonmatrix systems, so magnetized cells only need travel a short distance to be trapped. Nonmagnetized cells simply pass through the loosely packed matrix.

Molday and Molday (7) used similar filters for immunomagnetic cell separation. Coating the wires with a thin, biocompatible plastic polymer layer prevents corrosion and any possible damage to the cells through direct contact with the wire (8). The coated steel-wool matrix is ideally suited for cell depletion.

Ordered arrays of wire mesh (14) or iron spheres improve the uniformity of flow through the columns. At Miltenyi Biotec and AmCell, we use a closely packed matrix of ferromagnetic spheres coated with a biocompatible polymer for optimal positive selection. This permeable matrix affords uniform fluid flow throughout the column.

When the iron-sphere matrix column is placed between the poles of an external magnet, a high magnetic gradient is generated in the immediate vicinity of the spheres. Optimizing the size of the spheres is a matter of balancing the magnetic force, which is inversely proportional to the radius, with the need to allow cells to pass easily through the spaces between the spheres. Spheres of 0.05–2 mm in a field of 0.5–1 T are appropriate for cell separations. The maximum attractive potential is at the poles of each sphere, while there is a repelling force at the equator. Thus, magnetically labeled cells are focused in particular regions of the column and nonmagnetically labeled cells flow through.

Both wire and iron-sphere matrix columns can be prepared in a range of sizes. Current columns are appropriate for separating up to approximately 10^{11} total cells, with as many as 10^9 being magnetically labeled and retained on the column. This scaling feature makes it possible to process small (e.g., 1 mL blood) and large (e.g., a leukapheresis harvest) samples in the same amount of time, within 30 minutes after labeling. This is in striking contrast to fluorescence-activated cell sorting (FACS), where cells are processed one at a time. Assuming a rate of 4000 cells/second in a conventional sorter, it would require 0.7 hour to process 10^7 cells and 700 hours to process a leukapheresis pack with 10^{10} cells.

D. Cell-Isolation Strategies

High-gradient MACS with colloidal superparamagnetic particles is a flexible technique that affords both enrichment (i.e., positive selection) and depletion (i.e., negative selection) of specific cell populations. Like other cell-separation procedures, the strategy of choice is dependent upon the availability of specific reagents, the characteristics of the cell mixture to be separated, the relative proportion of desired cell population within the mixture, and full consideration of how the recovered cells are to be used, including any restrictions regarding purity and activation state.

Advantages and Disadvantages of Depletion and Enrichment

Depletion methods ensure target cells remain unlabeled, and this may be advantageous for functional studies such as T-cell activation. It is often cumbersome, however, to remove all nontarget cells, especially if the cells of interest, such as hematopoietic stem cells, represent only a minor fraction of the starting population. A depletion approach would require a complete panel of Abs directed against markers on all other cells. Moreover, as the separation column must be large enough to retain nearly the entire sample, the process is not well suited for bulky samples such as leukapheresis packs. In contrast to positive selection, the resultant cells are diluted during the separation process because of the extensive washing needed. Taken together, these factors suggest that depletion strategies are of limited use in isolating rare cell populations.

Positive enrichment is particularly well suited for isolating rare cell populations from bulky cell mixtures. The iron-sphere columns are optimized for this approach and need only be large enough to hold the desired cells. Consequently, as with other types of affinity chromatography, the selected cells can be effectively concentrated on the column. For the isolation of rare populations, nonspecific cell trapping may be equal or exceed the number of target cells. It is possible, however, to improve the cell purity by eluting the magnetically labeled cells and redepositing them on a magnetic column. Cells labeled with the small magnetic particles can be used in flow cytometry and other cellular and molecular assays. They can be readily transferred into animals without toxicity (15). In fact, dextran-stabilized iron oxide has been used since the 1960s for the treatment of iron deficiencies (16,17). The magnetically enriched cells have been shown to be suitable for cell therapy without removal of the microbeads.

Multiparameter Magnetic Separations

Combinations of positive and negative selection make isolation of cell populations not defined by a single Ag possible. There are several different strategies. First, magnetic separation can be used as a preenrichment for FACS sorting. The iron-dextran microbeads do not interfere with the fluorescent and scatter properties of the cells. Use of FACS for the second step allows for separations based on finer quantitative distinctions in Ag density than is practical in magnetic separations. Second, negative depletion can be used before positive selection. The nonretained cells from the first separation are simply labeled with a second magnetic microbead reagent and enriched on a second, usually smaller, column. Success depends on efficient retention of the labeled cells in the first separation. Finally, positive magnetic selection can be done before either a second positive selection or negative depletion, provided the magnetic labels can be released from the cells. This can be accomplished by specific competition or biochemical removal of the microbeads. Success depends on efficient release of the magnetic reagent after the first separation. Examples of multiparameter separations are given in the following applications section.

Cell-Labeling Strategies

Labeling the cells with immunomagnetic reagents is relatively straightforward. Both direct (one step) and indirect (two or more steps) methods are used with appropriate incubations and washes. A general advantage of the colloidal magnetic beads is the faster binding reaction compared with larger particles. Incubation times of 15–30 minutes are typical.

It is important to titrate each magnetic bead reagent for optimal cell separation. Cells typically are suspended at a concentration of 10^8/mL, and microbeads are typically used at a concentration of 0.05–5 OD_{450}. In general, the optimal microbead concentration depends on the affinity of the Ab, the level of Ab/microbead, and the nature of the Ag. Too much reagent may lead to retention of nonspecific cells on the column.

Titrations of two anti–T-cell magnetic bead reagents directed to different antigens are shown in Figure 2. The percentage of T cells not retained on the column is plotted as a function of microbead concentration. The anti-CD3 and anti-CD5 conjugates have approximately the same amount of Ab per microbead. The anti-CD5 reagent is substantially more effective in capturing T cells. In general, magnetic bead reagents are used below saturating levels, and it is still possible to label the target Ag with fluorescent reagents for flow cytometry. Higher amounts of reagent may lead to nonspecific and Fc receptor–mediated binding to irrelevant cells. We estimate that there are typically 10^2–10^4 microbeads/cell for efficient cell separation.

One-step reagents, where the mAb is directly coupled to the microbeads, are significantly more convenient to use than two-step reagents because the labeling time is consequently reduced. While substantial effort may be required to prepare several direct conjugates, their commercial availability for specific leukocyte Ags from humans and mice has made them the favored class of reagents. For many purposes, direct reagents provide good magnetic labeling.

Two-step staining is advantageous when a one-step magnetic bead

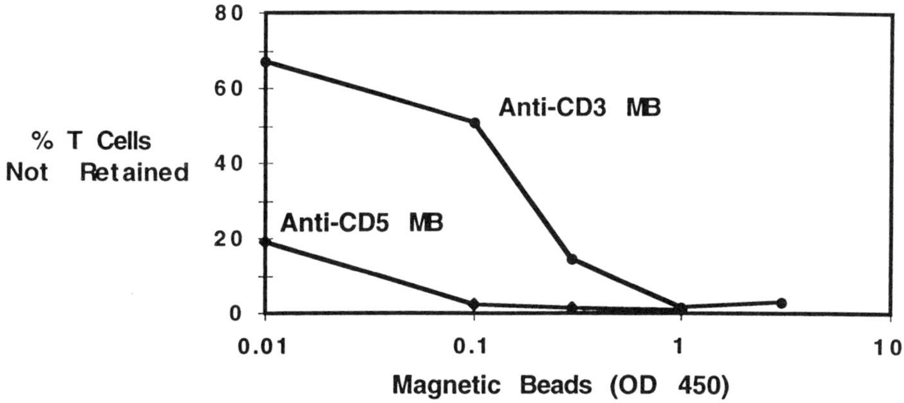

Figure 2 Titration of anti–T-cell reagents. Magnetic beads (MB) conjugated to either anti-CD3 or anti-CD5 mAbs were titrated using 10^7 peripheral blood mononuclear cells (PBMCs) at a concentration of 10^8 cells/mL. The beads have similar levels of antibody. The OD_{450} is a measure of the reagent concentration. The separation is monitored as the fraction of target cells not retained on the column.

reagent is unavailable, when multiple populations are targeted, and when the primary Ab does not work well directly conjugated to magnetic beads because of low affinity, difficulty with the coupling reaction, or steric hindrance. Four combinations of primary and secondary reagent pairs are in common use: unmodified primary Ab and anti-immunoglobulin–magnetic microbeads; primary tetrameric Ab complexes, consisting of an anti-dextran Ab coupled to an anti-target cell Ab, and unmodified iron-dextran microbeads (18); biotinylated primary Abs and avidin-magnetic beads; and hapten-conjugated (e.g., fluorescein or digoxigenin) primary Ab and anti-hapten magnetic microbeads.

Monitoring the Cell Separation

Magnetically labeled cells can be labeled with fluorescent Abs and analyzed by flow cytometry and/or microscopy. The small magnetic particles do not interfere with the light scattering or fluorescence properties of the cells. Analysis of the original and resultant fractions allows the separation process to be readily analyzed for yield, purity, and viability of the target cells. Since the magnetic bead reagents are usually used far below saturating concentrations, even fluorescent recognition of the same epitope can be used. Specific contaminating populations of interest can also be enumerated.

III. EXAMPLES OF MAGNETIC-CELL SEPARATIONS

The following examples focus on recent applications of MACS: multiparameter sorting, sorting based on an intracellular Ag, and instrument-mediated magnetic purification for potential therapeutic purposes.

A. Multiparameter Magnetic Cell Sorting

Negative Selection Followed by Positive Selection

There is no commonly accepted surface Ag to define human peripheral blood dendritic cells (19,20). However, a combination of positive and negative selection strategies can be used to isolate the small (0.5%) population of this cell type (Fig. 3). Peripheral blood dendritic cells express CD4 (21), which is also present on a subset of T cells, monocytes, and natural killer (NK) cells. These nondendritic cells are first removed from the blood peripheral blood mononuclear cells (PBMCs) by magnetic depletion using a two-step labeling process with hapten-conjugated anti-CD3, CD11b, and CD16 as the primary cocktail, followed by anti-hapten Ab conjugated to microbeads as the secondary reagent. The cells are depleted on a wire mesh column to enrich the dendritic cells to approximtely 16% of the resultant population. $CD4^+$ blood dendritic cells are then directly labeled using anti-CD4–conjugated microbeads and enriched using a iron-sphere column. Final dendritic cell purity of > 95% and yield of approximately 70% can be achieved with this approach.

Nucleated red blood cells from maternal blood are a potential source of fetal cells for the diagnosis of genetic disease (22). This population, which contains both maternal-derived and fetal-derived cells, is present at about 0.1% of peripheral blood nucleated cells. The cells can be enriched with a combination of negative and positive selection (23). An example is shown in Figure 4. The majority of leukocytes are depleted on a wire-mesh column using two-step magnetic staining with biotin–anti-CD45 as the primary reagent and avidin microbeads as the secondary reagent. The target cells, as identified by staining with anti-CD71 and glycophorin A, are enriched to 10–20% purity in this step with a yield of 50–70%. The $CD45^-$ cells are then labeled with anti-CD71 magnetic beads and the nucleated red blood cells further enriched on an iron sphere column. The resultant cells are typically 70–90% pure.

For double separations, it is important that the cells with the magnetic label are efficiently retained on the first column. Any labeled cells that pass through may be captured on the second column and contaminate the final population. This potential problem is minimized by using a wire matrix

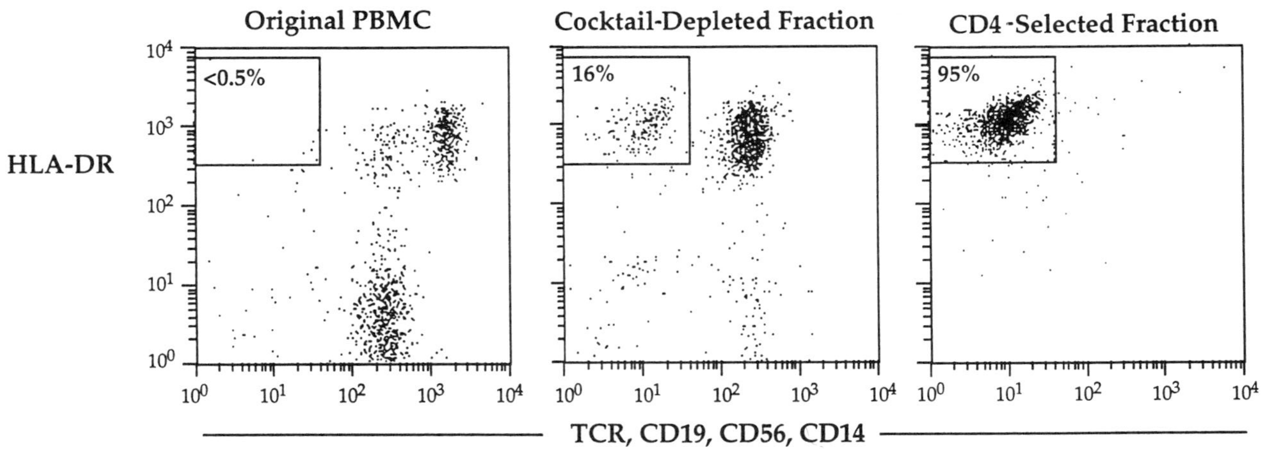

Figure 3 Enrichment of human peripheral blood dendritic cells. Dendritic cells were isolated from peripheral blood mononuclear cells (PBMCs) by depleting T cells, monocytes, and natural killer (NK) cells using a cocktail of hapten-conjugated anti-CD3, CD11b, and CD16 antibodies (Abs) followed by antihapten microbeads and a wire array column. Dendritic cells then were further enriched using anti-CD4 microbeads and an iron-sphere column. Fluorescence-activated cell sorting (FACS) analysis shows dendritic cells, which express the HLA class II DR molecule [but not markers for T cells (TCR), B cells (CD19), NK cells (CD56), or monocytes (CD14)] comprised < 0.5% of starting population. The depletion step increased the frequency to 16% and the positive selection further increased the purity to 95%.

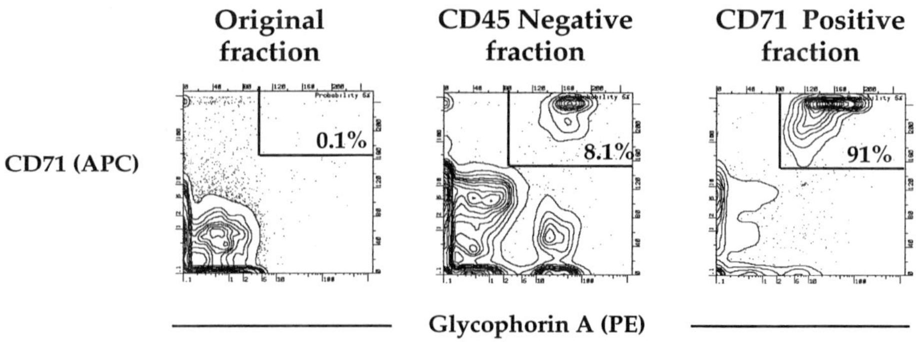

Figure 4 Isolation of human peripheral blood nucleated red blood cells. Leukocytes were removed using biotin anti-CD45, streptavidin microbeads, and a wire column. The CD45⁻ fraction was then labeled with anti-CD71 microbeads and enriched on an iron-sphere column.

column, designed to have a stronger magnetic field gradient, for the initial depletion step and an iron-sphere matrix for the subsequent positive selection.

Double-Positive Selection

Human B-1 cells express low levels of the "pan T cell" Ag CD5 and represent approximately 2% of all PBMCs (24). They can be isolated by two magnetic separation strategies: T-cell depletion followed by anti-CD5 positive selection or sequential positive selection with anti-CD19 and CD5 using bead-release technology. Data from the double-positive method are shown in Figure 5. B cells are first labeled with anti-CD19 microbeads and enriched to > 99% purity with an iron-sphere column. The magnetic microbeads then are released from the enriched B cells using a biochemical reaction. Since CD5 is expressed at low levels on B cells, the cells are labeled with a two-step procedure, hapten-conjugated anti-CD5 primary reagent and anti-hapten Ab conjugated to magnetic beads as the secondary reagent. The B-1 cells are positively selected on a second column to a purity of > 90%. The contaminating cells are mostly CD5⁻ B cells. This is a result of incomplete release of the first-step magnetic beads. It is often possible to remove cells with residual magnetic beads by passing the released cells over a magnetic column before labeling with the enrichment reagents.

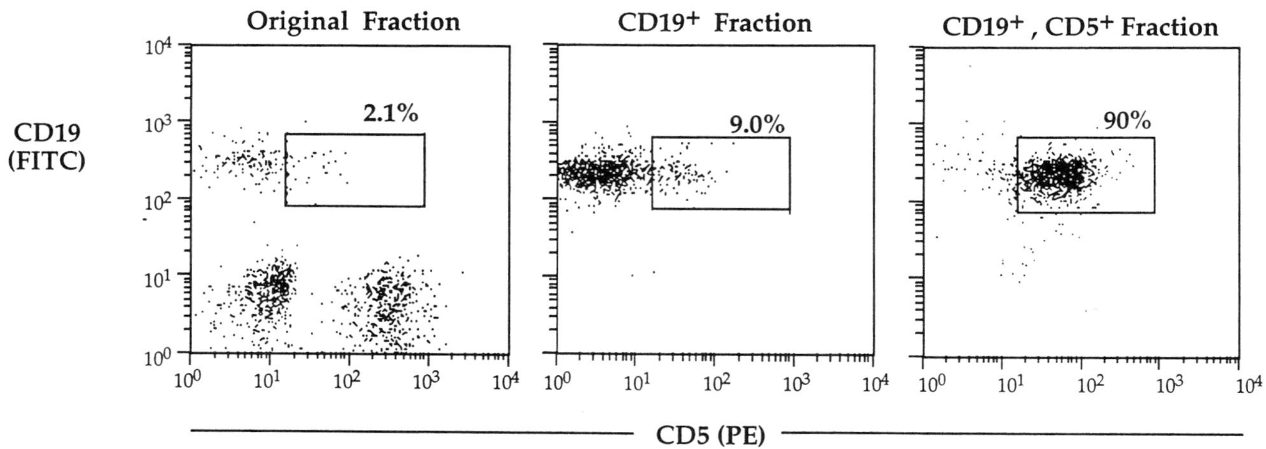

Figure 5 Isolation of peripheral blood B-1 cells. $CD19^+$, $CD5^+$ cells were isolated from peripheral blood mononuclear cells (PBMCs). Before separation, fluorescence-activated cell sorting (FACS) analyses showed that 2.1% of PBMCs were $CD19^+$ and $CD5^+$. PBMCs were labeled with anti-CD19 microbeads and the B-1 cells enriched to 9% purity using an iron-sphere selection column. The beads were released enzymatically, the cells magnetically labeled with fluorescein isothiocyanate (FITC)-anti-CD5 and anti-FITC microbeads, and $CD19^+$, $CD5^+$ cells purified to > 90% on a second column.

B. Positive Selection of Epithelial Tumor Cells Based on an Intracellular Marker

Detection and quantitation of epithelial tumor cells that have metastasized to the peripheral blood or marrow are potentially important for staging disease and evaluating cell preparations for autologous transplantation after myeloablative chemotherapy. The cytokeratins are well accepted as the most specific marker for epithelial cells, with essentially no background expression on blood leukocytes or erythrocytes (25). This is essential since the target cells are present at extremely low levels ($< 10^{-6}$) even in patients with late-stage disease. Permeabilization of the cells allows anticytokeratin magnetic beads to bind the Ag residing just below the plasma membrane. Griwatz et al. (26) have used preformed mouse anticytokeratin/rat-antimouse immunoglobulin magnetic beads and Siewert and coworkers (27,28) and Qian et al. (29) have used direct-conjugate anticytokeratin magnetic beads.

Briefly, blood samples are permeabilized with a mild detergent in a procedure that also lyses the red blood cells, fixed, reacted with the magnetic bead reagent, washed, and separated on iron-sphere columns. In contrast to separations based on cell-surface Ags, the cells are not viable after the permeabilization and fixation procedure. However, they are suitable for enumeration by flow cytometry, immunocytochemistry, and polymerase chain reaction (PCR).

An example is shown in Figure 6. T47D epithelial tumor cells were spiked into buffy coat preparations at 1000 cells/10^8 leukocytes. No tumor cells were detected by flow cytometry in the original fraction. After enrichment, a distinct cytokeratin positive, $CD45^-$ population is observed. The cells are enriched with a purity of 77% and a yield of 56%. Staining with anti-epithelial surface Ag-Ab confirms that these cells are epithelial.

C. Clinical Applications

The large-scale isolation of sterile, viable, clonogenic hematopoietic cells from human leukapheresis and bone marrow harvests is becoming increasingly important for therapeutic transplantation and gene therapy. Clinical-grade MACS are facilitated by the computer-controlled Cell Selector developed and manufactured by AmCell Corporation and introduced in clinical trials for autologous CD34 transplantation by Amgen Inc. (Thousand Oaks, CA).

The instrument and tubing set are shown in Figure 7. Bags containing magnetically labeled cells and buffer are connected to the separation column by sterile tubing. Two iron-sphere matrix columns are used. The first acts as a

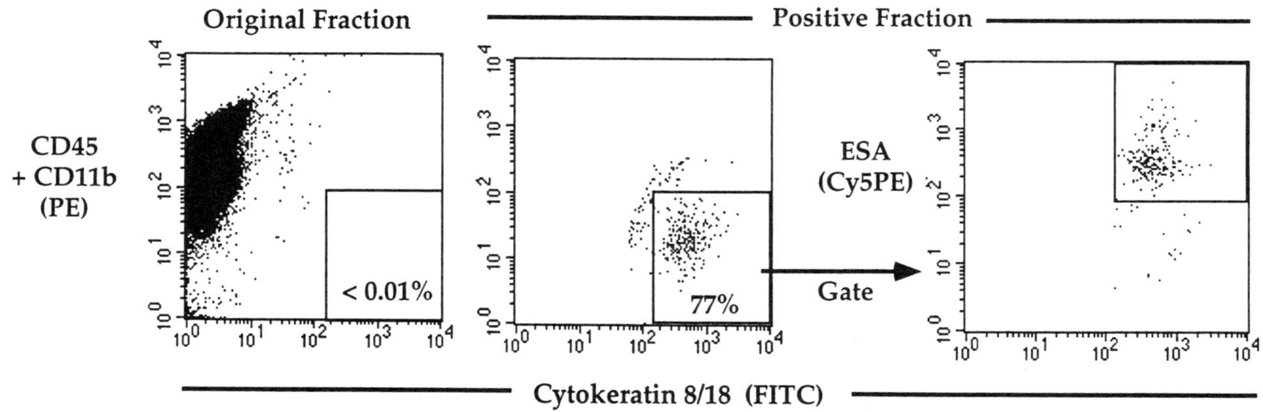

Figure 6 Isolation of epithelial tumor cells from peripheral blood. T47D epithelial tumor cells were spiked into peripheral blood mononuclear cells (PBMCs) at 1000 cells/10^8. Cells were permeabilized, fixed, and then incubated with anticytokeratin 8/18 magnetic beads. The magnetically enriched cells were analyzed by fluorescence-activated cell sorting (FACS) for cytokeratin, the epithelial marker ESA, and CD45 + CD11b.

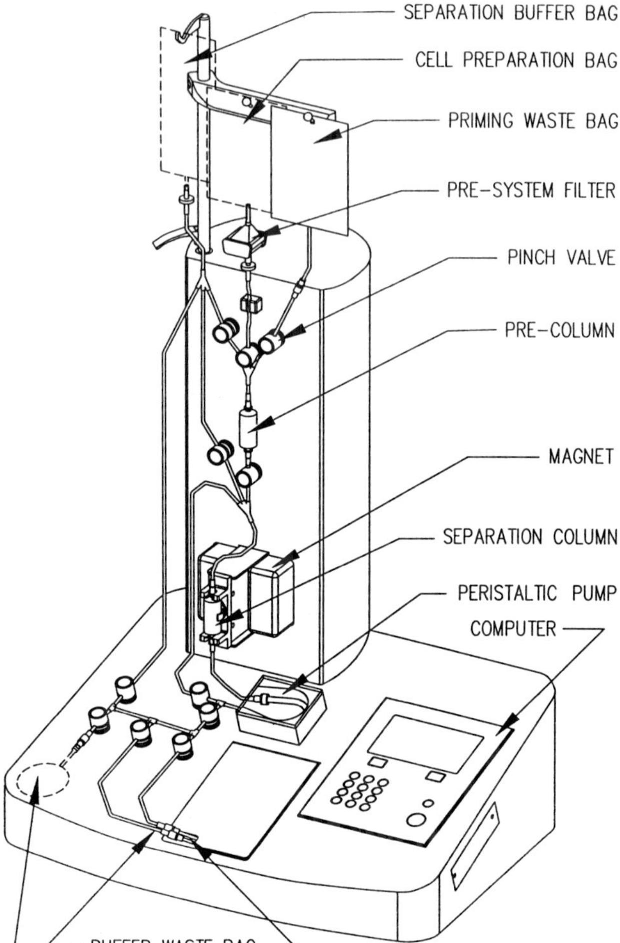

Figure 7 The Cell Selector instrument. The instrument is shown with the tubing set, including the iron-sphere ferromagnetic columns installed. Key features are identified in the diagram.

prefilter to remove nonspecific "sticky" cells, and the second is for magnetic retention of selected cells. The computer controls the placement of the column in the magnetic field and fluid flow through a pump and a series of valves. Each step in the cell-separation process is automated: cell deposition, washing, elution, redeposition, washing, and elution of the final product into a sterile collection bag. Two depositions on the magnetic separation column typically are used to improve cell purity.

The $CD34^+$ cell population contains progenitor cells required for hematopoietic reconstitution after myeloablation during high-dose cytotoxic chemotherapy. A number of groups have reported success using MACS with microbeads to purify $CD34^+$ cells (9,30–33), including some with the automated Cell Selector (34,35).

Figure 8A shows a sample FACS analysis of the enrichment of CD34 cells with the Cell Selector. The sample is a leukapheresis harvest from a donor who has been stimulated with Filgrastim (r-metHuG-CSF) to mobilize progenitor cells into the peripheral circulation. The preparations typically have $> 10^{10}$ leukocytes and a relatively high proportion of CD34 cells—typically 1%. Cells were labeled in the collection bag with a direct conjugate, anti-CD34 magnetic bead reagent, then washed in-bag by centrifugation, and attached to the tubing set. The resulting CD34 cells were 99% pure. Table 2 summarizes the results from 10 separations. Average purity and yield were 95 and 92%, respectively.

For autologous or allogenic transplantation, it is usually necessary to monitor the removal of specific contaminating populations, such as cancer cells or T cells. This removal is usually discussed in terms of log depletion. For example, a log depletion of 3 corresponds to a 1000-fold reduction in the number of a specific population. The value depends on the number of cells in the final fraction, the original fraction, and the nonspecific retention of the cell type in the Cell Selector. Table 2 includes log-depletion data from the CD34 cell isolations from the stimulated leukapheresis samples. T cells are depleted the very effectively, by an average of 4.5 logs. B cells and monocytes are removed by about 4 logs. These results suggest that cancer cells can be effectively removed from patient leukapheresis harvests, assuming they behave similarly to T cells.

T cells have been isolated with the Cell Selector. In an experiment designed to test the capacity and versatility of the system, T cells were isolated from a leukapheresis harvest with anti-CD5 direct-conjugate magnetic microbeads (Fig. 8B). The sample is from an unstimulated donor and contained 55% $CD3^+$ T cells among all the nucleated cells. A total of 2.4×10^9 T cells were retained on the column and isolated with excellent purity (99.6%) and good yield (78%).

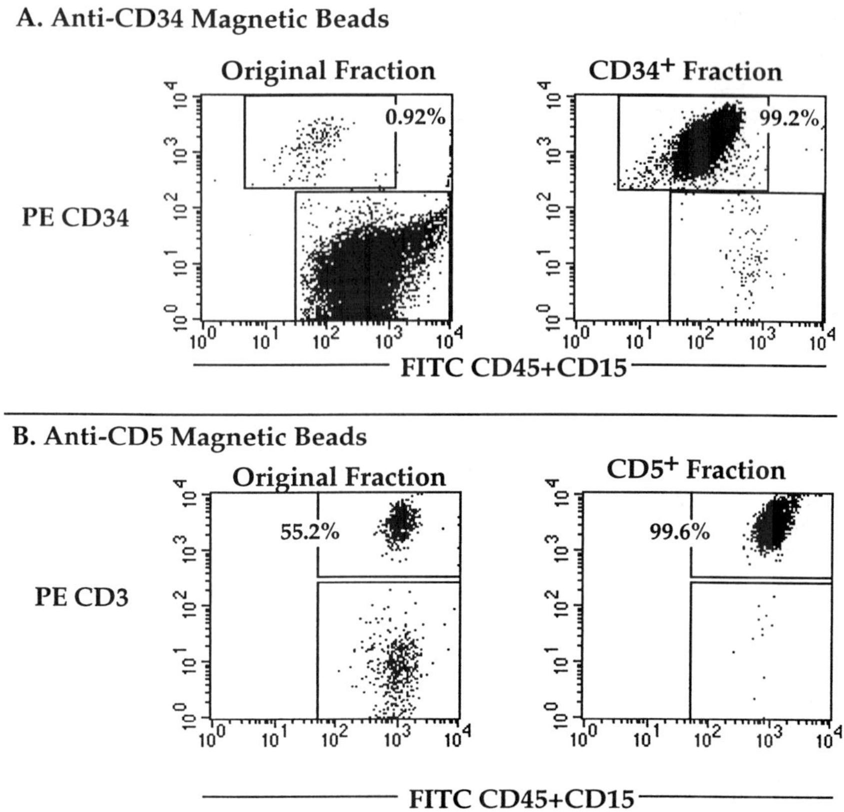

Figure 8 Clinical scale isolation of hematopoietic cells from leukapheresis harvests. (A) $CD34^+$ cells were isolated using a direct anti-CD34 microbead reagent and the Cell Selector. Fluorescence-activated cell sorting (FACS) analysis indicates the $CD34^+$ cells are enriched to high purity and yield. (B) T cells were enriched with anti-CD5 magnetic microbeads. A total of 2.4×10^9 T cells were enriched to high purity. The yield was only 78% because limitations in the column capacity were reached.

IV. SUMMARY

The applications of MACS using colloidal superparamagnetic particles are growing as both an exclusive cell-separation step or as a complement to other isolation protocols such as FACS. The most exciting areas include the isolation of human cells for therapeutic transplantation, such as with CD34 progenitor cells, and potential uses for clinical diagnosis such as with breast cancer cells.

Table 2 Therapeutic Scale Isolation of CD34 Progenitor Cells from Leukapheresis Harvest of Filgrastim-Stimulated Patients

	CD34 cells[a]			Log Depletion		
	Original	Purity	Yield	CD3 T cells	CD14 monocytes	CD20 B cells
Average	0.8%	95%	92%	4.5	3.9	4.1
SD	0.3	4	5	0.2	0.2	0.3
Range	0.4–1.3%	88–99%	84–99%	4.1–4.7	3.4–4.4	3.6–4.7

[a] $n = 10$ for CD34 cells data and 5 for log depletion data.

Future developments in MACS technology will almost certainly increase its utility in both the clinic and the research laboratory.

ACKNOWLEDGMENTS

We gratefully acknowledge the ongoing contributions of our colleagues at Miltenyi Biotec and AmCell Corporation. We especially thank the individuals who provided data or graphics for this paper: Janette Phi-Wilson, Kevin Sheehan, Doug Dority, Paul Raju, Henry Ajo, Kyung Kim, Wendy Chen, Simone Guth, Christiane Siewert, Verena Martin, Fawn Qian, Mathias Köster, Sabine Cremer, Anja Fuchs, Alexondar Horst, Susunne Höher, and Sandra Kosten.

REFERENCES

1. Kantor AB, Roederer M. FACS analysis of leukocytes. In: *Handbook of Experimental Immunology*, 5th ed., Edinburgh: Blackwell Scientific, 1996, 49.1–49.13.
2. Owen CS, Lindsay JG. Ferritin as a label for high-gradient magnetic separation. *Biophys J* 1983; 42:145–150.
3. Owen CS, Sykes NL. Magnetic labeling and cell sorting. *J Immunol Methods* 1984; 73:41–48.
4. Poynton CH, Dicke KA, Culbert S, Frankel LS, Jagannath S, Reading CL. Immunomagnetic removal of CALLA positive cells from human bone marrow. *Lancet* 1983; 1:524.
5. Molday RS, Yen SP, Rembaum A. Application of magnetic microspheres in labelling and separation of cells. *Nature* 1977; 268:437–438.
6. Molday RS, MacKenzie D. Immunospecific ferromagnetic iron-dextran reagents

for the labeling and magnetic separation of cells. *J Immunol Methods* 1982; 52:353–367.

7. Molday RS, Molday LL. Separation of cells labeled with immunospecific iron dextran microspheres using high gradient magnetic chromatography. *FEBS Lett* 1984; 170:232–238.
8. Miltenyi S, Muller W, Weichel W, Radbruch A. High gradient magnetic cell separation with MACS. *Cytometry* 1990; 11:231–238.
9. Kato K, Radbruch A. Isolation and characterization of CD34+ hematopoietic stem cells from human peripheral blood by high-gradient magnetic cell sorting. *Cytometry* 1993; 14:384–392.
10. Paul F, Roath S, Melville D. Differential blood cell separation using a high gradient magnetic field. *Br J Haematol* 1978; 38:273–280.
11. Melville D, Paul F, Roath S. Direct magnetic separation of red cells from whole blood. *Nature* 1975; 255:706–708.
12. Kolm HH. The large-scale manipulation of small particles. *IEEE Trans Magnet* 1975; 11:1567.
13. Oder RR. High gradient magnetic separation theory and applications. *IEEE Trans Magnet* 1976; 12:428.
14. Thomas TE, Lansdorp PM. Selective separation of cells using magnetic colloids. *Prog Clin Biol Res* 1994; 389:65–77.
15. Schmitz J, Thiel A, Kuhn R, et al. Induction of interleukin 4 (IL-4) expression in T helper (Th) cells is not dependent on IL-4 from non-Th cells. *J Exp Med* 1994; 179:1349–1353.
16. Groden BM, Whitelaw J, Will G. The treatment of iron-deficiency anemia by iron dextran infusion, with special reference to the effect on blood-grouping, coagulation, sedimentation and haemolysis. *Postgrad Med J* 1968; 44:433.
17. Slade IJ, Iosefa RN. Intravenous iron-dextran (Imferon) in the treatment of iron deficiency anemia. *J Med Assoc Georgia* 1968; 57:137–139.
18. Thomas TE, Abraham SJ, Otter AJ, Blackmore EW, Lansdorp PM. High gradient magnetic separation of cells on the basis of expression levels of cell surface antigens. *J Immunol Methods* 1992; 154:245–252.
19. Thomas R, Davis LS, Lipsky PE. Isolation and characterization of human peripheral blood dendritic cells. *J Immunol* 1993; 150:821.
20. Steinman RM. The dendritic cell system and its role in immunogenicity. *Annu Rev Immunol* 1991; 9:271–296.
21. O'Doherty U, Steinman RM, Peng M, et al. Dendritic cells freshly isolated from human blood express CD4 and mature into typical immunostimulatory dendritic cells after culture in monocyte-conditioned medium. *J Exp Med* 1993; 178:1067–1076.
22. Bianchi DW, Flint AF, Pizzimenti MF, Knoll JH, Latt SA. Isolation of fetal DNA from nucleated erythrocytes in maternal blood. *Proc Natl Acad Sci USA* 1990; 87:3279–3283.
23. Busch J, Huber P, Pfluger E, Miltenyi S, Holtz J, Radbruch A. Enrichment of fetal

cells from maternal blood by high gradient magnetic cell sorting (double MACS) for PCR-based genetic analysis. *Prenat Diagn* 1994; 14:1129–1140.
24. Bhat NM, Mithal A, Bieber MM, Herzenberg LA, Teng NN. Human CD5+ B lymphocytes (B-1 cells) decrease in peripheral blood during pregnancy. *J Reprod Immunol* 1995; 28:53–60.
25. Pantel K, Felber E, Schlimok G. Detection and characterization of residual disease in breast cancer. *J Hematother* 1994; 3:315–322.
26. Griwatz C, Brandt B, Assmann G, Zanker KS. An immunological enrichment method for epithelial cells from peripheral blood. *J Immunol Methods* 1995; 183:251–265.
27. Siewert C, Martin VM, Miltenyi S, Schmitz J. Enrichment of micrometastic carcinoma cells from hematopoietic cell samples by high-gradient magnetic cell sorting (MACS). *Mulhouse Manual* 1996.
28. Martin VM, Siewert C, Miltenyi S, Schmitz J. Enrichment of epithelial tumor cells from peripheral blood of breast and prostate carcinoma patients by high gradient magnetic cell sorting (MACS). ISHAGE Conference, 1996.
29. Qian F, Kantor AB, Gibbons I, Chen W, Recktenwald D. A sensitive carcinoma cell detection assay. ISHAGE Conference, 1996.
30. Martin-Henao GA, Ingles-Estevee J, Cancelas JA, Garcia J. Isolation of CD34+ hematopoietic progenitor cells in chronic myeloid leukemia by magnetic cell sorting (MACS). *Bone Marrow Transplant* 1996; 18:603–609.
31. Dinicola M, Siena S, Corradini P, et al. Elimination of follicular non-Hodgkin-lymphoma (F-NHL) cells and high recovery of $CD34^+$ cells from blood-cell transplants by the Miltenyi cell sorting system. *J Exp Hematol* 1996; 24:696.
32. Briddell R, Zilm K, Kern B, et al. Depletion of T-lymphocyte subsets in peripheral blood apheresis harvests from normal donors by purification of cells expressing the CD34 antigen. *Blood* 1995; 86:918.
33. Guth S, Koster M, Thiel A, et al. Immunomagnetic isolation of subpopulations of CD34+ human hematopoetic progenitor cells using releasable superparamagnetic microbeads. *Exp Hematol* 1995; 23:812.
34. McNiece I, Briddell R, Stoney G, Kern B, Recktenwald D, Miltenyi S. CD34 cell isolation using the Amgen Cell Selection Device. *Exp Hematol* 1996; 24:95.
35. McNiece I, Briddell R, Recktenwald D, et al. Large scale isolation of CD34 expressing cells using the Amgen Cell-Separation Device results in high levels of purity and recovery. *Exp Hematol* 1995; 85.

9

Immunomagnetic Cell Separation Using Antibodies and Superparamagnetic Microspheres

Adrian P. Gee

University of South Carolina School of Medicine, Columbia, South Carolina, and M. D. Anderson Cancer Center, Houston, Texas

I. INTRODUCTION

Selective cell separation is an essential component of many processes in biology and medicine. In the majority of cases, the aim is to achieve complete removal of the target population from a cell mixture, however, additional considerations can have a major impact on the method that is chosen to effect the separation. For some applications, a secondary aim may be to recover the selected cells for study for additional manipulation. In other cases, the purity of the separation may take precedence over the yield of selected cells, or vice versa. Factors such as processing time, viability of residual cells, and exposure to nonmedically approved reagents also must be considered when selecting the separation technique.

II. ANTIBODY-MEDIATED CELL SEPARATION

Immunomagnetic cell separation has become the method of choice for a wide variety of applications (1), since it combines the specificity of monoclonal antibodies (mAbs) to identify the target cell population, with a rapid, nontoxic method for their collection. Other methods that use mAbs include panning (2,3), immunotoxins (4,5), and complement-mediated cytotoxicity (6). Each approach has its limitations.

Immobilization of a fixed amount of mAb to a flat surface for panning, as opposed to a three-dimensional particle, results in a cumbersome configuration of surfaces that occupies a much larger volume to achieve the same efficiency of cell capture. For example, a 600 cm^2 area can be configured as eight 75 cm^2 tissue-culture flasks, or as 0.3 mL of 3.4-μm-diameter beads, or 10 μL of 100-nm-diameter nanoparticles (7). The preparation of immunotoxins requires handling of highly toxic reagents that must be conjugated to each Ab that is to be used to eliminate the target cells (4,5,8). The immunotoxin must then be translocated across the cell membrane. This may require the use of agents to prevent nonspecific binding and to facilitate transmembrane transport (9). Since the targeted cell is destroyed by these agents, it is no longer available for further study. This is also true of complement-mediated cell selection, however, unlike toxins, complement can be added as a generic agent to target cells that have been presensitized with multiple mAbs. These Abs must be of a complement-fixing isotype. In addition, we originally reported (10), and others have recently confirmed (11), that certain cells release a decay-accelerating factor that can interfere with the cytotoxic activity of complement. In the United States there is also encouragement from regulatory agencies to avoid the use of animal sera in human therapeutic procedures. These considerations have fueled the use of immunomagnetic separation for many applications.

III. MAGNETIC SEPARATION

A number of magnetic matrices have been developed for immunomagnetic cell separation (1,12–16). Commercial manufacturers include Polysciences (Warrington, PA), Bangs Laboratories (Carmel, IN), ICN Biomedicals Inc. (Cleveland, OH), Amersham International (Amersham, UK), and PerSeptive Diagnostics (Cambridge, MA). The most widely used matrices are the microspheres from Dynal Inc. (12) and nanoparticles from Miltenyi Biotec (13). The use of nanoparticles is discussed elsewhere in this volume; this chapter focuses on immunomagnetic separations using superparamagnetic microspheres.

Although Dynal manufactures several sizes of magnetic microspheres (2.8, 4.5, and 5.0 μm in diameter), the M450 polystyrene microspheres are the most widely used for cell separation (12,14). The beads are 4.5 μm in diameter, with a surface area of 1–4 $m^2/1.4 \times 10^{10}$ beads. They contain approximately

20% by weight of magnetite (γFe_2O_3) dispersed throughout their volume (Fig. 1). The particles are superparamagnetic, meaning that they can be collected in magnetic fields (Fig. 2), but do not retain any magnetism when removed from that field. The Dynal microspheres are available uncoated or precoated with a variety of anti-immunoglobulin Abs (antimouse pan-IgG, anti-IgG subclass specific, and anti-IgM, anti-rabbit IgG, and anti-rat IgM), or mAbs directed towards a variety of CD antigens (Ag) (anti-CD2, CD3, CD4, CD8, CD19, CD14, CD15, CD34, HLA class II) (17). Other coatings are available for molecular biology applications, for use with biotinylated Abs, and for chemical coupling of biologicals to the bead surface (17). The microspheres have found wide application in the selective separation of bacteria (18) and mammalian cell populations (1,12,19), subcellular components (20), nucleic acids (21), and proteins (22) for both basic science and clinical applications. Although not sold in the United States for clinical use, the microspheres have been tested under the investigational new drug approval mechanism. Here, their primary use has been in stem cell graft manipulation for the removal of tumor cells from autologous grafts (23–25), T lymphocytes from allogeneic grafts (26–28), and for the enrichment of hematopoletic stem cells for transplantation (29–31).

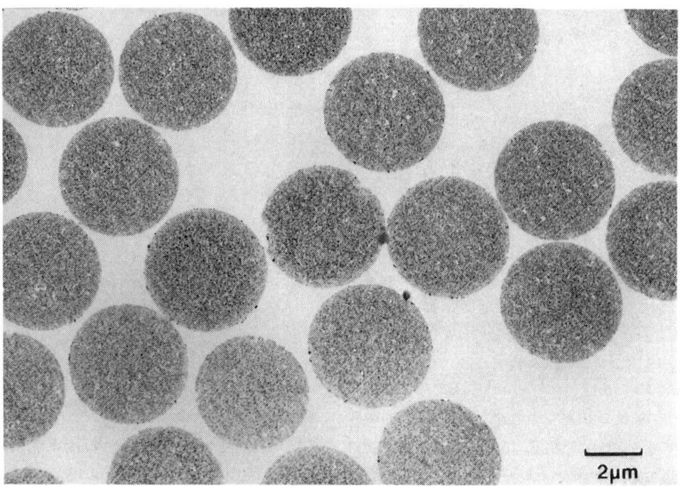

Figure 1 Transmission electronmicrograph of Dynal M450 microspheres in cross section. Note the uniformity of bead size and distribution of darkly staining magnetite within each bead. (Electronmicrograph courtesy of Baxter Biotech Immunotherapy.)

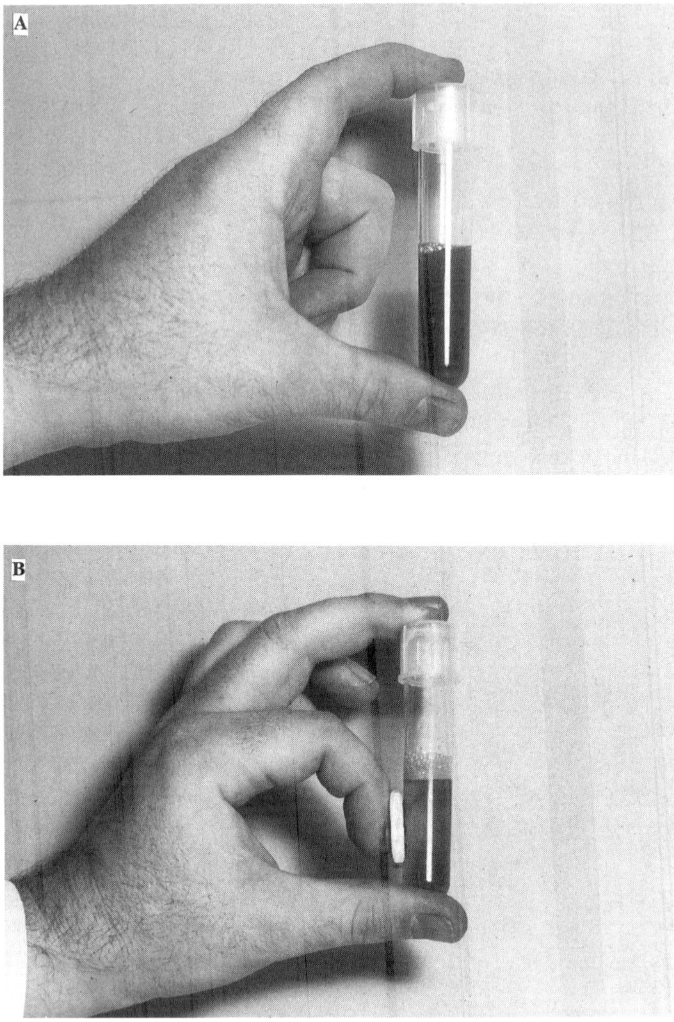

Figure 2 Collection of Dynal microspheres in a magnetic field. (A) Microspheres in suspension appear as a cloudy liquid. (B) A small permanent magnet held to the side of the tube starts to attract the microspheres within a few seconds. (C) Within 30 seconds, all of the microspheres are attracted to the magnet and held against the side of the tube.

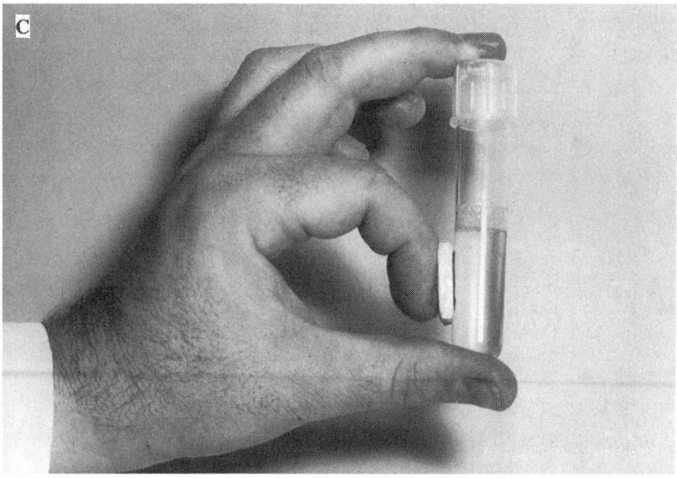

IV. PRINCIPLE OF IMMUNOMAGNETIC SEPARATION

Selective cell separation requires a method to identify the target cell and a means to effect its collection. In immunomagnetic separations, identification is achieved by use of a mAb directed against a cell-surface component that is not expressed by other cells in the mixture. This component need not, however, be biologically exclusive to the target cell population; for example, an Ab used for removing tumor cells from bone marrow may cross-react with Ags expressed on liver cells and still be suitable for the intended use. Other agents that bind to cell surface components have also been used, e.g., lectins (32,33); however, their specificity towards a particular cell type requires extensive confirmation if they are to achieve the performance that is characteristic of mAb. Once the target cell has been identified, it must be linked to the solid phase that will be used to separate it physically from the mixture. This consists of a superparamagnetic microsphere or bead that has been coated with an Ab that reacts with the mAb used to identify the target cell, i.e., an antiimmunoglobulin Ab. Alternatively, if biotinylated mAbs are used, an avidin- or streptavidin-coated bead can be used for capture (34). The coated microspheres are added to the cell mixture and mixed gently for 30–90 minutes, during which time the beads attach specifically to the target cells to form a rosette (Fig. 3). These rosettes, together with any remaining nonrosetted

Figure 3 Lymphoma cell surrounded by immunomagnetic microspheres. (Scanning electron micrograph provided by Dr. Gunnar Kvalheim, Norwegian Radium Hospital, Oslo.)

beads, can then be separated by exposing the cell mixture to a magnetic field, such as that generated by an array of permanent magnets. The nontarget cells remain in suspension and can be collected by aspiration or decanting, whereas the target cells and free beads are attracted to the magnet and are retained against the side of the separation vessel. The positively selected target cells can be recovered by removal of the magnetic field and resuspension of the rosetted beads in a suitable cell culture medium. Removal of the beads from the cells can be achieved in a number of ways, as discussed later.

V. ADVANTAGES AND DISADVANTAGES OF MICROSPHERES AS THE MAGNETIC SOLID PHASE

A number of paramagnetic solid phases have been used for cell separation. These vary widely in their size, shape, and magnetic properties (1,12–16,35,36). A major difference is in the method that is used for their collection. Larger particles, such as microspheres, can be easily collected in a magnetic field generated by a small permanent magnet (12,14). Rare earth magnets composed of samarium-cobalt or neodymium-iron-boron are particularly suitable for this purpose. Paramagnetic nanoparticles and magnetic colloids are generally too small to be affected by this type of field and must be collected using a high-gradient magnetic array (13,15,35,36). In this technique the particles are collected onto a column or array of metallic wires or particles that is magnetized by placing it into a field generated by externally located magnets. The particle-coated cells attach to the wires and can be recovered by removing the column or array from the field and eluting the cells with buffer or medium. Although efficient, this technique requires somewhat more sophisticated hardware than is needed when beads are used.

The larger size of the microspheres can, however, have disadvantages (37,38). They will tend to settle out of suspension within a few minutes and must therefore be continually mixed when reacting with the target cells. At high bead-to-cell ratios there may be damage to delicate cell types or even shearing of the beads from target cell projections or processes. As the beads are drawn to the magnets, their larger size can result in nonspecific trapping of bystander cells, which become sequestered in the bead mass at the magnet surface. The smaller M-280 Dynal beads (2.8 pm in diameter) have been found to be more effective when used with IgM mAbs. This was attributed to their lower weight and iron content and their ability to bind to the Fab fragment of IgM molecules, whereas the larger beads bind predominantly to the Fc fragment of IgG (39). The remarkable size uniformity of the Dynal beads probably contributes to their reproducibility of performance and ease of collection. It also facilitates easy counting of the beads in a hemacytometer or automated cell counter, which allows use of accurate bead-to-cell ratios in separations. Unlike nanoparticles (13), the beads have distinct light-scattering properties, which can interfere with analysis of rosetted target cells by flow cytometry. They have a long shelf life in both the coated and uncoated forms and are nontoxic, allowing rosetted cells to be tissue-cultured without difficulty.

VI. POSITIVE VERSUS NEGATIVE SELECTION

The procedure described above in which the target cells are rosetted with the beads is known as positive selection (Fig. 4). It has the advantage that a single mAb that is specific for the target cell can be used to mediate cell separation, but also the disadvantage that the separated cells are recovered rosetted with the beads. In negative selection (Fig. 4), the aim is to achieve the same end, but by rosetting and removing all nontarget cells so that a microsphere-free target cell population is obtained. In most cases, negative selection of a homogeneous target cell population from a complex cell mixture may require

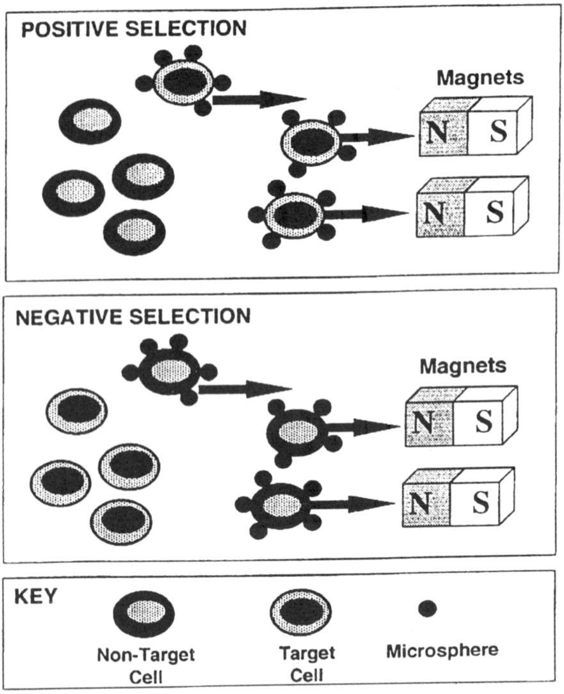

Figure 4 Positive versus negative immunomagnetic separation. In positive selection, the target cells are identified using the beads and collected onto the magnets. In negative selection the target cells are enriched by immunomagnetic removal of nontarget cells.

use of multiple mAbs to identify all of the nontarget cells (40,41) and several cycles of separation to achieve high target cell purities (42,43). Repeated treatments effectively increase the bead–to–target cell ratio on each cycle, with a resulting improvement in selection efficiency, however, this is usually accompanied by a parallel decrease in nontarget cell yields and increased expense. Negative selection is, however, the method of choice for applications in which the aim is to remove a specific contaminating cell population from a heterogeneous cell mixture required for further study. The best example is probably removal of metastatic cancer cells (neuroblastoma, breast cancer, lymphoma, and small-cell lung carcinoma) from bone marrow required for transplantation (39,42–45). Since, until relatively recently, no mAb was available that could identify and positively select the small population of hematopoietic stem cells in bone marrow, it was necessary to use negative selection to recover the heterogeneous cells of the normal marrow required for transplantation by positively selecting the cancer cells. The discovery of the CD34 antigen on early marrow progenitors (46), including pluripotent stem cells, has now allowed the use of anti-CD34 mAbs for positive selection of the primitive hematopoietic cells (29–31), but, this also necessitated the development of methods to remove the beads from these cells before transplantation (30,47).

VII. DIRECT VERSUS INDIRECT SEPARATION

Immunomagnetic separations can be classified as either direct or indirect (Fig. 5). Up to this point, the discussion has focused on indirect separations, in which the target cell is first sensitized with the mAb and subsequently reacted with beads coated with anti-immunoglobulin (Fig. 5). An alternative approach is direct separation using microspheres that have been directly coated with the antitarget mAb (48) (Fig. 5). This has the advantage of being a one-step procedure that eliminates both incubation of target cells with fluid phase mAb and extensive washing of the cells before the addition of the beads. It has been found to be particularly useful for separating cells that show uniform high expression of the target Ag (e.g., T cells) (49,50). There are, however, a number of important disadvantages that mandate careful consideration when choosing to use the direct or indirect methods.

Indirect separations allow the use of multiple mAbs to identify the target cells, followed by incubation with a single capture reagent (the anti-immunoglobulin coated beads). This can compensate for both target Ag heterogeneity (40,41,51,52) and variability in target Ag density (48,53–55), which can adversely affect the efficiency of separation when only a single mAb is used (Fig. 6). Where this may be a concern, the use of a direct technique would require preparation of either beads directly coated with more than one mAb, or the addition of mixtures of different types of beads, each of which is prepared using a single mAb. In either case, it is difficult to compensate effectively for variations in target Ag expression without extensive phenotyping of the target cells before the separation. In addition, the use of directly coated beads often results in slower reaction kinetics, since both the mAb and its Ag are immobilized. This can be overcome by increasing the bead–to–target cell ratio (38), but this is more complex and expensive when bead mixtures must be used.

A further complication of the direct procedure is the restricted choice of mAb that can be used. In most cases IgG Abs can be coated onto the beads, but do not achieve effective separations. It is thought that this is due, at least in part, to steric considerations (48), in that the Ag-combining site may be

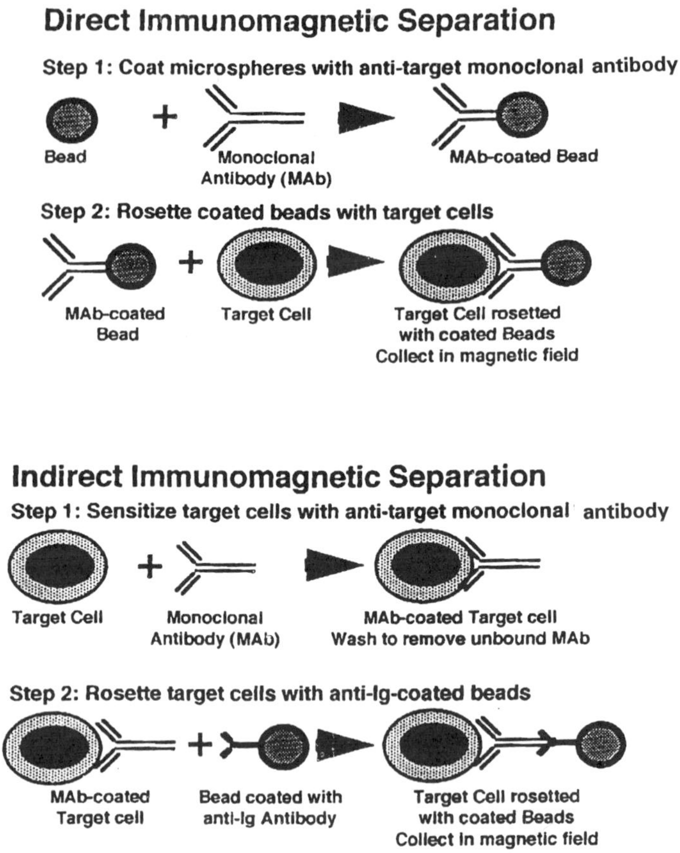

Figure 5 Direct and indirect immunomagnetic separation. (Top panel) Direct technique; (bottom panel) indirect technique.

presented to the cell in a configuration that is too close to the bead surface and does not permit stable attachment. This is supported by the finding that directly conjugated beads prepared with IgM mAbs usually work well, due to the availability of multiple Ag-binding sites away from the bead surface. The use of a semidirect method can overcome some of the problems associated with IgG mAbs (A. P. Gee, unpublished). In this procedure the IgG is reacted to anti-Ig–coated microspheres, which are then washed and added to non-sensitized cells. The anti-Ig Ab appears to act as a spacer and provides correct orientation for the IgG to bind to the target Ag.

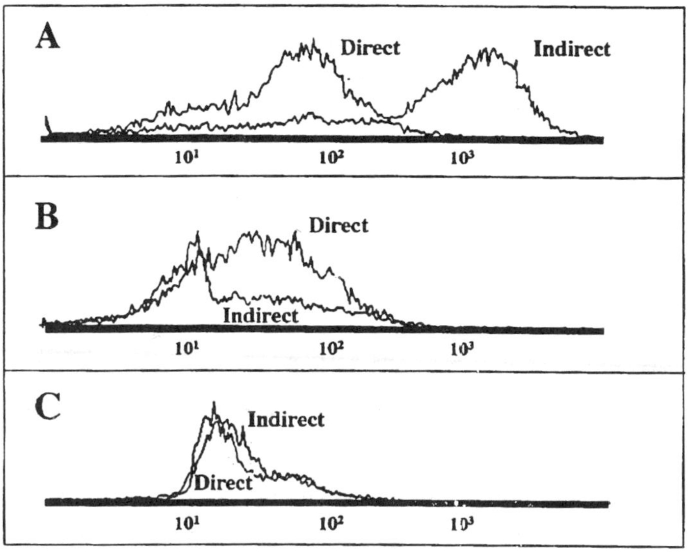

Figure 6 Changes in target antigen selection following direct and indirect immunomagnetic separation with a single monoclonal antibody. The figure shows flow cytometry histograms of residual tumor cells following immunomagnetic separation. (A) "Unusual" profile in which the largest decrease in antigen expression (most efficient separation) was seen using the direct technique. This occurred with an antigen showing very high expression which prevented efficient interaction with the microspheres. Limiting the antibody in the system by attaching it to the antimouse beads improved separation efficiency. (B) "Usual" profile, typical for most immunomagnetic separations. Here improved separation is achieved using the indirect technique, as shown by the decreased expression of the target antigen by the residual target cells. (C) Controls for panels A and B in which depletion was performed using an isotype control antibody and the residual cells were examined for staining with the control antibody.

In the indirect procedure, variability in target Ag expression can be overcome by using cocktails or panels of fluid-phase mAb to saturate all of the binding sites on the cells. Although separation conditions should be optimized for each application, effective rosetting of the sensitized cells and anti-lg–coated beads usually reaches the end point within 60–90 minutes (38). These kinetics are, however, slower than those associated with magnetic nanoparticles (35,56,57), and prolonged mixing of the beads and cells may result in physical damage to the nontarget population.

The major disadvantage of the indirect method is the potential for residual fluid-phase mAbs to block the binding sites on the beads and reduce separation efficiency. It is therefore necessary to remove unbound mAbs before adding the beads by performing three or four washes of the cells; this adds considerably to the length of the procedure and may increase nonspecific cell losses. In contrast, certain colloidal magnetic materials have such huge mAb-binding capacity that unbound Ab is easily absorbed by the colloid, which still retains the ability to bind to the mAb on the target cells (35,56).

VIII. SEPARATION PROCEDURES

A. Choice of Target Antigen and Antibody

Efficient immunomagnetic separation is primarily dependent on the availability of a good mAb to identify the target-cell population. As stated above, the target Ag need not be absolutely specific to the target cell, as long as other cells expressing that Ag are not present in the starting cell mixture. At first glance the selection of the target Ag may appear easy. It should ideally be expressed universally on the target cells, however, the density of the determinant may have a critical influence on separation efficiency (41,48,51,55). We have previously shown that immunomagnetic techniques can be used to separate cells that are resistant to other forms of mAb elimination by virtue of low target Ag expression (54). Low-Ag-density leukemia cells that are resistant to complement-mediated Iysis can be readily rosetted with beads and collected in a magnetic field. There is, however, a lower limit of Ag density beyond which it is not possible to effect magnetic separation using microspheres—presumably due to lack of adequate bonding between the bead and the target cell. In the case of breast cancer cells, we estimated that at least 5×10^4 binding sites per cell were required for effective separation (48), although this will depend on the Ag-Ab combination that is used. At lower densities, improved separation may be achieved by using nanoparticles and high gradient magnetic fields.

A surprising finding has been that efficient separation can also be compromised by high expression of the target Ag. In a breast cancer model system (Fig. 7), we were able to demonstrate that cells expressing high levels of a high molecular weight mucin target were not efficiently removed by magnetic beads (48). Reducing the Ag-Ab interactions by restricting the amount of mAb used by coupling it to the beads resulted in improved separation efficiency. The most likely explanation for the high-Ag-density effect is that overcrowding of mAb molecules on the cell surface causes steric hindrance of binding to the anti-lg on the bead.

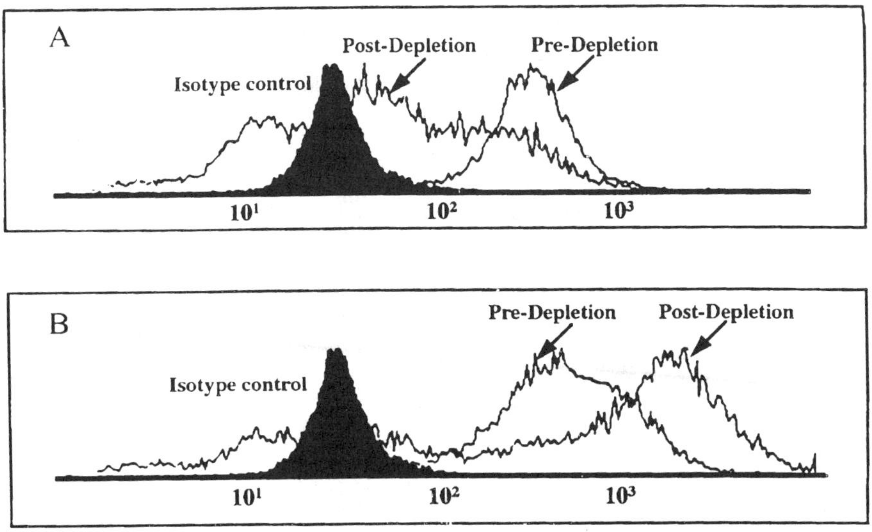

Figure 7 Effect of high target antigen density on the efficiency of immunomagnetic separation. The figure shows flow cytometry histograms of tumor cells before immunomagnetic separation and of the residual cells that were not captured by the beads. (A) Routine indirect immunomagnetic separation in which residual target cells display a reduced expression of the target antigen. (B) Indirect immunomagnetic separation of cells with high target antigen density shows a different pattern, in which the residual cells are those with higher antigen density. The high levels of expression may cause steric interference in the binding of the cell-bound monoclonal antibody to the anti-immunoglobulin on the beads.

Criteria used for the selection of the target Ag should not focus solely on the distribution and density of the Ag but also on its biological function. Highly mobile Ags that undergo rapid internalization upon cross-linking or binding are generally unsuitable. We have consistently found that CD3 is not a good target Ag for the separation of T lymphocytes when microspheres are used (27). Breakage of the various bonds between the Ag/mAb/anti-lg can be detected to varying extents following these separations. This is apparently dependent on a number of factors, since efficient separations can be achieved using anti-CD3 and magnetic nanoparticles collected by high gradient separation (A. P. Gee, unpublished). It is not clear whether this is due to the size of the particles, the reaction kinetics, the strength of the magnetic field, or the possibility that cells may still be susceptible to collection when they have internalized the Ab and attached nanoparticles.

These effects of target Ag density emphasize the need for careful analysis of the separation system. Uniform conditions cannot be successfully used to separate cells of different types. In our hands, one of the most powerful tools for optimizing separations has been residual cell analysis (28,37,48,58). In this approach, following a separation, the negative cell fraction is analyzed with respect to phenotype and Ag expression. As described later, careful selection of the Abs used to perform this analysis can result in generation of extremely useful data that can be used to further optimize the separation. These are then combined with mass balance calculations so that the fate of each cell subpopulation can be quantitatively determined. In the majority of cases, the residual cells that are detectable should show a marked decrease in expression target Ag (28,48), as determined using Ab directed to an epitope different to that used to mediate separation.

Following the identification of a target Ag, the choice of mAb is usually straightforward, being based on binding profile, availability, purity, cost, etc. Antibodies of almost any isotype may be used if an indirect method is being developed. For IgM monoclonals it is also possible to use the direct technique in which the mAb is chemically coupled or passively adsorbed to uncoated beads. If this is the choice, it is important to use Ab preparations of high purity with little extraneous added protein, since adsorption to the bead surface is nonspecific.

B. Coating of Microspheres

For separations, beads must be coated for direct procedures with the mAb directed against the target Ag or, for indirect separations, with a monoclonal or, more frequently, a polyclonal anti-Ig Ab. Streptavidin-coated M-280 beads are also available for capturing biotinylated mAb (17), but are less widely used. Anti-Ig beads can be prepared by passive adsorption or chemical coupling (59) of the Ab to the bead; however, the potential savings associated with using this approach must be offset against the convenience of using the commercially available precoated beads. These are prepared with reagents that show good reactivity with most mAbs of the specified isotype, and it is unusual to see failure of a separation directly attributable to this problem.

Coating with mAb for use in direct separations is subject to the limitations described earlier, in that it is often difficult to achieve good separations with IgG-coated beads, whereas IgM-coated microspheres generally work well (60). Coating instructions are provided by the manufacturer (17); however, a brief generic summary is provided for convenience.

Uncoated beads are provided in distilled water and may be sterilized before coating by washing with 70% alcohol or by gamma irradiation. The beads can be most easily collected between washes by using a small permanent magnet. If coating small numbers of beads, it is advisable to use glass or polypropylene tubes for handling to reduce adsorption of the coating Ab to the tube surface.

It is recommended that the concentration of beads be checked by counting an aliquot in a hemacytometer or automated cell counter.

The required number of beads should be washed with *protein-free* buffer (e.g., phosphate or borate buffer at pH 7.0–8.5 and resuspended in a volume of buffer that will result in a bead concentration of 4–6 $\times 10^8$/mL) *after addition of the Ab.*

The coating Ab should be highly purified. This reduces the amount of non-Ab protein binding in the preparation that will also bind to the beads. The Ab preparation should be in phosphate or borate buffer that is free of sugars or stabilizers. Approximately 2.5–4.0 μg Ab is required for each 1×10^7 beads, of which 40–80% will be adsorbed.

After addition of the Ab, the tube is capped and rotated or mixed continuously overnight at 4°C. Avoid "dead" spots in the tubes where there is inadequate mixing.

Following coating, the beads are collected using the magnet and washed three times in buffer. Phosphate buffer containing a small amount (2–5% v/v) of albumin or fetal bovine serum (FBS) is recommended for this purpose, as the added protein will block any noncoated surfaces on the microspheres. In our experience, three rapid washes and a 30-minute incubation at ambient temperature are sufficient to remove unbound Ab and block the microspheres. Dynal recommends an overnight incubation at 4°C.

The coated beads are usually stable for several months at 4°C, although addition of 0.02% sodium azide is recommended to inhibit bacterial growth.

Before use, beads should be washed in buffer used for the separation procedure.

C. Antibody Sensitization for Indirect Separations and Cell Preparation

For indirect separations, the mAb is offered to the target cell suspension (Fig. 4). The amount of Ab to be used should be determined empirically. In initial experiments this can be done by flow cytometry, although it is often possible to use less mAb than the saturating concentration determined by immunofluorescence measurements. As a rule of thumb, we generally start with a concentration of 2 μg mAb/1 × 10^6 target cells and then titrate downwards in serial threefold dilutions.

It is important to pay attention to the composition of the starting cell mixture. Careful preparation upfront may avoid subsequent problems with non-specific binding. For example, removal of macrophages and polymorphonuclear leukocytes prevents these cells attaching to, and attempting to phagocytose, the microspheres. Frustrated phagocytosis can result in death of the cell and release of DNA, which gels and further interferes with the separation. For separation of blood or marrow leukocytes, a mononuclear cell preparation is recommended, if necessary this may be depleted of monocytes by incubation in a plastic flask or on nylon wool columns (61), or by a "predepletion" step. In predepletions, the cell preparation is exposed to either uncoated beads (62) or beads that have been coated with a control, non-reactive Ab. Cells that would have attached non-specifically to the beads that are used for the separation step will adhere to the "mock" beads and can be easily removed. The extent to which these preseparation manipulations are of value will depend on the goal of the separation procedure. For example, 70% to 90% depletions of $CD4^+$ T cells can be achieved by adding directly conjugated CD4 beads to whole blood (A. P. Gee, unpublished). In contrast, positive selection of highly purified $CD34^+$ hematopoietic cells usually requires careful preparation of the starting cells and extensive optimization of experimental conditions.

Sensitization of the target cells is usually performed at high cell concentrations in media or buffer containing 2–10% v/v protein (e.g., serum, plasma protein fraction, or albumin), with occasional or continuous mixing. Incubations are normally performed at 4°C, although satisfactory results may be achieved at ambient temperature. The incubation time ranges from 15 minutes to > 1 hour, depending on the system. Unbound mAb is removed in a series of three or four washes with large volumes of cold buffer. When performing large-scale separations, the efficiency of washing can be increased by splitting the cells into more than one container. Expense is also reduced by washing with simple salt solutions containing 2% protein. Optimal mAb removal is achieved by increasing the number of washes, rather than the wash volume, at each step. We have found that four 100-mL washes of an average marrow graft produce satisfactory results (A. P. Gee, unpublished). For incubation with the microspheres, the cells should be returned to a buffered salt solution or culture medium containing protein.

D. Incubation with Microspheres

Incubations of mAb-treated target cells with anti-Ig coated-beads (indirect separation) or nonsensitized cells with mAb-coated beads (direct separation) (Fig. 4) are usually carried out under similar conditions. These must be determined empirically for each system, however, the following guidelines may help in optimizing a system:

For target cells at high proportions in the starting mixture (10 to ≥ 70%), bead-to-cell ratios of 4–10 beads/target cell are usually adequate.

For smaller proportions of target cells (10 to < 1%), the bead-to-target cell ratio should be increased to 10 to > 40 beads/cell. The upper limit is determined by the risk of physical damage to the cells, cost, increased possibility of nonspecific cell trapping, and capacity of the magnets to retain large numbers of beads. Where the target cell population is variable, but is usually < 2%, an alternative approach is to add 1–5 beads per total nucleated cell.

The incubation conditions should favor interaction between the microspheres and the target cells (e.g., low reaction volume, good mixing, and high cell concentrations). Generally, cell concentrations of $> 1 \times 10^7$/mL give good results. At very high total cell concentrations and low target cell proportions, it may be difficult to achieve good bead-cell interaction. For mixing purposes, slow (4 rpm) end-over-end rotation is satisfactory.

Mixing containers used should not have dead spaces (e.g., the access ports of transfer packs and caps of tubes) in which cells and/or beads can become trapped. Avoid containers with a capacity that greatly exceeds the reaction-mixture volume. This prevents excessive tumbling of the cells and beads during mixing that can lead to physical damage to the cells.

Incubations should be performed at 4°C. This reduces capping and internalization of cell-bound mAb and phagocytosis of the beads by neutrophils and monocytes.

The incubation time depends on the system and incubation conditions. Times of 30–90 minutes are usually satisfactory (38). Direct separations often benefit from longer incubation times (63).

Before collection of the beads, the volume of the reaction mixture can be increased by the addition of buffer or medium. This reduces the likelihood of cells being nonspecifically trapped as the beads are drawn to the magnets.

E. Magnetic Separation and Separators

For small-scale separations, the collection of rosetted cells and beads is easily accomplished by holding a permanent magnet to the outside of the mixing tube. The beads are drawn to the magnets within a few seconds, and after 2–3 minutes the nonrosetted cells may be aspirated or decanted. Magnetic holders are commercially available for 0.5- to 50-mL tubes (Dynal) (17), and for multiple separations, side-pull and bottom-pull magnetic racks are available for 12 × 75 mm tubes (Corning).

Large-scale separations require more sophisticated hardware (64–67) . Two approaches have been described. The first uses a large array of permanent magnets onto which the container of cells and beads is placed. The magnetic array should be engineered to maximize magnetic reach, so that rosetted cells at some distance from the magnet surface can be effectively captured. Following the collection period, the nonrosetted cells are pumped or drained from the container, taking care not to disturb the target cells and beads held in contact with the sides of the container by the magnets. This system has the advantage of simplicity of both design and use, however, there is the potential for high nonspecific cell losses due to settling of the nontarget cells during the collection period, trapping as the beads are collected onto the magnets, and failure to recover all of the ceils when the container is emptied. This can be overcome by performing the collection in a nonhorizontal orientation (e.g., in a column), by collecting the beads against gravity (upwards collection) and by angling the container so that the exit port is at the lowest point.

The second approach is to pass the cell-bead mixture over the magnetic array as a continuous flow (66). In this case, the array should be designed to provide maximum holding force at the surface, to avoid the collected cells being swept through the system by the flow of oncoming cells. This approach requires a starting container for the bead-cell mixture, a separation vessel, and a collection container. Fluid flow must be controlled either by careful use of gravity or by a peristaltic pump, which should be placed downstream to avoid traumatic injury to the rosetted cells by passage through the pump rollers. This design is more complex, but can result in higher cell yields by minimizing cell settling. The geometry of the collection vessel must be such that the entire flow path is exposed to the magnetic array and that build-up of rosettes and beads does not occur at the mouth of the container. This can obscure the flow path and result in a sieving effect on incoming beads and cells.

In most cases a combination of the two collection systems has been used (64,65,68) (Fig. 8). Static collection is first performed to debulk the suspension of the majority of free beads and rosetted target cells. The effluent from this

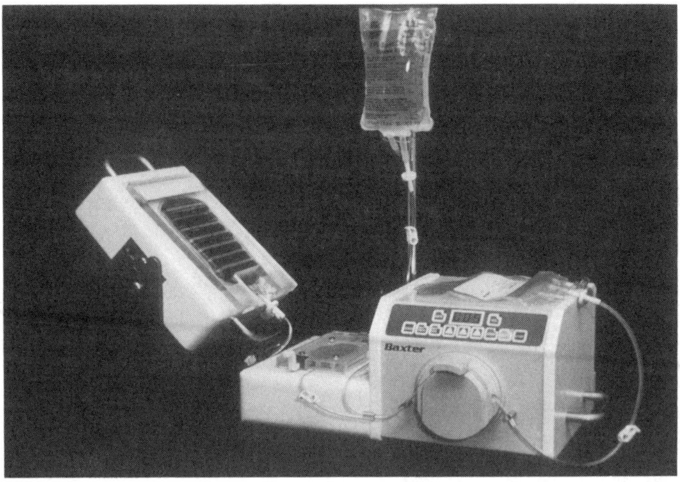

Figure 8 Baxter MaxSep™ large-scale magnetic separator for research applications. A large magnetic array with high reach-out is at the left of the separator and is angled for improved recovery of the nontarget cells. These are pulled from the transfer pack by use of a peristaltic pump built into the control component of the separator, shown at the right of the picture. The smaller, horizontal magnetic array, shown at the center, has high holding force and serves to capture any beads that escape from the primary magnet. The fluid in the hanging transfer pack is used to prime the system. This device is not approved for clinical use in the United States. (Photograph courtesy of Baxter Biotech Immunotherapy.)

primary collection is then passed through a continuous-flow device, since the high holding force in these systems provides an excellent means to capture magnetic particles that may have escaped the first separation. A number of separators using this combination of capture systems have been described (64,65,68) for negative immunomagnetic separation applications.

For positive selections the considerations are different (30,63,69). Separators designed for negative selection generally emphasize capture and retention of target cells no longer required. For positive selection, the rosetted cells must be recovered from the separator with high yields and viabilities. This necessitates a different approach, in which the collection is usually performed over a large surface area using a weaker magnetic field (68). This reduces nonspecific trapping and dense packing of the beads and target cells onto the magnetic array. Large, densely packed bead-cell complexes are often impossible to disaggregate without lethal damage to the cells and/or substantially reduced yields and purities.

Most positive separators employ relatively simple arrays of columns or rows of identical permanent magnets spaced along the length of the container (30,62) (Fig. 9). These magnets must have a combination of reach-out and holding properties that ensures highly efficient capture of the target cells and their retention while the nontarget population is removed. For this reason, the

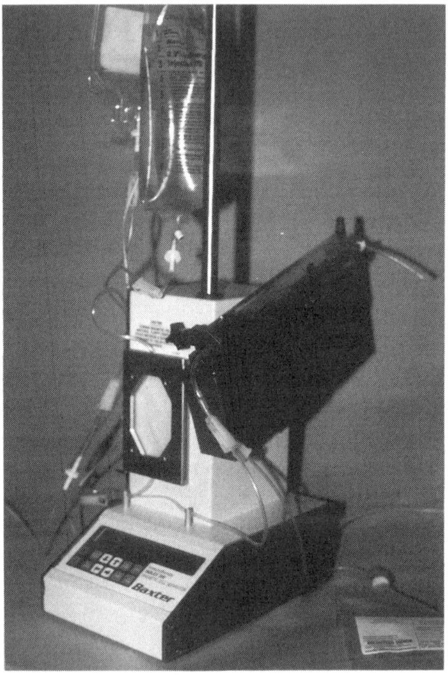

Figure 9 Baxter Isolex™ 300 magnetic separator for positive selection applications. The separation is performed in the disposable column shown on top of the tilted magnetic array. Target cells are mixed with microspheres in this chamber using an in-built rocker. The column is then exposed to the magnets, the target cells collected, and nontarget cells are drained from the column. The target cell–bead rosettes are then washed in the separation chamber, to which is added chymopapain to release the beads (the fluid in the hanging transfer packs is used to prime the system and to wash the rosettes). The free beads are then retained by exposure to the magnetic array and the released cells are drained from the column. This device is not yet approved for clinical use in the United States. (Photograph courtesy of Baxter Biotech Immunotherapy.)

arrays are usually placed on more than one surface of the separation vessel and collection is performed in multiple directions. This results in capture of the target cells as a thin layer onto the walls of the container. After removal of the noncaptured cells, the container is refilled with buffer and the magnetic array removed. Gentle mixing is used to resuspend the rosetted cells and any trapped nontarget cells. The vessel is then reexposed to the magnets. The aim of the second (and sometimes third) collection step is to increase the purity of the separation by releasing any nontarget cells that were trapped during the initial exposure to the magnets.

IX. REMOVAL OF MICROSPHERES FROM SELECTED CELLS

Depending upon the final application, it may be necessary to remove the rosetted beads from the target cell. A number of methods have been used with varying success (30,47,62,69-72) (Fig. 10). The simplest, but least generally successful, approach is to place the rosetted cells into culture at 37°C (72). In some cases the beads will spontaneously detach in 24–72 hours. A more rapid technique is to use enzymes to digest the proteins linking the mAb and the bead. Trypsin, dispase, and a number of other proteases have been tested with limited success. Detachment may occur, but often at the expense of cell viability or unwanted alterations to the cell surface. The most notable exceptions have been the use of chymopapain (30,62) (Fig. 11) and O-sialoglycoprotease (69) for the release of beads from $CD34^+$ cells. These enzymes attack epitopes on the CD34 Ag with which the anti-CD34 mAbs react (73,74). Removal of these epitopes releases the Ab-bead complex from the cell surface; however, these enzymes may digest epitopes on other Ags (30). This can interfere with both immunophenotypic characterization of the selected population and subsequent Ab-mediated manipulations. Culture of the released cells overnight usually results in reexpression of the intact Ag (30).

An alternative approach is to use specific peptides to displace the bound epitope from its combining site on the mAb (Fig. 10). This results in the release of the target cell with unaltered surface properties. The availability of these peptides is very limited, and their cost is likely to be high if the application is limited. A similar effect has been achieved by using an anti-Fab polyclonal Ab (47,71) (commercially available as DETACHaBEAD® reagent from Dynal) to attach to and distort the combining site on the mAb attached to the target cell. This results in the release of the bead-antibody complex without the use of

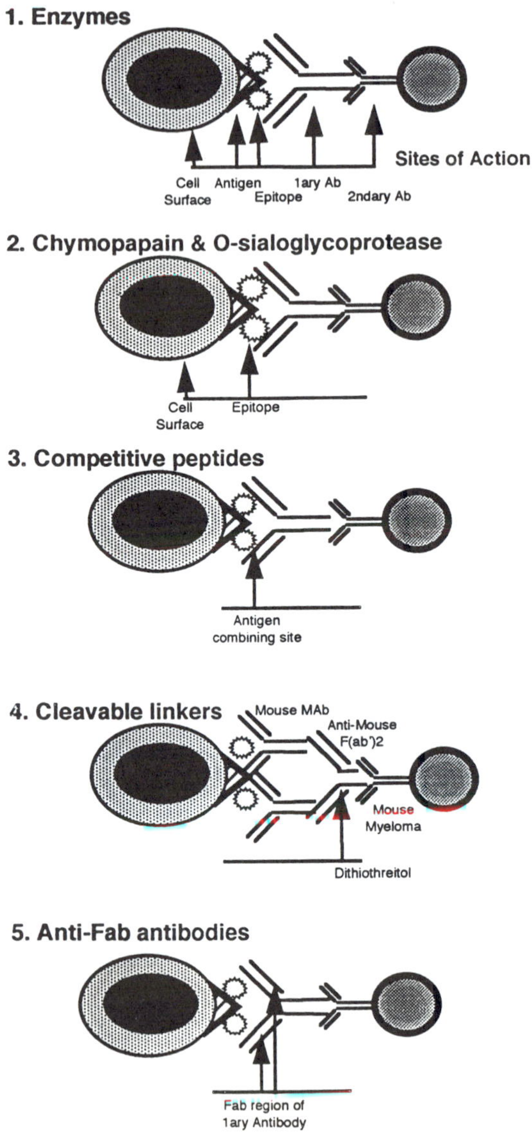

Figure 10 Methods for the release of positively selected cells from magnetic matrices. (For details see text.)

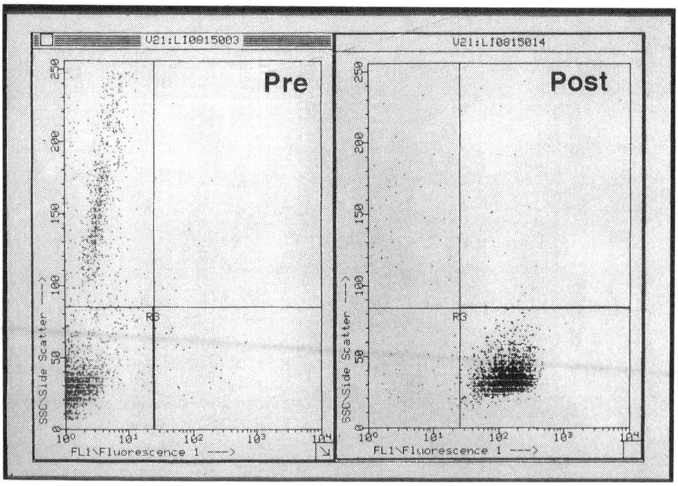

Figure 11 Enrichment of CD34$^+$ bone marrow cells using positive selection with magnetic microspheres and chymopapain for cell release. (Pre) CD34$^+$ cells are shown in the lower right quadrant and represent $< 2\%$ of total nucleated cells in the sample. (Post) After positive indirect selection using magnetic microspheres, and chymopapain to release the cells from the beads, the proportion of CD34$^+$ cells in the sample postenrichment has increased to $> 93\%$.

enzymes. It appears that this approach requires the generation of specific anti-Fab Abs for each mAb (47,71), as attempts to use a generic Ab have been unsuccessful. This, in turn, restricts the choice of mAb that can be used for positive selection. Successful target cell attachment has also been achieved by introducing a cleavable link between the bead and the cell and then releasing the cell by treatment of the complex with 10 μM dithiothreitol (70). When nanoparticles are used for positive selection, the target cells remain coated with the magnetic particles that are invisible by conventional light microscopy and flow cytometry. Although their presence does not appear to affect the cells' viability or biological properties, it is not yet known whether coated cells will traffic normally when infused into a living host. The approaches described above to release microspheres from target cells can be applied to nanoparticles. A new enzymatic procedure (using MACS MutliSort release reagent from Miltenyi Biotec) has been recently described (75), which does not affect other cell surface molecules. Verification of release will be more complex than with beads due to the very small size of the nanoparticles.

X. RESIDUAL CELL ANALYSIS

Immunophenotypic analysis of the cells remaining after immunomagnetic separation can provide extremely useful information that can be used to optimize separation efficiency (28,37,48,58) (Figs. 6 and 7). Residual target cells should be detected by their ability to bind an Ab to a different epitope on the Ag used to effect separation. In addition, wherever possible, confirmation should be obtained using a second independently expressed Ag present on all target cells. Adequate sensitization of the target cells can be verified by incubating the residual population with labeled Ab to the target Ag. Positive staining is an indication of lack of mAb in the system.

Other explanations for system failure can be achieved by staining residual cells with anti-immunoglobulin Abs. Where a murine mAb was used for separation, residual cells staining positively with labeled antimouse Ig indicates failure of the beads to react with adequately sensitized target cells, or detachment of the beads. Positive staining with antisheep Ig suggests that sheep antimouse polyclonal Ab has detached from the microspheres during the procedure.

As the separation conditions are further optimized, the expression of target Ag by residual cells generally decreases to a baseline level (48), in which case consideration should be given to adding additional mAbs with different specificity. Eventually the number of residual cells will be too small to permit detection reliably by flow cytometry (< 1 in 10,000–25,000). At this point, limiting dilution analysis (28,77) and/or the polymerase chain reaction (PCR) may be used if the target cells express a specific gene marker (78,79); alternatively, in model systems, it is possible to introduce a gene that can be used for tracking. PCR sensitivities of 1 target cell in > 1,000,000 cells have been reported; however, reliable quantitation may be difficult to achieve without very careful optimization of the reaction conditions and immaculate experimental technique. In the case of positive selection, it may also be important to assess the integrity and functional activity of the selected cells. The nature of these assays will depend upon the identity of the target cells, however, dye exclusion viability tests and clonogenic, proliferative, or cytotoxicity assays of function are the most widely used (23,30).

XI. EVALUATION OF NONTARGET CELLS

When negative selection is used, assays of cell function similar to those described above for positively selected cells can be used to assess cell function.

In addition, phenotyping of the negatively selected population can provide useful information. For example, we were able to show that nontarget cells took up fragments of an IgM mAb used for T-cell depletion nonspecifically (80). This Ab fragmented after attachment to the beads and during the separation procedure. Residual cell analysis revealed a large number of cells binding labeled anti-IgM but few cells co-staining with a mAb directed against a second T-cell–specific Ag. Although the T cells had been effectively removed, since this separation technique was under development for clinical use, the presence of IgM fragments on the nontarget was still of concern.

XII. VALIDATION OF SEPARATION SYSTEMS

The assays described above are useful in determining the efficiency of cell separation and may be adequate for optimizing small-scale procedures. If the goal is to develop a large-scale technique for ultimate clinical use, optimization should be more rigorous. In the United States, regulatory approval will be required to test the clinical performance of the system. In general, such approval will require extensive preclinical data demonstrating the performance of the system (23,81) (e.g., cell separation efficiency, cell purities, yields, viabilities, functional assay data, toxicity of the various components). Information on the manufacture of the hardware and biologicals is necessary, together with detailed results of testing for potential viral contamination of the mAb. Collection of these data usually requires the evolution of an experimental model system (81). Depending on the identity of the target cell, initial experiments focus on small-scale depletion of naturally occurring targets (e.g., leukocyte subpopulations in blood) from the cell mixture. Alternatively, cultured human cell lines may be used as the target population. This has the advantage that the cells may be prelabeled with either a fluorochrome (82) or a marker gene (83) to aid in their subsequent detection. If cell lines are used, it is important to verify, as early as possible, that they express the target Ag at levels comparable to those seen on naturally occurring cells of the same histology. This is frequently not the case for cells that have been grown in culture for a prolonged period. Since low-Ag-density targets are usually the most difficult to separate (48,53–55), a cell line that stably expresses low levels of the Ag can be a useful target in optimization experiments (55). Such lines can be generated by repeated exposure of the parental line to IgG Ab and complement, which selectively kills the cells with high target Ag densities.

Although it may be difficult to obtain human material for system development, animal cells are usually a poor alternative. Most nonhuman mammalian cells behave differently from their human counterparts with respect to size, sedimentation, and surface characteristics (81), and many human-directed mAb do not cross-react with the analogous Ag on animal cells. This can make it difficult to scale-up procedures to clinically appropriate cell numbers and volumes. For separations involving bone marrow, it is possible to substitute peripheral blood and/or bovine bone marrow (81), however, data generated using these cells must ultimately be confirmed using human marrow. For this purpose, some investigators have had success using thawed cryopreserved marrows that are no longer required for transplantation. Various techniques have been described for thawing and diluting this material in the presence of additives to prevent cell clumping and death (84). It should also be remembered that cryopreservation per se may alter the cellular composition of the marrow that should be reflective of the starting material used for the separation (85). Scale-up often presents the biggest single challenge in developing a separation technique (23,81). Minor problems of nonspecific binding, reduced yields and purities, and separator geometry can become major challenges on a large scale. In addition, each application may have characteristic properties that preclude the development of a generic approach. As with most biological systems, careful optimization of each component is the secret to success.

XIII. COMBINATION SEPARATIONS

If extremely high purity is the most important goal in a separation, it may be necessary to use a combination of negative and positive immunomagnetic techniques (86). Negative selections frequently require the use of multiple mAbs, which can be expensive. It is therefore logical to use a large-scale, up-front positive selection to debulk the starting cell population, followed by a small-scale negative selection to remove any contaminating cells. Historically this sequence has been difficult to use with magnetic microsphere separations, since proteolytic enzymes were used to detach the beads from the target cells after positive selection (74). These enzymes frequently destroyed other Ags on contaminating nontarget cells required to effect their removal by negative selection (30). This limitation has been overcome by the availability of nonenzymatic bead-removal techniques (47,70,71), which do not affect the antigenic properties of the cells.

In two-step separations, careful attention must be paid to the conditions used in the negative-selection procedure. The yield of target cells from the positive-selection step is frequently about 50% of an initial population that may initially be < 1% of the total nucleated cells. A subsequent yield of 50% from the negative selection may result in target cell recoveries that are unacceptable. Where permissible, a compromise between purity and yield may have to be reached to achieve adequate cell numbers for further use. Prolonged manipulation of the cells during combined separation can also adversely affect viability and cell function and increase the likelihood of microbial contamination.

XIV. ANTICIPATED RESULTS

Immunomagnetic separation is widely regarded as one of the most specific and effective methods for selective cell separation. The success of the procedure is critically dependent on the nature of the components and the careful use of optimization. Performance characteristics are specific to each system, rather than to the generic approach. Single-step separations may give > 95% target cell purities with yields of 60–80%; however, the same system in different hands with similar cells obtained from a different source may perform very poorly. In many cases this can be improved by both uniform up-front preparation of the cells and further experience with the technique. In large-scale clinical marrow purging, depletions of < 3 to > 5 logs of target cells have been reported, with approximately 50% recovery of nucleated nontarget marrow cells (31,44). Improved efficiency has been reported by using multiple cycles of purging (66,87,88) and by combining immunomagnetic separations with cytotoxic agents (89). Comparative studies have shown that use of beads can give superior results to those obtained with complement (54,90); however, two rounds of beads were required to produce results that were comparable to purging with immunotoxins (91). Positive-selection efficiency depends largely on the starting material, but separation purities of > 95% and yields of > 60% can be achieved. For higher purities, combinations of positive and negative selection should be considered, although improvements are usually achieved at the expense of yield.

Biological variation is, however, inherent in these systems and can be moderated, but probably not eliminated, by a thorough knowledge of the effects of modifying separation condition. In spite of these limitations, immunomagnetic separation using paramagnetic microspheres remains one of the most efficient and simple methods for eliminating or enriching target cell populations.

ACKNOWLEDGMENTS

The author would like to acknowledge the excellent secretarial assistance of Mrs. Jean Winter and Ms. Gloria Todd and financial support from the South Carolina Cancer Center and the Board of Directors of the Center for Cancer Treatment and Research, Richland Memorial Hospital, Columbia, South Carolina, for some of the studies described in this review.

REFERENCES

1. Kemshead JT. *Magnetic Separation Techniques Applied to Cellular and Molecular Biology.* Oxford, UK: Wordsmiths Conference Publications, 1991.
2. Wysocki LJ, Sato VL. "Panning" for lymphocytes: a method for cell separation. *Proc Natl Acad Sci USA* 1978; 75:2844–2848.
3. Lebkowski J, Schain L, Harvey M, Moseley A, Okarma TB. Isolation and culture of human and CD34+ hematopoietic stem cells using AIS CELLectors™. In: Wunder E, Sovalat H, Henon PR, Serke S, eds. *Hematopoietic Stem Cells, The Mulhouse Manual.* Dayton, OH: Alpha Med Press, 1994:215.
4. Uckun FM, Myers DE. Allograft and autograft purging using immunotoxins in clinical bone marrow transplantation for hematologic malignancies. *J Hematother* 1993; 2:155–163.
5. Antin JH, Weinstein HL, Bouloux C, Bierer BE. Imunotoxinmediated depletion of CD5+ T cells from bone marrow for graft-vs-host disease prophylaxis. In: Gee AP, ed. *Bone Marrow Processing and Purging: A Practical Guide.* Boca Raton, FL: CRC Press, 1991:213.
6. Gee AP, van Hilten J, Graham-Pole J, Boyle MD. Antibody-complement killing of neuroblastoma cells. *Prog Clin Biol Res* 1985; 175:471–483.
7. Lansdorp PM, Thomas TE, Schmitt CR, Eaves CJ. Marrow contamination: positive selection. In: Armitage JO, Antman KH, eds. *High-Dose Cancer Therapy.* Baltimore: Williams and Wilkins, 1992:276.
8. Siegall CB. Targeted toxins as anticancer agents. *Cancer* 1994; 74:1006–1012.
9. Raso V, Lawrence J. Carboxylic ionophores enhance the cytotoxic potency of ligand- and antibody-delivered ricin A chain. *J Exp Med* 1984; 160:1234–1240.
10. Gee AP, Bruce KM, Morris TD, Boyle MD. Evidence for an anticomplementary associated with human bone marrow cells. *J Natl Cancer Inst* 1985; 75:441–445.
11. Zhong RK, Kozii R, Ball ED. Homologous restriction of complement-mediated cell lysis can be markedly enhanced by blocking decay-accelerating factor. *Br J Hematol* 1995; 91:269–274.
12. Ugelstad J, Stenstad P, Kilaas L, et al. Monodisperse magnetic polymer particles. New biochemical and biomedical applications. *Blood Purif* 1993; 11:349–369.

13. Miltenyi S, Müller W, Weichel W, Radbruch A. High gradient magnetic cell separation with MACS. *Cytometry* 1990; 11:231–238.
14. Ugelstad J, Mork PC, Kaggerud KD, Ellingsen T, Berge A. Swelling of oligomer particles: new methods of preparation of emulsions and polymer dispersions. *Adv Colloid Interface Sci* 1980; 13:101.
15. Thomas TE, Lansdorp PM. Selective separation of cells using magnetic colloids. *Prog Clin Biol Res* 1993; 389:65–77.
16. Trickett AE, Ford DJ, Lam-Po-Tang PR, Vowels MR. Immunomagnetic bone marrow purging of common acute lymphoblastic leukemia cells: suitability of BioMag particles. *Bone Marrow Transplant* 1991; 7:199–203.
17. Dynal Inc. *Dynabead*® Biomagnetic Separation Product Catalog. Lake Success, NY: Dynal Inc., 1996.
18. Patel PD, Blackburn CW. Detection of food-poisoning agents using immunomagnetic particles. In: Kemshead JT, ed. *Magnetic Separation Techniques Applied to Cellular and Molecular Biology.* Oxford: Wordsmiths Conference Publications, 1991:93.
19. Jackson CJ, Gallery ED, Schrieber L. Human endothelial cells bind to, and can be purified by Ulex europaeus I coated Dynabeads. In: Kemshead JT, ed. *Magnetic Separation Techniques Applied to Cellular and Molecular Biology.* Oxford: Wordsmiths Conference Publications, 1991:63.
20. Saucan L, Palade GE. Membrane and secretory proteins are transported from the Golgi complex to the sinusoidal plasmalemma of hepatocytes by distinct vesicular carriers. *J Cell Biol* 1994; 125:733–741.
21. Morgan JG, Dolganov GM, Robbins SE, Hinton LM, Lovett M. The selective isolation of novel cDNAs encoded by the regions surrounding the human interleukin 4 and 5 genes. *Nucleic Acids Res* 1992; 20:5173–5179.
22. Karlsson GB, Platt FM. Analysis and isolation of human transferrin receptor using the OKT-9 monoclonal antibody covalently crosslinked to magnetic beads. *Anal Biochem* 1991; 199:219–222.
23. Gee AP. Malignant cell purging: immunophysical techniques. In: Lasky L, Warkentin P, eds. *Marrow and Stem Cell Processing for Transplantation.* Bethesda, MD: American Association of Blood Banks, 1995:83.
24. Gee AP. Immunologically based methods for the elimination of tumor cells from autologous stem cell grafts. *Immunomethods* 1994; 5:232–242.
25. Kemshead JT. The immunomagnetic manipulation of bone marrow. In: Gee AP, ed. *Bone Marrow Processing and Purging: A Practical Guide.* Boca Raton, FL: CRC Press, 1991:307.
26. Vartdal F, Kvalheim G, Lea TE, et al. Depletion of T lymphocytes from human bone marrow. *Transplantation* 1987; 43:366–371.
27. Gee AP, Mansour V, Weiler M. T-cell depletion of human bone marrow. *J Immunogen* 1989; 16:103–115.
28. Collins NH, Gee AP, Henslee-Downey PJ. T cell depletion of allogeneic bone marrow transplants by immunologic and physical techniques. In: Lasky L,

Warkentin P, eds. *Marrow and Stem Cell Processing for Transplantation.* Bethesda, MD: American Association of Blood Banks, 1995:149.

29. Egeland T, Gaudernack G. Immunomagnetic isolation of CD34+ cells: methodology and monoclonal antibodies. In: Wunder E, Sovalat H, Henon PR, Serke S, eds. *Hematopoietic Stem Cells, The Mulhouse Manual.* Dayton, OH: Alpha Med Press, 1994:141.
30. Ishizawa L, Burgess J, Hardwick A, Hangoc G, Gee AP, Law P. Selection of CD34+ cells using indirect immunomagnetic procedures and a magnetic cell separation system. In: Wunder E, Sovalat H, Henon PR, Serke S, eds. *Hematopoietic Stem Cells, The Mulhouse Manual.* Dayton, OH: Alpha Med Press, 1994:171.
31. Shpall EJ, Gehling U, Cagnoni P, Purdy M, Hami L, Gee AP. Isolation of CD34-positive hematopoietic progenitor cells. *Immunomethods* 1995; 5:197–203.
32. Rhodes EG, Baker PK, Rhodes JM, Davies JM, Duguid JK. Autologous bone marrow transplantation for myeloma patients using PNA and CD19-purged marrow rescue. *Bone Marrow Transplant* 1994; 13:795–799.
33. Reisner Y, Gan J. Differential binding of soybean agglutinin to human neuroblastoma cell lines; potential application to autologous bone marrow transplantation. *Cancer Res* 1985; 45:4026–4031.
34. Qin S, Cobbold SP, Pope H, et al. "Infectious" transplantation tolerance. *Science* 1993; 259:974–977.
35. Fletcher RC, Picolli SP. Capacity of colloidal goat anti-mouse IgG ferrofluid to negatively deplete cells. *Prog Clin Biol Res* 1994; 389:79–87.
36. Roath S, Thomas TE, Watson JH, Lansdorp PM, Smith RJ, Richards AJ. Specific capture of targeted hematopoietic cells by high gradient magnetic separation by the use of ordered wire array filters and tetrameric antibody complexes linked to a dextran iron particle. *Prog Clin Biol Res* 1994; 389:155–163.
37. Mansour V, Weiler M, Gee AP. Factors limiting the efficiency of immunomagnetic cell separation. *Prog Clin Biol Res* 1992; 377:169–179.
38. Pilling D, Kitas GD, Salmon M, Bacon PA. The kinetics of interaction between lymphocytes and magnetic polymer particles. *J Immunol Methods* 1989; 122:235–241.
39. Kvalheim G, Feld JG, Pihl A, et al. Immunomagnetic removal of B-lymphoma cells using novel mon-sized magnetizable polymer bead, M-280, in conjunction with primary IgM and IgG antibodies. *Bone Marrow Transplant* 1989; 4:567–574.
40. Malpas J, Kemshead JT, Pritchard J, Greaves M. Variability of surface antigen expression on neuroblastoma cells as revealed by monoclonal antibodies. In: Raybard C, Clement R, Lebreuil G, Benard J, eds. *Proceedings of the Thirteenth Meeting of the International Society of Pediatric Oncologists.* Amsterdam: Exceprta Medica, 1981:90.
41. Taupier MA, Kierney JF, Leibson PJ, Loken MR, Schreiber H. Nonrandom escape of tumor cells from immune lysis due to intraclonal fluctuation in antigen expression. *Cancer Res* 1983; 43:4050–4056.

42. Kvalheim G, Fodstad Ø, Pihl A, et al. Elimination of B lymphoma cells from human bone marrow, I. The use of monodisperse magnetic particles with attached monoclonal antibodies. *Cancer Res* 1987; 47:846–851.
43. Reynolds CP, Seeger R, Vo DD, Black AT, Wells J, Ugelstad J. Model system for removing neuroblastoma cells from bone marrow with monoclonal antibodies using monoclonal antibodies and magnetic immunobeads. *Cancer Res* 1986; 46:5882–5886.
44. Shpall EJ, Johnston C, Hami L. Bone marrow purging. In: Armitage JO, Antman KH, eds. *High-Dose Cancer Therapy* Baltimore, MD: Williams and Wilkins, 1992:249.
45. Vredenburgh JJ, Ball ED. Elimination of small cell carcinoma of the lung from human bone marrow by monoclonal antibodies and immunomagnetic beads. *Cancer Res* 1990; 50:7216–7220.
46. Civin CI, Strauss LC, Brovall C, Fackler MJ, Schwartz JF, Shaper JH. Antigenic analysis ot hematopoiesis. III. A hematopoietic progenitor cell surface antigen defined by a monolconal antibody raised against KG-1a cell. *J Immunol* 1984; 133:157–165.
47. Herikstad BV, Lien E. Immunomagnetic separation of hematopoietic progenitor cells using Dynabeads® and DEATCHaBEAD@. In: Wunder E, Sovalat H, Henon PR, Serke S, eds. *Hematopoietic Stem Cells, The Mulhouse Manual.* Dayton, OH: Alpha Med Press, 1994:149.
48. Gee AP, Mansour VH, Weiler MB. Effects of target antigen density on the efficacy of immunomagnetic cell separation. *J Immunol Methods* 1991; 142:127–136.
49. Gee AP, Lee C, Sleasman JW, Madden M, Ugelstad J, Barrett DJ. T lymphocyte depletion of human peripheral blood and bone marrow using monoclonal antibodies and magnetic microspheres. *Bone Marrow Transplant* 1987; 2:155–163
50. Jansen J, Hanks S, Akard L, et al. Selective T cell depletion with CD8-conjugated magnetic beads in the prevention of graft-versus-host disease after allogeneic transplantation. *Leukemia* 1995; 9:271–278.
51. Shapiro SJ, Leibson PJ, Loken MR, Schreiber H. Changes in susceptibility to cytotoxic antibody among tumor cells surviving exposure to chemotherapeutic agents. *Cancer Res* 1982; 42:2622–2627.
52. Treichel RS. Autologous bone marrow transplantation for leukemia: monoclonal antibody-mediated purging of multi-drug resistant leukemia. *Leuk Res* 1993; 17:491–499.
53. Gee AP, Boyle MD. Purging tumor cells from bone marrow by antibody and complement: a critical appraisal. *J Natl Cancer Inst* 1988; 80:154–159.
54. Janssen WE, Lee C, Gross S, Gee AP. Low antigen density leukemia cells: selection and comparative resistance to antibody-mediated marrow purging. *Exp Hematol* 1989; 17:252–257.
55. Gee AP, Bruce K, van Hilten J, et al. Selective loss of expression of a tumor-associated antigen on a human leukemia cell line induced by treatment with monoclonal antibody and complement. *J Natl Cancer Inst* 1987; 78:29–35.

56. Kemshead JT, Hancock J, Liberti P. Immunomagnetic colloids for the enrichment of tumor cells from peripheral blood and bone marrow: a model system. *J Hematother* 1994; 3:51–57.
57. Miltenyi S, Guth S, Radbruch A, Pflüger E, Thiel A. Isolation of CD34+ hematopoietic progenitor cells by high gradient magnetic cell sorting (MACS). In: Wunder E, Sovalat H, Henon PR, Serke S, eds. *Hematopoietic Stem Cells, The Mulhouse Manual.* Dayton, OH: Alpha Med Press, 1994:201.
58. Gee AP. Monoclonal antibodies as agents to purge tumor cells from bone marrow: laboratory and clinical experiences. In: Kemshead JT, ed. *Pediatric Tumors: Immunolonical and Molecular Markers.* Boca Raton, FL: CRC Press, 1989:105.
59. Jackson CJ, Garbett PK, Nissen B, Schreiber L. Binding of human endothelium to Ulex europaeus I-coated Dynabeads: application to the isolation of microvascular endothelium. *J Cell Sci* 1990; 96:257–262.
60. Manyonda IT, Soltys AJ, Hay FC. A critical evaluation of the magnetic cell sorter and its use in the positive and negative selection of CD45RO+ cells. *J Immunol Methods* 1992; 149:1–10.
61. Nakamura Y, Noma M, Kidokoro M, et al. Isolation of CD34+ peripheral blood stem cells using an immunomagnetic positive selection system: prior monocyte depletion using a nylon-wool column. *J Hematother* 1994; 3:129–133.
62. Burgess J, Traycoff C, Hardwick A, et al. Separation of clinical quantities of $CD34^+$ cells from human bone marrow using immunomagnetic procedures. *Prog Clin Biol Res* 1994; 389:293–302.
63. Gee AP, Bruce K, Lee C, Bauer PC, Janssen WE, Gross S. Depletion of cALLa-positive leukemia cells from bone marrow using monoclonal antibodies and magnetic microspheres. *Proc Am Assoc Cancer Res* 1986; 27:367.
64. Gee AP, Moss T, Mansour V, et al. Large scale immunomagnetic separation system for the removal of tumor cells from bone marrow. *Prog Clin Biol Res* 1992; 377:181.
65. Kvalheim G, Sorensen O, Fodstad Ø. Immunomagnetic removal of B Iymphoma cells from human bone marrow: a procedure for clinical use. *Bone Marrow Transplant* 1988; 3:31–41.
66. Ball ED, Powers FJ, Vredenburgh JJ, Heath CA, Converse AO. Purging of small cell lung cancer cells from bone marrow using immunomagnetic beads and a flow-through device. *Bone Marrow Transplant* 1991; 8:35–40.
67. Hardwick RA, Prisco MR, Shah DO. A large-scale magnetic separator for selective cell separations with paramagnetic microbeads. *Artific Organs* 1990; 14:342–347.
68. Hardwick RA, Kulcinski D, Mansour V, Ishizawa L, Law P, Gee AP. Design of large-scale separation systems for positive and negative immunomagnetic selection of cells using superparamagnetic microspheres. *J Hematother* 1992; 1:379–386.
69. Marsh JC, Sutherland DR, Davidson J, Mellors A, Keating A. Retention of

progenitor cell function in CD34+ cells purified using a novel O-sialoglycoprotease. *Leukemia* 1992; 6:926–934.

70. Grimsley PG, Amos TA, Gordon MY, Greaves MF. Rapid positive selection of CD34+ cells using magnetic microspheres coated with monoclonal antibody QBEND/10 linked via a cleavable disulphide bond. *Leukemia* 1993; 7:898–908.
71. Rasmussen AM, Smeland EB, Erikstein BK, Caignault L, Funderud S. A new method for detachment of Dynabeads from positively selected B Lymphocytes. *J Immunol Methods* 1992; 146:195–202.
72. Lea T, Smeland E, Funderud S, et al. Characterization of human mononuclear cells after positive selection with immunomagnetic particles. *Scand J Immunol* 1986; 23:509–519.
73. Sutherland DR, Keating A. The CD34 antigen: structure, biology, and potential clinical applications. *J Hematother* 1992; 1:115–129.
74. Civin CI, Strauss LC, Fackler MJ, Trischmann TM, Wiley JM. Positive stem cell selection—basic science. *Prog Clin Biol Res* 1990; 333:387–401.
75. Guth S, Köster M, Thiel A, et al. Multiparameter magnetic cell sorting using releasable super-paramagnetic microbeads (MACS MultiSort) for rapid isolation of subsets of CD34+ human hematopoietic cells (abstract #2912). 9th International Congress of Immunology, San Francisco, 1995.
76. Scouten WH, Konecny P. Reversible immobilization of antibodies on magnetic beads. *Anal Biochem* 1992; 205:313–318.
77. Lee C, Brouillette M, Lamb L, et al. Use of a closed system for Vαβ-positive T cell depletion of marrow for use in partially-mismatched related donor (PMRD) transplantation. *Prog Clin Biol Res* 1994; 389:523–532.
78. Martin H, Atta J, Zumpe P, et al. Purging of peripheral blood stem cells yields BCR-ABL-negative autografts in patients with BCR-ABL-positive acute lymphoblastic leukemia. *Exp Hematol* 1995; 23:1612–1618.
79. Straka C, Kroner C, Dorken B, Kvalheim G. Polymerase chain reaction monitoring shows a high efficacy of clinical immunomagnetic purging in patients with centroblastic-centrocytic non-Hodgkin's lymphoma. *Blood* 1992; 80:2688–2690.
80. Gee AP, Silverman D, Lamb LS, et al. Immunomagnetic T cell depletion for allogeneic bone marrow transplantation—factors affecting separation efficiency. *Blood* 1994; 84:2924.
81. Moss T, Xu ZJ, Mansour VH, et al. Quantitation of tumor cell removal from bone marrow: a pre-clinical model. *J Hematother* 1992; 1:65.
82. Reynolds CP, Black AT, Woody JN. Sensitive method for detecting viable cells seeded into bone marrow. *Cancer Res* 1986; 46:5878–5881.
83. Brenner MK, Rill D, Moen RC. Gene marking to trace origin of relapse after autologous bone marrow transplantation. *Lancet* 1993; 341:85–86.
84. Rowley SD. Recombinant human deoxyribonuclease for hematopoietic stem cell processing. *J Hematother* 1995; 4:99–104.
85. Rosillo MC, Ortuno F, Rivera J, Moraleda JM, Vicente V. Cryopreservation modifies flow-cytometric analysis of hemopoietic cells. *Vox Sang* 1995; 68:210–214.

86. Gee A. Purging of peripheral blood stem cell grafts. *Stem Cells* 1995; 13:52.
87. DeRosa L, Montuoro A, Pandolfi A, et al. Immunomagnetic purging procedure for autologous bone marrow transplantation in Iymphoid malignancies. *Hematologica* 1991; 76:37–40.
88. Wang MY, Kvalheim G, Kvaloy S, et al. An effective immunomagnetic method for bone marrow purging in T cell malignancies. *Bone Marrow Transplant* 1992; 9:319–323.
89. Anderson IC, Shpall EJ, Leslie DS, et al. Elimination of clonogenic breast cancer cells from human bone marrow. *Cancer Res* 1989; 48:4659–4664.
90. Kiesel S, Haas R, Moldenhauer G, Kvalheim G, Pezzutto A, Dorken B. Removal of.cells from a malignant B-cell line from bone marrow with complement and immunoglobulin switch variant mediated cytolysis. *Leuk Res* 1987; 11:1119 1125.
91. Myklebust AT, Godal A, Juell S, Pharo A, Fodstad Ø. Comparison of two antibody-based methods for elimination of breast cancer cells from human bone marrow. *Cancer Res* 1994; 54:209–214.

10

Free-Flow Magnetophoresis: Continuous Immunomagnetic Sorting of Cells and Organelles by Magnetic Deviation and Focusing

Michael Hausmann and Christoph Cremer
Institute of Applied Physics, University of Heidelberg, Heidelberg, Germany

Roland Hartig
Max-Planck Institute for Cell Biology, Ladenburg, Germany

Hans-George Liebich
Institute for Veterinary Anatomy, University of Munich, Munich, Germany

Georg H. Lüers
Institute of Anatomy and Cell Biology, University of Heidelberg, Heidelberg, Germany

Armin Saalmüller
Federal Research Centre for Virus Diseases of Animals, Tübingen, Germany

Reinhard Teichmann
University Clinic of Tübingen, Tübingen, Germany

I. INTRODUCTION

Magnetic separation is a widely used technique that has been applied in commercial metal extraction as well as in life sciences. Different sorting devices have been developed according to the respective requirements of each field (1–4). These techniques appear to be able to handle large quantities of particles. The differences in magnetic properties, however, of native biological particles usually are not large or specific enough to separate subpopulations. Thus, magnetic solid supports with specific affinity couples have become a commonly used method for separating cell organelles and microorganisms in biology, microbiology, medicine, and environmental sciences (5–10). Magnetic microspheres and other magnetizable carriers can be used and labeled by antibodies (Abs). These microspheres are often polystyrene beads (typically 0.5–5 μm in diameter) containing iron oxide (11,12). The biological particles are immunochemically fixed to these magnetic beads by a highly specific antigen (Ag)-Ab bridge. Several companies offer magnetic beads in different sizes coated with Abs and other affinity couples ready to use for a broad spectrum of applications.

The most often used magnetic beads are produced by Dynal A/S (Oslo, Norway). These "Dynabeads" are superparamagnetic, polystyrene-based particles 2.8 μm (M-280) or 4.5 μm (M-450) in diameter. They are well-shaped, spherical beads with iron oxide homogeneously distributed throughout the entire particle. They have a density of 1.5 g/cm^3 with an iron content of about 20% (w/v) and a maximum variation in diameter of $\pm$ 5% (13). These beads are well suited for high-purity bulk sorting by the separation technique described in this chapter.

In principle, separation of the magnetic beads with biological particles from nonmagnetic material can be performed by several appropriate devices; in some cases, a simple hand-held permanent magnet may be used (13). This method appears to be useful with a low density of magnetic beads and less stringent requirements for sorting purity. Both components of the sample, magnetically labeled and nonlabeled particles, are homogeneously suspended in a separation vial. When the sample is exposed to the magnet, the strongly magnetizable beads form chains, oriented into the direction of the magnetic field. These chains are diverted into the direction of the magnetic force, i.e., in the direction of the magnetic field of the permanent magnet. Because nonlabeled particles are randomly dispersed in the separation volume, some of these particles are in the region of the highest magnetic forces at the beginning of the separation process. Magnetically labeled particles are diverted into this area, forming aggregates with the nonlabeled particles, reducing the sorting purity. Collective effects can induce aggregation of particles (1) that cannot be resuspended by washing steps without altering or impairing the biological particles.

Large numbers of particles can also be separated in magnetic filters

(14,15). The filter material consists of magnetic stainless steel wool in a magnetic field of 0.1–1.0 Tesla. Nonlabeled particles pass through the filter chamber, while magnetic microspheres and magnetically labeled particles are retained. These can be flushed out when the magnetic field is switched off. Although magnetic filtration has an indisputable success in various separation technologies in biology and medicine, there are still several basic aspects, especially in cell separation, that require further research. One requirement of magnetic filtration is that the capacity of the filter chamber must be coordinated with the number of particles retained in the chamber. Using too many particles may block the filter system or lead to unwanted contamination in the separated fractions. In the case of rare particles to be sorted, there may be a deficit of particles. Another feature of the magnetic filtration technique is that the retained particles can only be eluted by switching off the magnetic field; continuous separation of large quantities of particles cannot be performed. Additionally, both components of the sample are randomly dispersed in the entire filter chamber; the disadvantage of this was discussed earlier in this chapter. Retained cells rosetted by a few magnetic beads may be destroyed by forces of shear stress along the steel wool. This treatment can influence the viability of the cells to be sorted.

Although different magnetic sorting devices are commercially available (13,14,16,17), several applications in medical research and routine application still require sorting techniques of a high throughput rate and a high purity combined with good visibility of the biological particles that are sorted. Many magnetic sorting devices do not fulfill these requirements simultaneously; this may be one reason why they are often used in combination with other sorting techniques, e.g., flow cytometry (16). Here, free-flow magnetophoresis (18,19), also called continuous immunomagnetic sorting (CIMS) (17,20) is described. A fast, gentle diversion of the specifically labeled sample from the nonlabeled sample allows depletion, as well as concentration (21), and CIMS is expected to contribute substantially to a widening of the range of applications of magnetic sorting in biology and medicine.

II. CONTINUOUS IMMUNOMAGNETIC SORTING

A. Separation Principles

Aggregate-free suspensions of cells or organelles are prepared and specific labeling is performed with commercially available magnetic beads using appropriate affinity couples according to product information. For sorting, the suspensions containing magnetically labeled and nonlabeled cells or organelles are injected by an electronically controlled perfusor system into a separation chamber through two sample inlets. In the perfusor system, the specimen is protected against magnetic fields to avoid the formation of aggregates during long-term sorting. In contrast to many other sorting techniques, the separation medium can be freely chosen according to the biological requirements only, a highly useful advantage for viable cell sorting. The separation medium, e.g., a cell buffer, flows inside the sorting chamber (in vertical direction upwards) as a thin film. The conditions of strict laminar flow must be preserved. The velocity of this flow is controlled by a peristaltic pump. The sample, being transported in two narrow streams by the separation medium, passes a well-defined inhomogeneous magnetic field perpendicularly oriented to the direction of the flow. According to the magnetic moments induced by the magnetic field, the strongly magnetizable particles, magnetic beads. and magnetically labeled particles are diverted into the direction of the magnetic forces, while the nonlabeled, weakly magnetizable particles pass through the magnetic field with negligible interaction. Depending on the shape of the poles of the electromagnet, the diverted particles are focused into an area where the diverting magnetic forces are equal to zero. Magnetic beads and the magnetically labeled particles leave the area of the magnetic field in a stream that is completely separated from the nondiverted streams of nonlabeled particles. The stream of the separation medium is split into several fractions by a fraction line composed of 60 Teflon® tubes. These fractions are continuously collected into separate vials. The labeled and the nonlabeled particles remain in suspension during the entire separation process.

For long-term operation, a computer-controlled optical scanning system can be adapted to the separation chamber. This scanning system detects absorbed as well as scattered light from the diverted and nondiverted particle streams. This allows the on-line detection of disturbances in the laminar flow conditions influencing the sorting process.

Two different magnetic sorting techniques can be performed: concentration or depletion of magnetically labeled particles (18). The concentration-sorting technique means that the *target* population is specifically labeled by magnetic beads. This allows the rapid isolation of rare cells present in a suspension. The disadvantage of this technique is that the desired target cells carry the magnetic beads. Thus, for further experiments with the sorted material, the beads may need to be detached, e.g., in a procedure where the Abs are kept on the cell surface. Where the concentration technique is not indicated, the depletion technique can be applied; *unwanted* particles are labeled by magnetic beads and sorted. In this case, the desired target cells are not influenced by Abs. They can pass the magnetic field without interaction, whereas the labeled cells are diverted. In practice, the sample suspension consists of more magnetically labeled particles than unlabeled ones. Isolation of rare cell subpopulations becomes difficult because suitable Abs to label all the other populations need to be found. However, in cases of strict demands for viability and preservation of shape and morphology, depletion may be more advantageous than concentration.

B. Apparatus

Configuration of the Magnet

An emerging technique to perform continuous magnetic sorting is the so-called open-gradient separation. The magnet configuration of CIMS is based on one of the earliest devices using open-gradient technology, the Frantz Isodynamic Magnetic Separator (1). To realize an appropriate field configuration by a less complex shape than a Frantz magnet, a pair of wedge-type magnetic pole pieces is used, producing appropriate separation forces in the gap between the pole pieces. An air-cooled coil induces the magnetic field (Fig. 1). The magnetic force can be controlled by the coil current (0–9 A). A schematic section through the configuration shows the separation chamber and the magnetic field lines between the pole pieces (Fig. 2) calculated by a computer program ("Poisson"; access kindly provided by the Max-Planck-Institute for Nuclear Physics, Heidelberg).

Particle Forces. The deviation force on a single magnetizable particle in a magnetic field with magnetic induction $B = \mu_0 H$ is given by (22):

$$F = -\frac{1}{2}\mu_o \chi V_\rho \,\mathrm{grad}\,(H^2) \tag{1}$$

where V_ρ is the volume and χ is the susceptibility of the particle. The direction of the magnetic force is determined by the direction of the gradient of the square of the magnetic field. If a particle is suspended in a fluid of susceptibility χ_f, the force on the particle depends on the susceptibility difference χ of the particle χ_ρ and the surrounding fluid ($\chi = \chi_\rho - \chi_f$). In the separation chamber, approximately the following conditions are fulfilled:

$$H_x = H_y = 0; \quad H_z \neq 0$$

$$\frac{\partial H_z}{\partial y} = \frac{\partial H_z}{\partial z} = 0; \quad \frac{\partial H_z}{\partial x} \neq 0$$

where H_x, H_y, H_z are the components of the magnetic field in x, y, z directions.

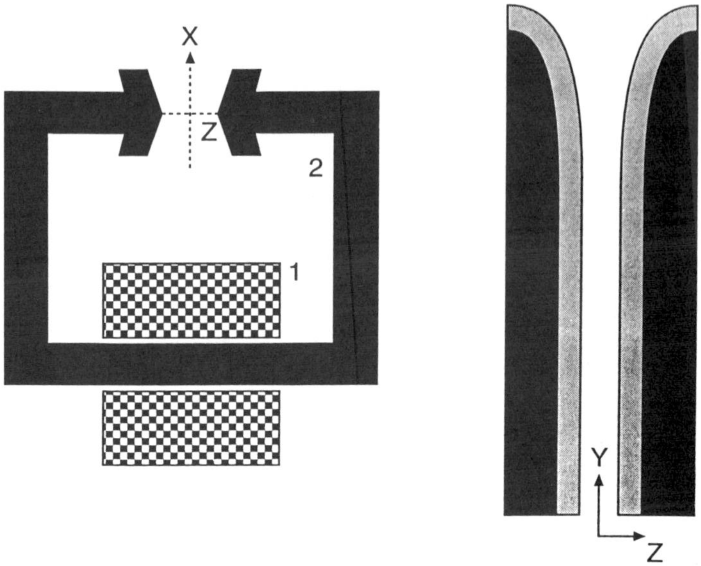

Figure 1 Cross section of the magnet. The left figure shows the z-x cross section of the whole magnet. The right figure shows the z-y view of the magnetic pole pieces to illustrate their shape. These have a width of 40 mm in the x direction, a length of 300 mm in the y direction, and an air gap of 30 mm in the z direction. The angles of the pole tips are 120° (see also Fig. 2). 1, Coil; 2, steel core. (Adapted from Ref. 20.)

With respect to Eq. (1), the force on a magnetizable particle in the pole axis x between two wedge-type pole pieces is given by (23,24):

$$Fx = \frac{1}{4\mu_0} \frac{\cot^2\vartheta}{(x + 0.5d \cot \vartheta)^3} \chi V_p B^2 d^2 \qquad (2)$$

where d is the distance between the two pole pieces and ϑ an angle [$\vartheta = 0.5$ (180° – Θ); Θ is the angle as shown in Fig. 2]. If the separation chamber is positioned in the middle between both tips of the pole pieces, F_z is negligible so that the magnetic particles do not attach to the walls of the flow chamber. Particle displacement in the x direction is due only to the magnetic forces, the viscous drag forces of the separation medium in the chamber, and the field exposure time. Figure 3 is an example for the product of the magnetic induction and its gradient (BdB/dx), which is proportional to the magnetic force as a function of the coordinate x. The figure indicates that the vectors of the deviating magnetic forces are directed antiparallel around the tip of the

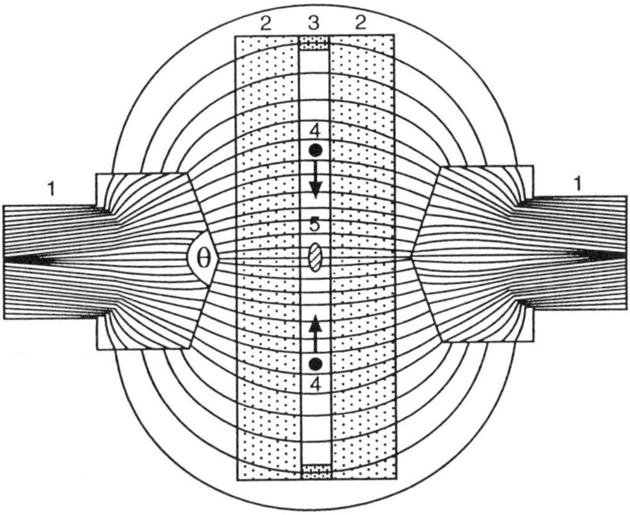

Figure 2 Field lines in the z-x plane and schematic diagram of the flowing chamber between the two wedge-type pole pieces of the magnet. The arrows illustrate the directions of the deviating magnetic forces: 1, magnetic pole pieces; 2, safety glass plate; 3, spacer gasket; 4, stream of the probe suspension; 5, stream of the focused particles (streams 4 and 5 perpendicular to drawing plane). (Adapted from Ref. 20.)

wedges of the two pole pieces. The sample inlets are located in the region of the two extremes of the diversion forces. Strongly magnetizable particles are diverted into the direction of the gradient of the magnetic field. In the region of the tips of the magnetic-pole pieces, the diversion forces are equal to zero. The strongly magnetizable particles and magnetic beads are focused into one stream located around the tips of the pole pieces of the magnet. Nonmagnetic particles pass the magnetic field in two parallel streams without significant interaction. Therefore, the stream of the diverted, strongly magnetizable particles and the magnetically labeled biological particles is completely separated from the streams of the nonlabeled particles.

It is anticipated that if the magnetizable particles are leaving the magnetic field, an additional gradient of the magnetic field that is directed antiparallel to the flow direction (y direction) may stop the streaming particles. This would inhibit continuous separation. To overcome this disadvantage, this gradient must be as small as possible. A hyperbolic shape of the pole pieces parallel to the direction of the flow can make this problem negligible (Fig. 1).

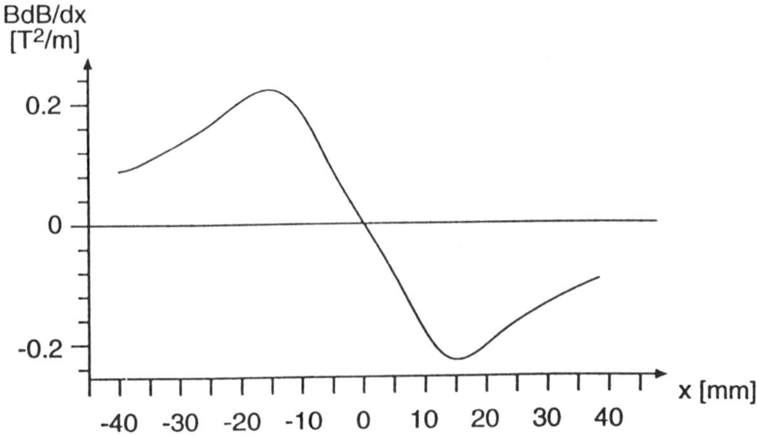

Figure 3 Plot of the product of the magnetic induction B and the gradient of the induction dB/dx along the x-axis using an air gap of 30 mm. The current of the coil was fixed at 8 A during the measurement of the magnetic parameters.

Electromagnet. The two pole pieces are directly connected to the core of an electromagnet (Fig. 1) with a coil ($160 \times 160 \times 280$ mm^3) of 1000 turns of an isolated copper wire (2×3 mm^2). The magnetic induction can be varied by the current of the coil. Up to a maximum current of 9 A, air cooling by means of 10-mm-thick air spaces between each 7 turns of the coil is sufficient. The temperature increase of the coil (measured on the surface relative to the surrounding air) can be approximated by:

$$\Delta T_o = \frac{N}{A(\alpha_r + \alpha_c)} \tag{3}$$

where N is the electric power (135 W), A the free surface area (0.33 m^2), α_r the coefficient of thermal radiation (7 $Wm^{-2}K^{-1}$), and α_c the coefficient of thermal convection (7.7 $Wm^{-2}K^{-1}$) (25). Assuming an isothermic coil, this would result in $\Delta T_0 = 29K$. The parabolic temperature profile inside the winding leads to a difference of temperature from coil surface to the center (assuming cylinder geometry):

$$\Delta T_1 = \frac{N}{2\lambda V} a^2 \tag{4}$$

where V is the total volume of the copper wire, a is half of the radius of the assumed cylinder, and λ the effective coefficient of thermal conductivity of the copper wire and the isolation of the wire. For the coil used here, ΔT_1 was approximated to 12.5K. In summary, the estimated temperature increase is: $\Delta T = \Delta T_0 + \Delta T_1 \approx 41.5K$.

Separation Chamber

In principle, any free-flow electrophoresis chamber, e.g., elphor VAP 21, can be used as the separation chamber for magnetophoresis (18,19). Since the fluid usually streams from the top to the bottom, this has the disadvantage that the flow velocity of the chamber buffer must be controlled by a pump reducing the flow through the fraction tubes. This procedure is not very gentle for the cells and is contrary to the requirements of viable cell sorting.

To overcome these problems, a counterflow sorting chamber (Fig. 4) was developed (20,26). This chamber consists of two parallel safety glass plates (100×640 mm^2) spaced 0.5 mm apart. This distance is required by the gasket and forms a leak-free seal. The separation volume is 50 mm wide and 600 mm high. The buffer, injected through the buffer inlet by a peristaltic pump, passes continuously through this chamber against the gravity force, i.e., from the bottom to the top. The sample inlets, 40 mm above the buffer inlet, are made of two Teflon© tubes fixed into the front safety glass plate. Convection and diffusion are negligible under typical experimental conditions. The separated streams of particles are collected in fractions by counterflow technique as shown in Figure 4. This means that simultaneous to the chamber buffer, a second buffer stream is injected by the counterflow inlet at the top of the separation chamber. This stream flows antiparallel to the chamber buffer with a given velocity, which is controlled by a peristaltic pump. At the array of the fraction tubes, the two buffer streams are arranged in layers and the fluid is collected into 60 fractions. Thus, the sorted sample does not pass any further pump.

It is necessary to have an equal flow in all fraction tubes to fractionate the different streams of particles with appropriate selectivity. These flow conditions can be achieved by a flow velocity ν in a single tube of:

$$\nu \geq \frac{\Delta P R^2}{8 \eta l} \tag{5}$$

where R is the radius of the tubes (0.15 mm), l the length of the tubes (500 mm), η the dynamic viscosity of the fluid, and ΔP the difference of pressure given by the distance between the fraction line and the fraction tubes. For example, using water as the buffer medium and an altitude difference of 300 mm between fraction line and fraction tube, these parameters result in $\geq 1.7 \times 10^{-2}$ m/s. This value defines the ratio of the flow velocity between the chamber buffer and the buffer of the counterflow. A very small velocity of the

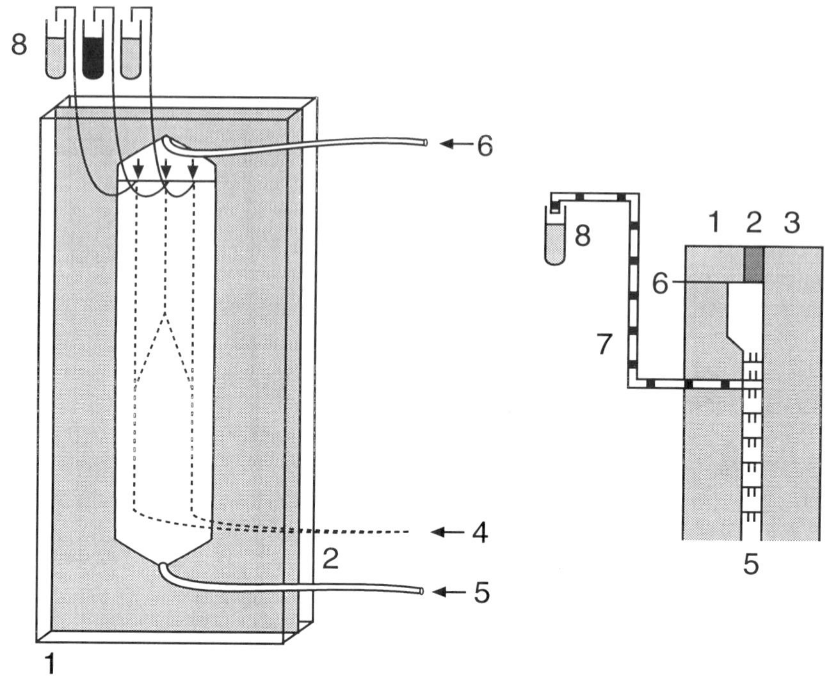

Figure 4 Schematic representation of the sorting chamber. The left figure presents the whole chamber; the right figure shows the side view of the chamber and illustrates the principle of counterflow technique. The arrows indicate the flow directions. Additionally, in the left figure the traces of the streams of the different particles are shown. The streams of the deviated particles are indicated by the dashed line. 1, Front plate; 2, back plate; 3, spacer gasket; 4, sample inlet; 5, chamber buffer; 6, counterflow; 7, fraction tube; 8, fraction. (Adapted from Ref. 20.)

chamber buffer must be compensated by a very high velocity of the counterflow. A minimum throughput volume of about 260 mL/h fluid is necessary. Using safety glass as chamber material and Teflon© tubes, the complete apparatus may be easily sterilized for clinical applications of viable cell sorting.

Sample Injection

The principle of continuous diversion by magnetophoresis allows the sorting of large amounts of material in long runs. Therefore, the specimen remains in the sample supply relatively long.This favors aggregation and sedimentation of the relatively heavy magnetic beads. To overcome this problem, a device for sample storage and injection (Fig. 5) was developed (27). It consists of an adapter for one-way (sterile) syringes (1, 5, and 10 mL) protected against static magnetic and electromagnetic fields. This set-up can be rotated (± 360°) with variable velocity. The sample can be injected by an electronically controlled, motor-driven precision mechanism via a Teflon© inlet tube (Fig. 5A). The injection rate can be varied in steps of 0.1 mL/h up to a range of 58 mL/h. To

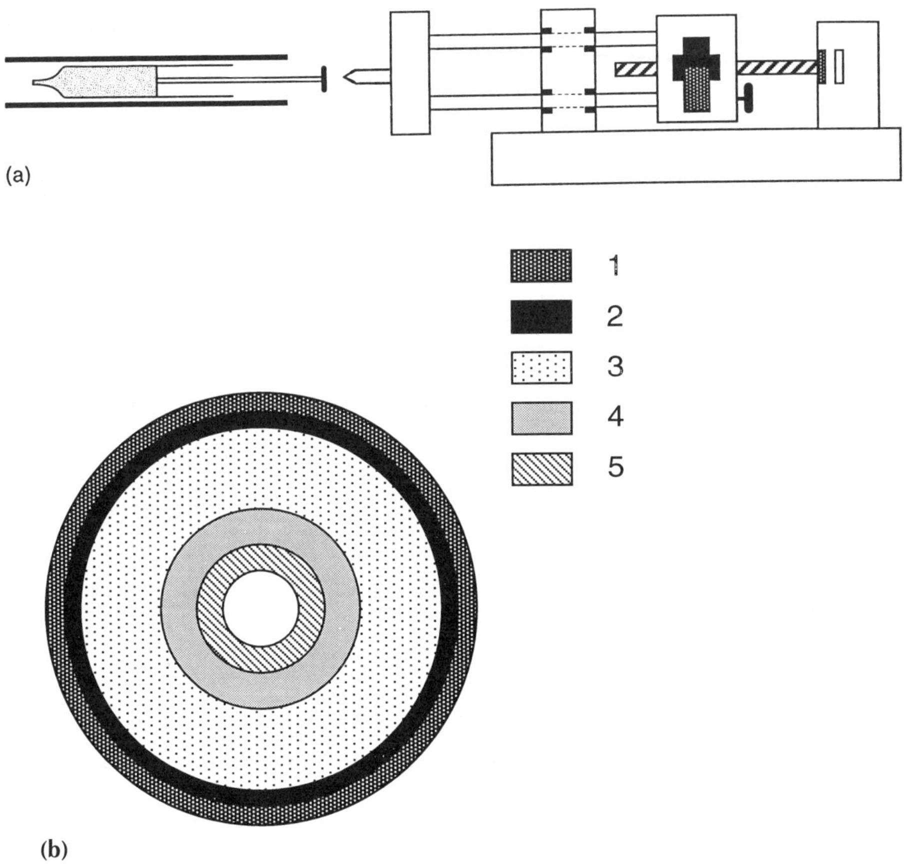

Figure 5 (A) Schematic representation of the sample perfusor. (B) Schematic section through the magnetostatically protected syringe adapter. 1, Aluminium cylinder; 2, Hyperm protection; 3, 10-mL syringe cylinder; 4, 5-mL syringe adapter; 5, 1-mL syringe adapter. (Adapted from Ref. 28.)

maintain the laminar flow conditions in the sorting chamber, the system works absolutely pulse-free. The magnetostatic protection of the sample supply is obtained by a cylinder of a FeNi material of high magnetic permeability ("Hyperm 766") (Fig. 5B). Inside the cylinder, a magnetic field strength less than the earth's magnetic field (50 μT) was measured during an exposure of 100 mT outside the perfusor system.

Optical Scanning System

CIMS can be used for long runs. The laminar flow conditions need to be maintained for hours. For on-line visualization of the particle flow streams, a computer-controlled optical scanning system can be positioned at the front of the separation chamber just below the fraction line. The scanning system consists of a step motor-driven stage that moves along the sorting chamber perpendicularly to the flow direction of the chamber buffer. The focusing and detection optics are mounted on a stage. Light from a mercury lamp can be focused into the separation volume to a light spot ≤ 0.9 mm in diameter. When this spot moves along the particle separation volume, scattered light as well as absorbed light can be detected by separate photodiodes. Along the width of the sorting chamber, as many as 1500 values can be recorded with reference to the light signal of the lamp not passing through the separation volume. A computer program can be used to evaluate the detected signal intensities. The result can be shown as a plot of the relative intensity versus the relative spot position of the focus. From the peak positions in the plot, the position of the particle streams and relative particle concentrations can be estimated. This allows on-line control of the sorting process and the detection of possible disturbances in the laminar flow conditions.

C. Physical Capacity

To test the physical capacity of the sorting device, weakly magnetizable particles "Latex-Beads" ($\chi_p = 1.2 \times 10^{-6}$; diameter = 4.3 μm) and strongly magnetizable particles "Magnetic-Beads" ($\chi_p = 0.13$; diameter = 4.5 μm) were mixed in different concentrations and sorted: the particles were injected into the sorting chamber by the two sample inlets at a rate of 5.0 mL/h each and passed the magnetic field in two parallel streams. The flow rate of the separation medium [0.15 M NaCl, 0.05% bovine serum albumin (BSA), pH 7.4] was fixed at 2.1×10^{-3} m/s. The current of the electromagnet was adjusted to 5.0 A, inducing a magnetic field of 75 mT in the air gap between the two wedge-type pole pieces. Magnetic beads were diverted and focused into a stream centrally located between the two parallel particle traces of the weakly magnetizable latex beads.

Suspensions of latex beads and magnetic beads in different concentrations and ratios were used. The ratio between the number of the weakly magnetizable latex beads and the strongly magnetizable magnetic beads was varied from 1000:1 to 1:400. The concentration of the suspension ranged between 5×10^6 and 5×10^7 particles/mL. In all experiments, sorting purities of approximately 99% were obtained under the conditions of a high throughput rate (Table 1). Within 5 minutes, as many as 8×10^7 particles were sorted corresponding to a throughput rate of about 1×10^9 particles/h. Counting the number of the particles in aliquots of the nondiverted as well as of the diverted fractions and comparing this number with the number of particles in the suspension before magnetic sorting, a recovery of more than 98.5% was detected.

Table 1 Microscopic Evaluation of the Sorting Purity of Magnetic and Nonmagnetic (Latex) Beads in Different Concentrations and Ratios

Particle concentration (mL^{-1})	Latex bead: Magnetic bead ratio	Sorting purity of the diverted fraction (%)	Sorting purity of the nondiverted fraction (%)
5×10^6	1:1	99.90	99.75
5×10^6	2:1	99.82	99.75
5×10^6	1:100	99.70	99.75
1×10^7	1:1	99.86	99.95
1×10^7	2:1	99.82	99.87
1×10^7	10:1	99.81	99.79
1×10^7	100:1	99.95	99.75
1×10^7	150:1	99.91	99.80
1×10^7	1:2	99.82	99.75
5×10^7	1:1	99.92	99.84
5×10^7	2:1	99.81	99.84
5×10^7	1:10	99.75	99.84
5×10^7	1:50	99.30	99.78
5×10^7	1:80	99.68	99.80
5×10^7	1:400	98.50	99.90

Source: Ref. 17.

III. APPLICATIONS

From the physical point of view, high-purity sorting under high-throughput conditions is possible. Consequently, the biological (e.g., surface properties of cells) or biochemical (e.g., specificity of the Ab coupling of the magnetic beads) parameters become the only limiting factors of magnetophoresis. In the following sections, some examples are described that indicate the broad spectrum of applications for CIMS.

A. Sorting in Immunology

The standard problem of present cell sorting is the specific sorting of lymphocytes in the field of immunology. Worldwide, flow cytometers are routinely used for immunological analysis and sorting of cells after fluorescent labeling by specific Abs (29,30). Typically, about 10^7 particles/h can be processed, from which subpopulations can be sorted with typical sorting purities of ≥ 80%.

To test magnetophoresis for applications in immunology, B and T lymphocytes were separated by concentration and depletion, respectively (see also Refs. 18,20). White blood cells were enriched from human peripheral blood by Ficoll density gradient. The cells were washed twice in phosphate-buffered saline (PBS). An aliquot of this sample was analyzed by flow cytometry using a FACSCAN. Forward and side light scatter were measured to define the lymphocyte gate. The B cells were labeled with phycoerythrin (red fluorescence) using a CD19-Ab, and the T cells with FITC (green fluorescence) using a CD2-Ab. The flow histogram of red versus green fluorescence revealed 8.9% B cells and 80.8% T cells before magnetic sorting.

For sorting, cell suspensions of 10^7 cells/mL were prepared at 4°C. In aliquot I, Dynabeads of 4.5 μm diameter coupled to CD19 Ab were washed in PBS/bovine serum albumin (BSA) five times. The beads were added to the cell suspension (beads: B cells, approximately 4:1) for a 25-minute incubation time. In aliquot II, Dynabeads 4.5 μm in diameter coupled to CD2-Ab were washed in PBS/BSA and added to the cell suspension (beads: T cells, approximately 4:1) for a 35-minute incubation time. Both aliquots were subjected to CIMS with a sample rate of 2 × 4.5 mL/h corresponding to a particle processing rate of about 10^8 cells/h. Hanks buffer solution with 2% fetal bovine serum (FBS) was used as chamber buffer. After magnetic sorting, the sorted fractions were analyzed by flow cytometry as described above (20).

B-cell depletion revealed high-purity T-cell sorting (≤ 0.2% B cells in the T-cell fraction) (Table 2). T-cell depletion, however, showed about 12.6% T cells in the B-cell fraction. Since microscopic observations found that neither

Table 2 Results of Magnetic Sorting of T and B Cells Analyzed by Flow Cytometry[a] of the Sorted Fractions

	Diverted fraction (%)		Nondiverted fraction (%)	
Aliquot	B cells	T cells	B cells	T cells
I (T-cell depletion) (CD19 labeling)	85.0	9.2	≈ 0.1	93.3
II (T-cell concentration) (CD2 labeling)	4.5	91.4	51.1	12.6[b]

[a]Typically 5000 particles were measured.
[b]Without magnetic beads as judged by microscopy.

T cells with magnetic beads nor isolated magnetic beads were in the nondiverted B-cell fraction, this result may be due to a minor specificity of the CD2 Ab. The relatively high amount of magnetically unlabeled cells in the diverted fractions (9.2% T cells for aliquot I and 4.5% B cells for aliquot II) may be due to small cell aggregates induced by cell-surface properties. This effect can be overcome by improved preparation. These results together with other experiments published elsewhere (18) indicate that in immunology cell sorting by CIMS is possible with sorting purities compatible to flow cytometry but with a throughput rate approximately one order of magnitude greater. Moreover, we tested the viability of the cells after CIMS by propidium iodide staining and flow cytometry. In all cases, < 1% of the cells showed a high propidium iodide signal indicating cell death, and 99% were still in viable conditions (17,18,20).

B. Sorting of Rare Events

To test the potential of CIMS for high-purity separation of rare events, B lymphocytes were directly sorted from human peripheral blood without any preenrichment step (17). One mL of blood contains approximately 5×10^9 erythrocytes, $0.5–1 \times 10^7$ leukocytes (i.e., about 67% granulocytes, 27% lymphocytes, and 6% monocytes), and $1.5–3 \times 10^8$ thrombocytes (31). Approximately 10% of all lymphocytes are B cells or, in absolute numbers of this estimate, about 2×10^5. Thus, direct sorting of B cells by the concentration technique corresponds to a particle ratio of 1:25,000 (B cells to all blood cells). For the experiments, 2 mL blood were diluted with PBS (1:1, v/v) and incubated 5 hours (experiment I) or 14 hours (experiment II), respectively, with magnetic Dynabeads (4.5 μm diameter) that were precoated with CD19 Abs.

Magnetic sorting was performed with a flow velocity of 2.1×10^{-3} m/s in a chamber buffer of PBS containing 0.05% BSA. The sample throughput rate was about 2 mL/h corresponding to a particle rate of 5×10^9 cells/h. Five fractions around the diverted particle trace were collected and pelleted. Microscopic evaluation (Table 3) showed a sorting purity of 95–98% of the magnetically labeled lymphocytes (presumptive B cells due to the specificity of the CD19 Ab detected in the experiment described above; see Table 2, aliquot I).

C. Sorting of Viable Gastrin Cells

Gastrin cells are parts of the mucous membrane of the antrum (32). They produce a hormone that controls the gastric acid (33). It is of high interest to investigate the hormone production of the gastrin cells in vitro to study digestion processes under different food conditions (34). This requires sorting of vital gastrin cells. Viable gastrin cell sorting by magnetic filter techniques and/or flow cytometry has, however, not been successful until now because of the permanent mucous secretion of these cells, which produces aggregates that block the sorting process. Here, preliminary results of gastrin cell sorting under viable conditions by CIMS are described that show that an enrichment of gastrin cells to about 30% under good viable conditions is feasible.

As a model system, the stomach of *Sus scrofa domestica* was used. Antral cell suspensions were prepared by enzymatic treatment with pronase and collagenase (35). The cell suspensions were diluted with PBS to a

Table 3 Microscopic Evaluation of Rare Blood Cell Sorting

Experiment	Incubation period (h)	Counted cells attached to beads	Counted cells: free lymphocytes	Counted cells: erythrocytes	Sorting purity (%)[a]
I	5	500	12	0	97.7
II	14	823	10	1	98.7
II[b]	14	950	50	0	95.0

[a]Sorting purity = number of cells attached to beads/total number of cells × 100%.
[b]Second evaluation after storage overnight.
Source: Ref. 17.

concentration of 1–2 × 10 cells/mL to avoid aggregate production by mucous secretion of the gastrin cells. Five-mL cell suspensions were sorted in each magnetophoresis run. Before sorting, about 16% dead cells were found in the suspension. By flow-cytometry measurements of scattered light, the cells of an aliquot were classified according to their size, and approximately 82% of the cell suspensions were lymphocytes.

Another 15% contained a mixture of epithelial cells and endocrine cells, among which the gastrin cells were a subpopulation comprising 0.4–1% of the cells. The depletion-sorting technique was done to keep the endocrine cells uninfluenced by Ab binding. Indirect magnetic labeling via an Ab bridge using MSA3/Abs was performed. After sorting, the nondiverted fraction was fixed with glutaraldehyde. A histological examination by light (Fig. 6) and electron microscopy revealed a depletion of 90% of the lymphocytic cells. Since cells with magnetic beads, or isolated magnetic beads, were not found in the sorted fraction, the purity of this depletion sorting was determined by the affinity of the Ab used. From the remaining cells in the sorted fraction, 60–70% were endocrine cells. Approximately 40% of these cells were gastrin cells corresponding to an overall percentage of 25–30% of the cells: i.e., an enrichment by a factor of 25–75 was achieved. A trypan blue test revealed a viability of 85% of the cells analyzed in this fraction.

In further experiments, cell suspensions from biopsies of human antrums were enzymatically prepared. As many as 10^5 cells were available per experiment. For magnetic labeling, beads precoated with CD45 Ab were added to the suspension. After incubation at 4°C for 40 minutes, 25–30% of the cells were found to be dead as judged microscopically by the trypan blue method. After sorting by CIMS (sorting rate 4–5 mL/h), the nondiverted fraction was incubated by anti-HEA 125 Ab directed against epithelial cells (kindly provided by Progen, Heidelberg) for 40 minutes at 4°C. The cells were washed in PBS and magnetic beads precoated with an antimouse Ab were added. After 40 minutes at 4°C, the cell suspension was again subjected to CIMS. After fixation with 2% gluturaldehyde, histological examinations of the nondiverted fractions were performed by light and electron microscopy. Twenty to 24% of the analyzed cells were found to be gastrin cells. Neither magnetically labeled cells nor isolated magnetic beads were found in this fraction. Visual morphological inspection of the cells revealed no considerable increase in the percentage of dead cells compared with the cell suspension before magnetophoresis. To our knowledge, no other sorting technique has been able to produce compatible results of enrichment of viable gastin cells from human biopsies.

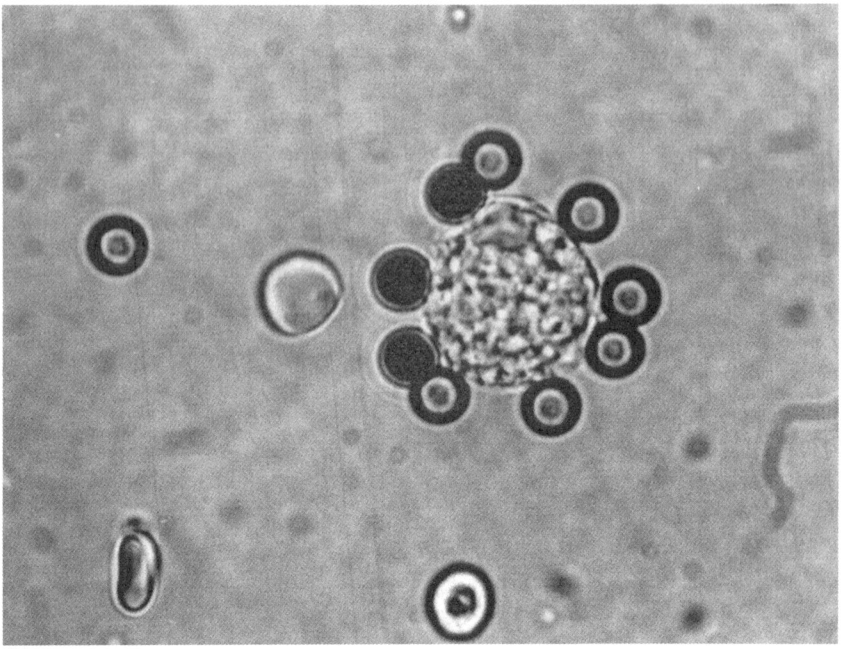

(a)

Figure 6 (A) Phase-contrast microscopic image of a lymphoid cell labeled with magnetic beads obtained from the deviated fraction after CIMS (magnification 100x, 40x). (B) Phase-contrast microscopic image of a glandular epithelial cell (large, spherical cell) together with endocrine cells (smaller cells with granula) after deviation sorting by CIMS obtained from the nondeviated fraction (magnification 100x, 40x).

D. Chromosome Sorting

Sorting of isolated metaphase chromosomes has contributed substantially to the construction of chromosome specific DNA libraries (36). For this purpose, flow cytometry was the most commonly used sorting technique (37). However, larger amounts of sorted chromosomes may be necessary for other biological investigations, e.g., the significance and specifically of nonhistone proteins associated with metaphase chromosomes. Rough estimates of the amounts of chromosomes possibly necessary to establish "chromosome-specific protein libraries" show that sorting of 10^{12}–10^{14} chromosomes may be required. Under these conditions, new methods of bulk sorting may be helpful.

For the application of chromosome sorting, magnetic beads must be

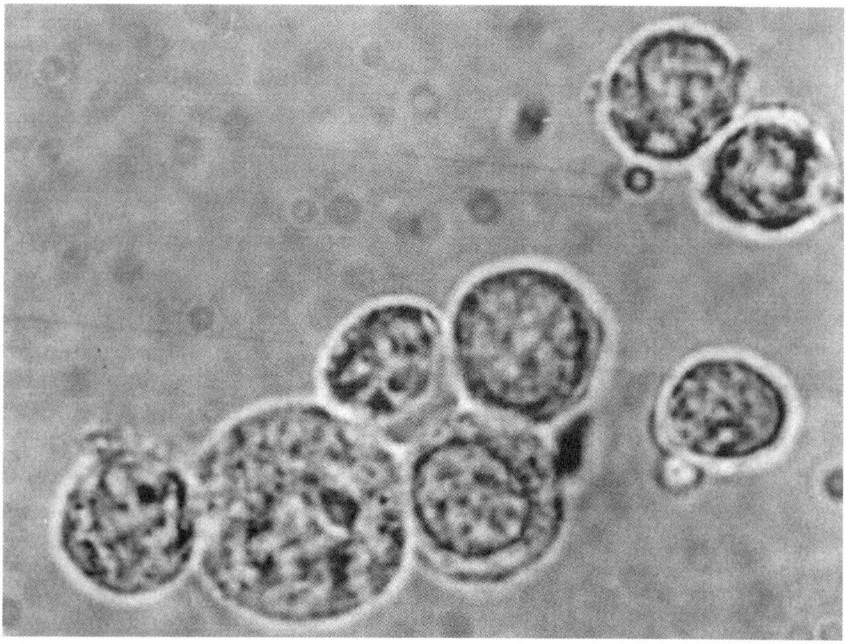

(b)

specifically bound to the chromosomes, e.g., by means of in situ hybridization, a routine method for fluorescence labeling of chromosomes fixed on microscope slides (38). Although basic experiments were successful, the feasibility of in situ hybridization for magnetic labeling of individual chromosomes isolated in suspension, however, is still in the state of development (39,40). The principle of magnetic in situ hybridization can briefly be described.

Single-stranded DNA sequences (probe) representing the sequence of a certain chromosome region can be labeled by a haptene (e.g., biotin) or a steroid (e.g., digoxigenin). After a chemophysical treatment (for details see Ref. 40) of the probe DNA and the chromosomal target DNA, the probe DNA binds to the chromosome region of complementary DNA sequence. After appropriate washing and Ab incubation, Dynabeads can be coupled to the chromosomes by a secondary Ab bridge. For example, if the DNA probe was labeled by digoxigenin, a mouse antidigoxigenin Ab and Dynabeads precoated with an antimouse Ab were used.

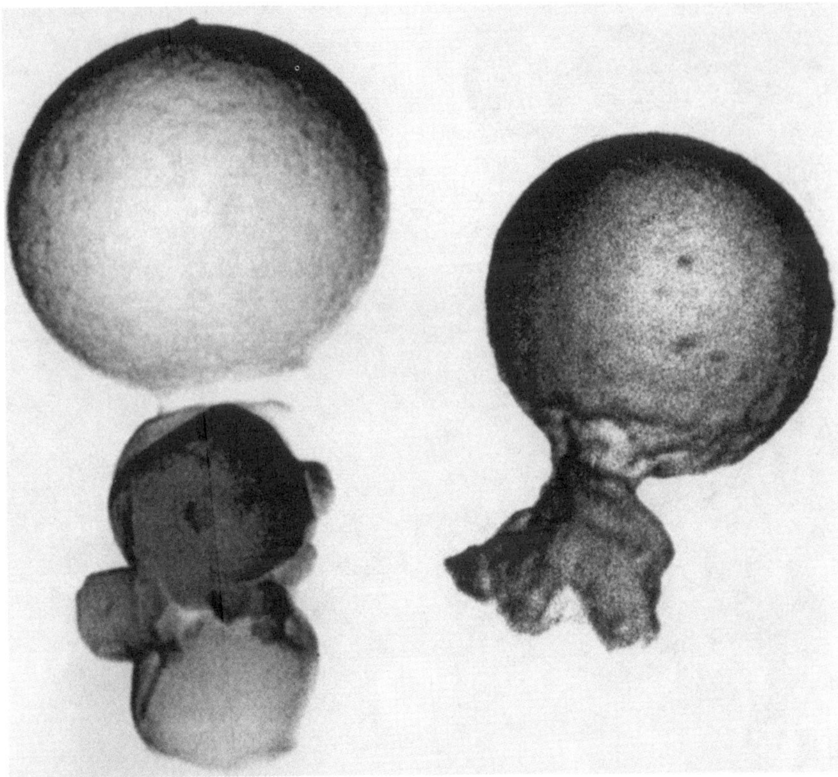

Figure 7 Scanning electron microscopy image of human chromosomes coupled to 4.5-μm Dynabeads. Right: Centromere coupling; left: telomere coupling.

Figure 7 shows a scanning electron microscopy image of human chromosomes coupled to magnetic beads in the centromeric region and in the telomeric region, respectively. In both cases, the magnetic bead had a diameter of 4.5 μm. Here, the beads were coated with streptavidin and coupled directly to the biotin-labeled DNA probe (for details see Ref. 39). To test the potential of CIMS for chromosome sorting, different modifications of in situ hybridization were done. Chromosomes of the Chinese hamster cell line DON were used. For probe DNA, genomic hamster DNA labeled with digoxigenin was used. Such experimental approaches have two advantages. For application, they can be used to purify chromosome suspensions from cell debris. For methodological developments, they have the advantage that the efficiency of the labeling and sorting process can be easily judged by microscopy. Chromosomes attached or not attached to magnetic beads can be counted without any further classification of the chromosome type. The sorting conditions were chosen in such a way that the diversion forces on magnetically labeled chromosomes were approximately 3×10^{-12} N. Under these conditions, chromosomes unspecifically attached to magnetic beads were detached during the sorting process and collected in separate fractions between the magnetically and nonmagnetically labeled particles (Fig. 8). Therefore, the chromosomes in the "diverted" fraction in Table 4 can be classified as magnetically labeled by in situ hybridization.

Depending on the chromosome buffer and hybridization parameters chosen, 9–26% of all chromosomes were magnetically labeled before magnetophoresis (Table 4). After CIMS, 70–80% (except in one experiment) of the chromosomes in the diverted fraction were labeled by magnetic beads. In the nondiverted fraction, ≤ 3% (again with the exception of one experiment) of the chromosomes evaluated were attached to a magnetic bead (for further details see Ref. 40). In these experiments, the sorting conditions were chosen in such a way that in the diverted fraction, the beads were bound to the chromosomes, whereas unspecifically attached beads were disrupted from the chromosomes and collected separately.

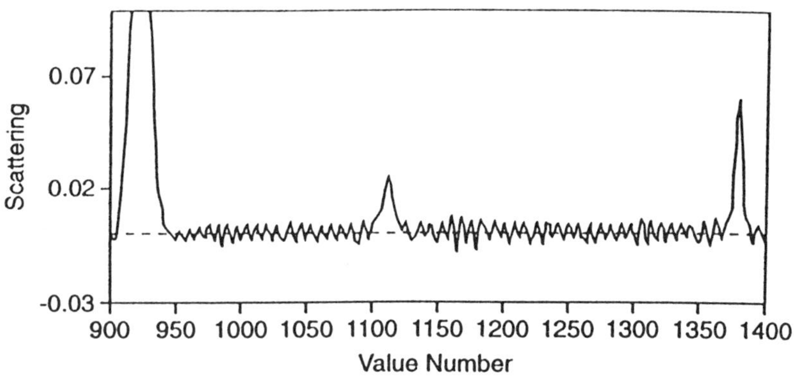

Figure 8 Example of optical scanning during chromosome sorting. The figure shows the relative intensity of scattered light (ordinate) along a part of the separation chamber. The abscissa shows the relative position of the light focus. The left peak results from the nondeviated, nonmagnetically labeled particles, the right one from the magnetic beads and chromosomes coupled to magnetic beads by in situ hybridization. The peak around position 1110–1120 results from chromosomes detached from magnetic beads during the sorting process.

Table 4 Sorting of Metaphase Chromosomes After Different Modifications of In Situ Hybridization

	Before CIMS		After CIMS	
Chromosome buffer	c (mL^{-1})	r (%)	r (%) diverted	r (%) nondiverted
$1 \times$ SSC	2×10^7	16.3 ± 2.4	71.2 ± 15.3	0.1 ± 0.1
$1 \times$ SSC + $MgCl_2$	1.5×10^7	18.7 ± 2.4	79.7 ± 3.0	0.3 ± 0.1
NaCl + $MgCl_2$	8×10^6	26.6 ± 6.4	66.7 ± 13.9	3.4 ± 2.0
$1 \times$ SSC + $MgCl_2$[a]	4.8×10^6	27.6 ± 4.6	26.6 ± 10.6	12.2 ± 3.4
$1 \times$ SSC[a]	5.3×10^6	9.2 ± 2.6	75.0 ± 17.3	1.6 ± 1.2

[a]Experiment without thermal denaturation of the target.
c = Chromosome concentration; r = ratio of chromosomes attached to magnetic beads to the total number of chromosomes.
Source: Ref. 40.

E. Sorting of Peroxisomes

Experiments on the immunomagnetic isolation of submicron-sized cell organelles (peroxisomes of rat liver, 150–500 nm diameter) were performed. After coating magnetic beads that were 2.8 and 4.5 μm in diameter (Dynabeads M-280 and M-450) with polyclonal peptide Abs against a peroxisomal membrane protein (PMP 70), as many as 20 organelles were bound to one magnetic bead. Scanning electron microscopy revealed the decoration of the beads (Fig. 9A). After separation, the organelles still showed a well-preserved morphology as detected by transmission electron microscopy (Fig. 9B). The purities of the diverted fractions were only limited by the specificity of the Abs used in these experiments. For biochemical analysis, the separated organelles were detached from the magnetic beads by adding a glycin buffer (0.2 M, pH 2.8) to the pelleted particles of the diverted fractions. The results of these experiments indicated that magnetic separation by specific diversion is an alternative approach for a reliable and specific isolation of highly purified, small-cell organelles (20,41).

IV. DISCUSSION

The free-flow magnetophoresis technique for CIMS has been described in detail. Specific deviation and focusing allows the separation of biological

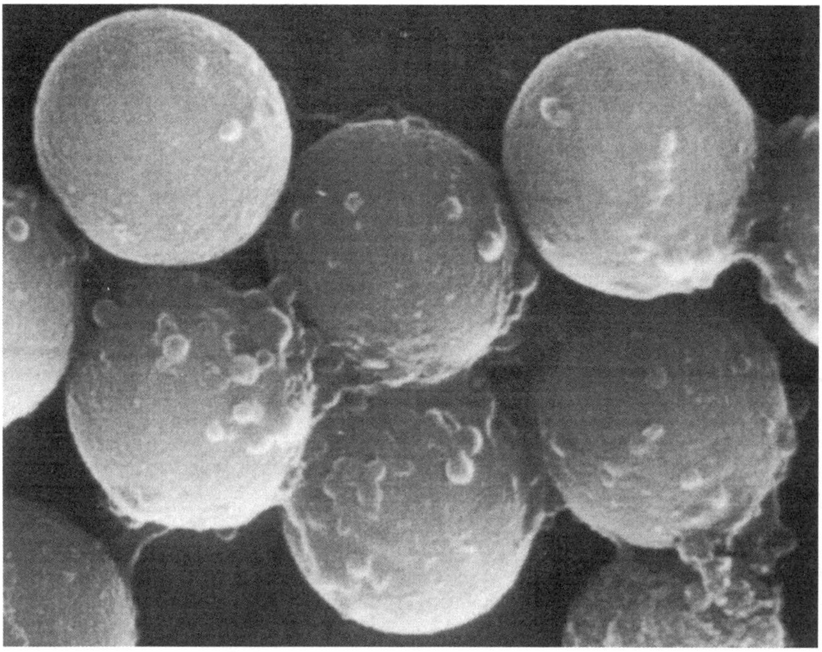

(A)

Figure 9 (A) Scanning electron microscopy image of peroxisomes coupled to 4.5-μm Dynabeads. (B) Transmission electron microscopy image of peroxisomes coupled to the surface of a magnetic bead. The peroxisome (PO) showed a well-preserved membrane and the uradoxidase crystal core.

particles that ranged from several hundreds of nm to several tens of μm in diameter. The major simultaneous advantages of CIMS are:

1. A *high throughput rate* of up to 5×10^9 particles/h. This can be further increased by multiplication the particle trajectories and magnetic pole pieces.
2. A *sorting purity* of physically ≥ 99%; only biological or biochemical parameters limit the sorting capacity.
3. A preservation of the *viability* of the cells, since the biochemical composition of chamber buffer is independent from the sorting process.
4. A potential to *sterilize* the complete apparatus, since only safety glass and Teflon© material may have contact with the biological material.

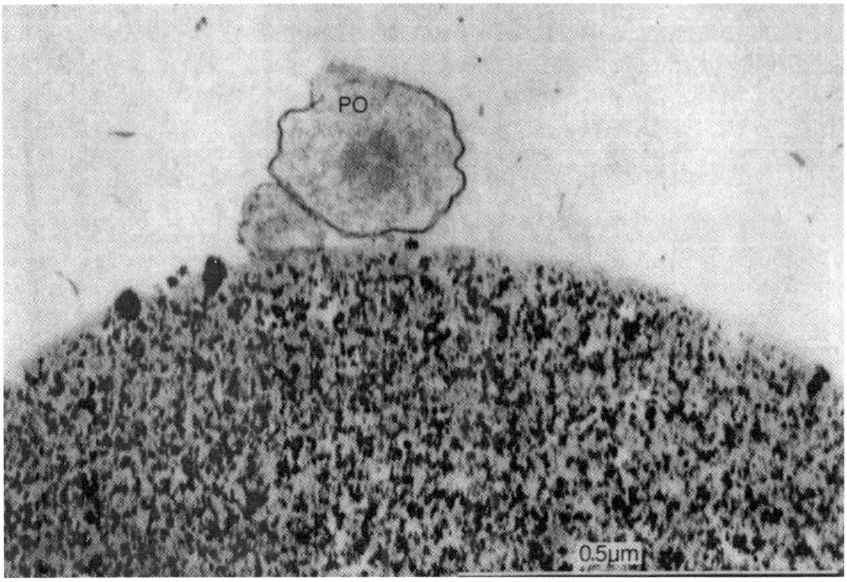

(b)

Figure 9 Continued

These advantages are combined with a high reproducibility of the sorting variables. They can be easily and quantitatively adjusted by the current of the magnet coil (magnetic force), by the flow and counterflow velocity (exposure time to the magnetic field), and by the electronically controlled sample injection unit. This supports the suitability of the system for routine use.

The system has been optimized for the application of Dynabeads 4.5 and 2.8 μm in diameter. However, any other magnetic beads having an appropriate magnetic susceptibility and appropriate geometrical conditions can be used. It is anticipated that the complete system is expensive if compared with permanent hand-held magnet or filter systems as described in Ref. 16. However, due to the advantages of the system, it is not necessary to combine it with flow cytometry or any other sorting technique for high-purity sorting. This is also true for rare event detection.

Additionally, the principle of sorting by magnetic deviation allows the separation of specifically labeled cells from cells unspecifically attached to magnetic beads. The well-defined variable deviation forces from femto N up to nano N allow to study the stability of Ab bindings.

Although it has been shown in several applications that magnetophoresis and CIMS is a versatile sorting device that offers new aspects in medicine [e.g., transplantation medicine (42)], its use in clinics is still not routine for several reasons that will not be discussed here. Therefore, one goal of this chapter was to argue that due to its merits, the magnetophoresis technique (CIMS) should find a more widespread application.

ACKNOWLEDGMENTS

The authors thank Michael Riedmiller, Michael Kraus, Dr. Markus Durm (Institute of Applied Physics, University of Heidelberg), Professor Dr. E. Weiss (Institute for Veterinary Pathology, University of Gieβen), Dr. J. Schmitt, Dr. D. B. J. Herrmann (Boehringer Mannheim), and Dr. B. Bartsch Sandhoff (Bad Tölz) for their support in the experiments and many helpful discussions. The authors also thank Hans Krieger (Institute of Applied Physics, University of Heidelberg) for his support in the preparation of the figures. The financial and hardware support of the Dr. Weber GmbH, Ismaning, the Deutsche Forschungsgemeinschaft, and the Heidelberg-Tübingen-Munich collaboration are gratefully acknowledged. The counterflow separation chamber and the magnetophoresis technique are patented. CIMS was commercially produced by Dr Weber GmbH, Am Englischen Garten 6, D-85737 Ismaning and is available via the authors (R. H., M. H.).

REFERENCES

1. Oberteuffer J. Magnetic separation: a review of principles, devices and applications. *IEEE Trans Magn* 1974; 10:223–238.
2. Kolm H, Oberteuffer J, Kelland D. High gradient magnetic separation. *Sci Am* 1975; 11:46–54.
3. Jeong KM, Petrakis L, Takayasn M, Friedlaender FJ. High gradient magnetic separation I: the removal of solids from shale oils. *IEEE Trans Magn* 1982; 18:1692–1694.
4. Emroy BB. Nuclear power reactor applications of high gradient magnetic filtration. *IEEE Trans Magn* 1982; 18:1686–1688.
5. Kemshead JT, Ugelstad J. Magnetic separation techniques: their application to medicine. *Mol Cell Biochem* 1985; 67:11–118.
6. Kronik P, Gilpin RW. Use of superparamagnetic particles for isolation of cells. *J Biochem Biophys Meth* 1986; 12:73–80.
7. Lea T, Vartdal F, Nustad K, et al. Monosized, magnetic polymer particles: their

use in separation of cells and subcellular components, and in the study of lymphocyte function in vitro. *J Mol Recogn* 1988; 1:9–18.

8. Bolt BA, Spurling TH. Water purification with magnetic particles. *Environ Monit Assess* 1991; 19:139–143.
9. Olsvik O, Popovic T, Skjerve E, et al. Magnetic separation technique in diagnostic microbiology. *Clin Microbiol Rev* 1994; 7:43–54.
10. Safarik I, Safarikovà M, Forsythe SJ. The applications of magnetic separations in applied microbiology. *J Appl Bact* 1995; 78:575–585.
11. Lea T, Vartdal F, Davies C, Ugelstad J. Magnetic monosized polymer particles for fast and specific fractionation of human mononuclear cells. *Scand J Immunol* 1985; 22:207–216.
12. Ugelstad J, Söderberg L, Berge A, Bergström J. Monodisperse polymer particles—a step forward for chromatography. *Nature* 1983; 303:95–96.
13. *General Cell Separation, Dynabeads M-450 Technical Handbook.* Oslo: Dynal A/S 1992.
14. Groman EV, Sullivan JA, Markinac JE, et al. BioMag—new support for magnetic affinity chromatography. *Bio Techniques* 1985; 3/4:156–160.
15. Hardwick RA, Prisco MR, Shak DO. A large-scale magnetic separator for selective cell separations with paramagnetic microbeads. *Artif Org* 1990; 14:342–347.
16. Miltenyi S, Müller W, Weichel W, Radbruch A. High gradient magnetic cell separation with MACS. *Cytometry* 1990; 11:231–238.
17. Hartig R, Hausmann M, Cremer C. Continuous focusing of biological particles by continuous immuno magnetic sorter: technique and applications. *Electrophoresis* 1995; 16:789–792.
18. Hartig R, Hausmann M, Schmitt J, Herrmann DB, Riedmiller M, Cremer C. Preparative continuous separation of biological particles by means of free-flow magnetophoresis in a free-flow electrophoresis chamber. *Electrophoresis* 1992; 13:674–676.
19. Hartig R, Bettag U, Bier F, Cremer C, Hausmann M, Weber G. *Vorrichtung zur Trennung von an magnetischen Partikeln (Beads) gebundenem biologischen Material in einem Magnetfeld.* Patent DE 39 25 093 C2, Deutsches Patentamt (1995).
20. Hartig R, Hausmann M, Lüers G, Kraus M, Weber G, Cremer C. Continuous sorting of magnetizable particles by means of specific deviation. *Rev Sci Instrum* 1995; 66:3289–3295.
21. Basch RS, Berman JW, Lakow E. Cell separation using positive immunoselective techniques. *J Immunol Meth* 1983; 56:269–280.
22. Jackson JD. *Classical Electrodynamics* New York: Wiley, 1975.
23. Zimmels Y. Magnetic forces and magnetic diffusion in magnetizable dispersions. *J Magnetism Magn Mat* 1990; 83:439–441.
24. Yaniv I, Lin IJ, Zimmels Y. Hydrostatic levitation of particles in electric and magnetic fields. *Sep Sci Technol* 1979; 14:557–570.

25. Schnell G. *Magnete: Grundlagen—Aufbau—Anwendungen*, Vol. 49. Munich: Verlag Karl Thiemig, 1973.
26. Weber G. *Verfahren und Vorrichtung zur kontinuierlichen, trägerfreien Ablenkungselektrophorese.* Patent DE 41 39 472 C1, Deutsches Patentamt (1993).
27. Hartig R, Riedmiller M, Hausmann M, Cremer C. *Probebevorratungs—und—zuführungs-System für Partikel in Suspension.* Gebrauchsmuster, G9401–778.6, Deutsches Patentamt, 1994.
28. Hartig R. *Die Free-Flow-Magnetophorese: Magnetische Ablenkung als Sortierverfahren.* Ph.D. thesis. University of Heidelberg, 1993.
29. Ault KA. Applications in immunology and lymphocyte analysis. In: Melamed MR, Lindmo T, Mendelsohn ML, eds. *Flow Cytometry and Sorting.* New York: Wiley-Liss, 1990:685–696.
30. Loken MR. Immunofluorescence techniques. In: Melamed MR, Lindmo T, Mendelsohn ML, eds. *Flow Cytometry and Sorting.* New York: Wiley-Liss, 1990:341–354.
31. Silbernagel S, Despopoulos A. *Taschenatlas der Physiologie.* Stuttgart: Thieme Verlag, 1983:58–59.
32. Arnold R, Creutzfeldt C, Track NS, Creutzfeldt W. Gastrin und Duodenalulcus. *Verh Dtsch Ges Inn Med* 1976; 83:1002.
33. Johnson R. Gastrointestinal hormones and their functions. *Ann Rev Physiol* 1977; 39:135–158.
34. Lichtenberger LM. Importance of food in the regulation of gastrin release and formation. *Am J Physiol* 1982; 243:429–441.
35. Lewin MJ, Cheret AM. Cell isolation techniques: use of enzymes and chelators. *Methods Enzymol* 1989; 171:444–461.
36. Van Dilla MA, Deaven LL. Construction of gene libraries for each human chromosome. *Cytometry* 1990; 11:208–218.
37. Cremer C, Hausmann M, Zuse P, Aten JA, Barths J, Bühring H-J. Flow cytometry of chromosomes: principles and applications in medicine and molecular biology. *Optik* 1989; 82:9–18.
38. Lichter P, Cremer T. Chromosome analysis by non-isotopic in situ hybridization. In: Rooney DE, Czepulkowski, eds. *Human Cytogenetics—A Practical Approach*, Vol. F. Oxford: IRL Press, 1992:157–192.
39. Dudin G, Steegmayer EW, Vogt P, et al. Sorting of chromosomes by magnetic separation. *Hum Genet* 1988; 80:111–116.
40. Kraus M, Hausmann M, Hartig R, Durm M, Haar F-M, Cremer C. Nonenzymatic, low temperature in situ hybridization of metaphase chromosomes for magnetic labelling and sorting. *Exp Techn Phys* 139–153.
41. Lüers G, Hartig R, Hausmann M, Fahimi HD, Cremer C, Völkl A. Immunomagnetic isolation of peroxisomes. *Mol Biol Cell Suppl* 1994; 5:476a (abstract).
42. Collins RH Jr. $CD34^+$ selected cells in clinical transplantation. *Stem Cells* 1993; 12:577–585.

11

Cell Separation Using Flow Cytometric Cell Sorting

Robert A. Hoffman and David W. Houck
Becton Dickinson Immunocytometry Systems, San Jose, California

I. INTRODUCTION

Flow cytometric cell sorting rapidly separates cells one at a time based primarily on optical and fluorescent properties. Fluorescent dyes to specific cellular targets or antibodies (Abs) tagged with fluorochromes are used to identify particular cells. Cells in liquid suspension flow past a sensing zone and are analyzed for the optical signal they emit. Analysis rates can be greater than 10,000 cells/s. Cells whose optical characteristics fall within preselected ranges or "gates" generate an electronic signal that initiates sorting of the cell. Less than one millisecond after a wanted cell is analyzed, it can be physically separated. Some early researchers coined the acronym FACS (fluorescence-activated cell sorting) for this technique, and FACS™ is currently a trademark of Becton Dickinson Immunocytometry Systems. Since characteristics other than fluorescence can be used to analyze the cells, this chapter will use a more general term, flow sorting, for this technique.

This chapter is intended as an introduction to flow sorting for the nonspecialist. It will provide an overview of the technology and survey a large, but not comprehensive, range of applications. Enough detail is provided so anyone considering using flow sorting will be able to determine if it is a suitable method for a particular application. It is assumed that if the user is new to the field of flow sorting, a knowledgable, trained instrument operator will run the sample. This will often be done in a central flow cytometry facility at an institution. This chapter should provide a new user with enough information to clearly communicate with the flow cytometry facility personnel and, with them, to determine the best approach for separating the desired cells.

There are several excellent books that the reader can refer to for more details on flow cytometric analysis and sorting (1–10). Shapiro (1) and Melamed et al. (2) provide comprehensive references for both the instrumentation and applications of flow cytometry. The books by Melamed et al. (2) and Van Dilla et al. (3) have the most details about various types of flow instrumentation. The smaller texts by Ormerod (5) and Given (6) are very good introductions to the field. Shapiro's book (1) is particulary good at describing the entire field of flow cytometry and sorting in a simple manner.

An article by Herzenberg et al. (11) describes the early history, application to immunology, and essential technology used in droplet sorting. Review articles by Horan et al. (12) and Arndt-Jovin and Jovin (13) provide additional perspective on the early instrumentation and applications. The journal *Cytometry*, which is cited in many of the references for this chapter, is a major primary resource for information on the technology and applications.

An increasingly useful source of information is the Internet. Many flow cytometry companies and laboratories using flow cytometry have Internet sites. A useful starting point for finding information about flow cytometry is the home page for the International Society for Analytical Cytology. Their Internet address at the time of this writting is http://nucleus.immunol.washington.edu/ISAC.html.

II. OVERVIEW OF FLOW CYTOMETRY AND CELL SORTING

The name "flow cytometry" is very descriptive of the process. "Cytometry" means cell measurements. Flow cytometry moves a suspension of cells in a liquid stream through a sensing zone, which in commercial cell sorters is usually a focused beam of light. Light scattered by cells provides information about their size, structure, or granularity. In addition, cells are usually stained with fluorescent probes used to identify more specific cell characteristics. The most commonly used probe is an Ab labeled with a fluorochrome. A wide range of fluorochromes, each providing a distinct fluorescence color, is available for Ab labeling. The use of two to four different Abs for staining a sample is common, and each Ab is labeled with a different color fluorochrome. Oligonucleotides, cytokines, and other site-specific molecules can be labeled with fluorochromes and used to stain cells. Fluorescent probes for nearly every cell constituent are also avaliable. Examples are probes for DNA, RNA, protein, mitochondria, plasma membrane, and calcium. In general, the number of different fluoroescent probes that can be used to simultaneously stain a sample is limited only by the number of detection channels on the flow cytometer.

A. Basic Technology

The flowing stream in a flow cytometer is composed of two fluids. A stream of cell suspension is injected into and surrounded by a second, cell-free sheath stream. A hydrodynamic focusing process controls the position and cross-sectional area of the sample stream within the sheath (Fig. 1). Cells maintained near the center of the sheath stream can be uniformly illuminated in the sensing zone and thus provide accurate information about the amount of fluorochrome labeling. This uniformity of illumination is especially important when measuring the amount of DNA per cell.

Signals given off by the cells as they pass through the optical sensing zone are detected and compared with the criteria for the cells to be sorted. Data are usually represented in either histograms representing the number of events as a function of signal value (Fig. 2A) or as dot plots showing the correlated data for two parameters (Fig. 2B). Linear scales are used to display data when the signal range is not too large (less than 100-fold variation) (Fig. 2A). Logarithmic scales are used when signals vary over 3 to 4 orders of magnitude (Fig. 2B). Peaks in the histogram or clusters of dots in the dot plot represent populations of cells with similar characteristics. If measurements from a cell are all within the range for desired cells, then the individual cell is separated from the flowing stream and collected in a container. Three different methods for separating individual cells from the sample stream are used in commercial instrumentation and are described in some detail below.

B. Multiparameter Analysis

A major advantage of flow cytometric sorting is the simultaneous use of multiple cell characteristics to identify the cells to be separated. The instrumentation to perform these measurements can have six or more optical detectors and multiple lasers. The basic optical layout of a flexible, multiparameter droplet cell sorter is shown in Figure 3. In this example, three lasers can be used to excite fluorescence or light scatter. The laser beams intercept a stream of cells (Fig. 4). Light emitted by the cells as they pass through the laser beam is collected by lenses in the direction of laser light (forward

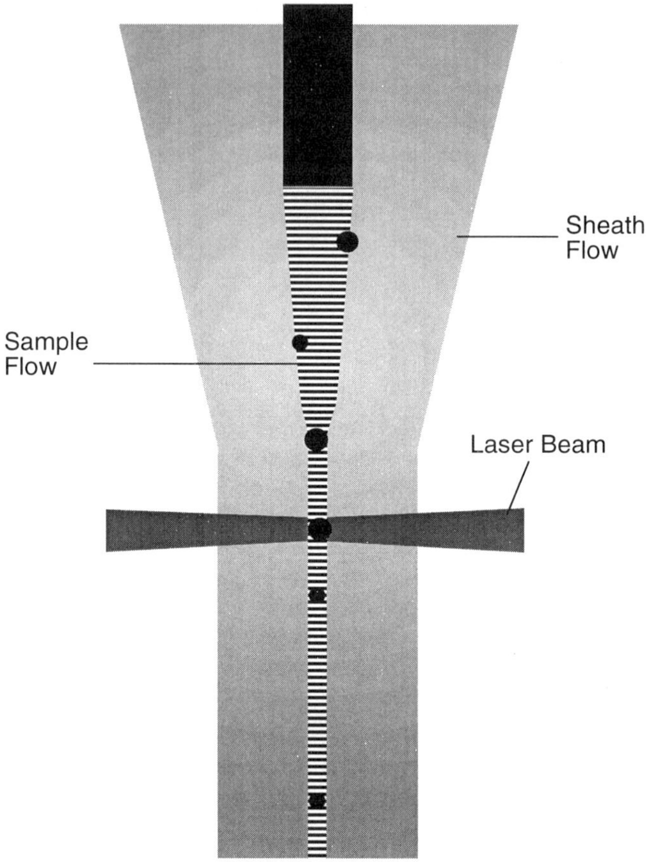

Figure 1 Schematic representation of hydrodynamic focusing of sample stream through the center of a laser beam. (By courtesy of Becton Dickinson Immunocytometry Systems, San Jose, CA.)

direction) and perpendicular to the direction of the laser beam. The signal measured in the forward direction is usually forward light scatter (FSC), which depends on cell size and other factors. Perpendicular to the laser beam, a video camera (shown) or alternatively a microscope (not used for this design) allows the sample stream to be viewed, using a stroboscopic illuminator, and the process of droplet generation can be monitored. Opposite the viewing camera (or microscope) is a lens that collects fluorescent light and light scattered in the side direction (side scatter, or SSC). SSC depends on the internal contents

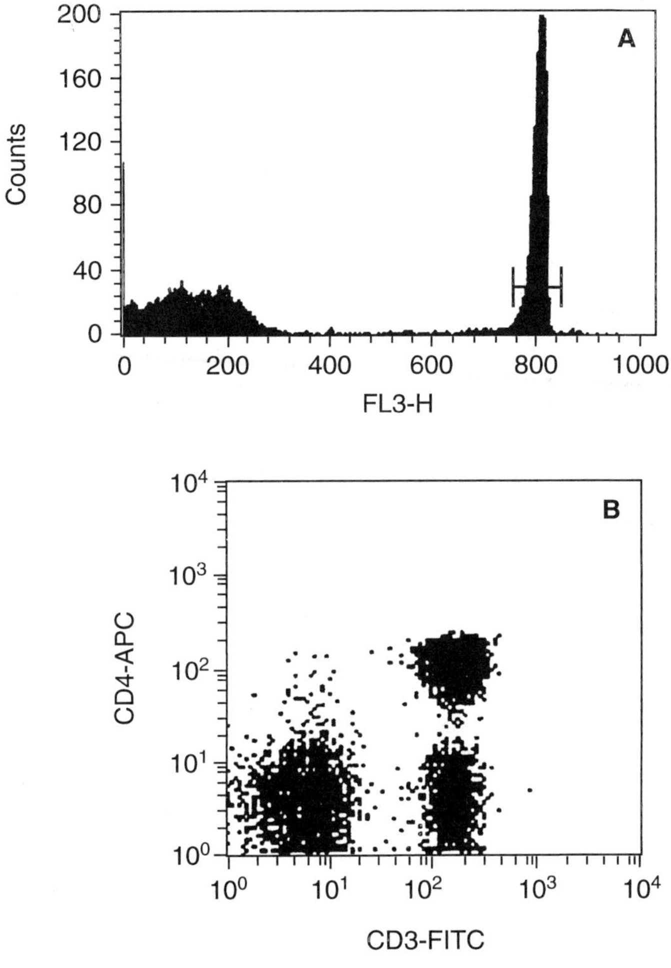

Figure 2 Examples of flow cytometry data representations. (A) Histogram showing value of measured parameter versus number of events. The x-axis is in linear units. (B) Dot plot of green fluorescence (FITC-CD3) signal versus red fluorescence signal (APC-CD4). The x and y axes are logarithmic and cover a four-decade range.

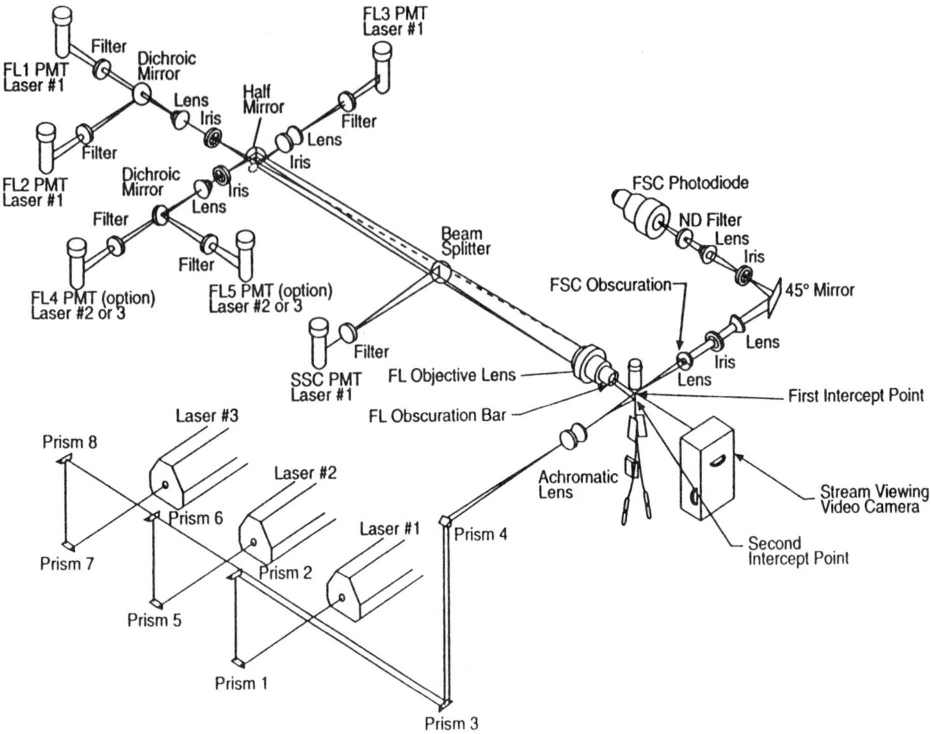

Figure 3 Optical diagram of a multilaser droplet sorter showing excitation and emission optical paths and major optical elements. (By courtesy of Becton Dickinson Immunocytometry Systems, San Jose, CA.)

and structue of the cell as well as cell size. For many practical purposes, SSC can be considered to depend on cell granularity. Fluorescence detectors are numbered FL1, FL2, etc.

Fluorescent light from different excitation beams is split off by mirrors, and different colors of fluorescence are selected by dichroic mirrors and filters. Each detector can be configured to respond to a different cell stain or probe. In the instrument illustrated in Figure 3, five different fluorescence detectors can measure five different probes for cell properties. An example of measurements made on a single cell as it passes through the laser beams could be the following: FSC and SSC may be measured from a blue laser; FL1, FL2, and FL3 may measure green, yellow, and red fluorescence, respectively, from a blue laser; FL4 may measure red or infrared fluorescence excited by a red laser, and FL5 may measure blue fluorescence excited by an ultraviolet laser.

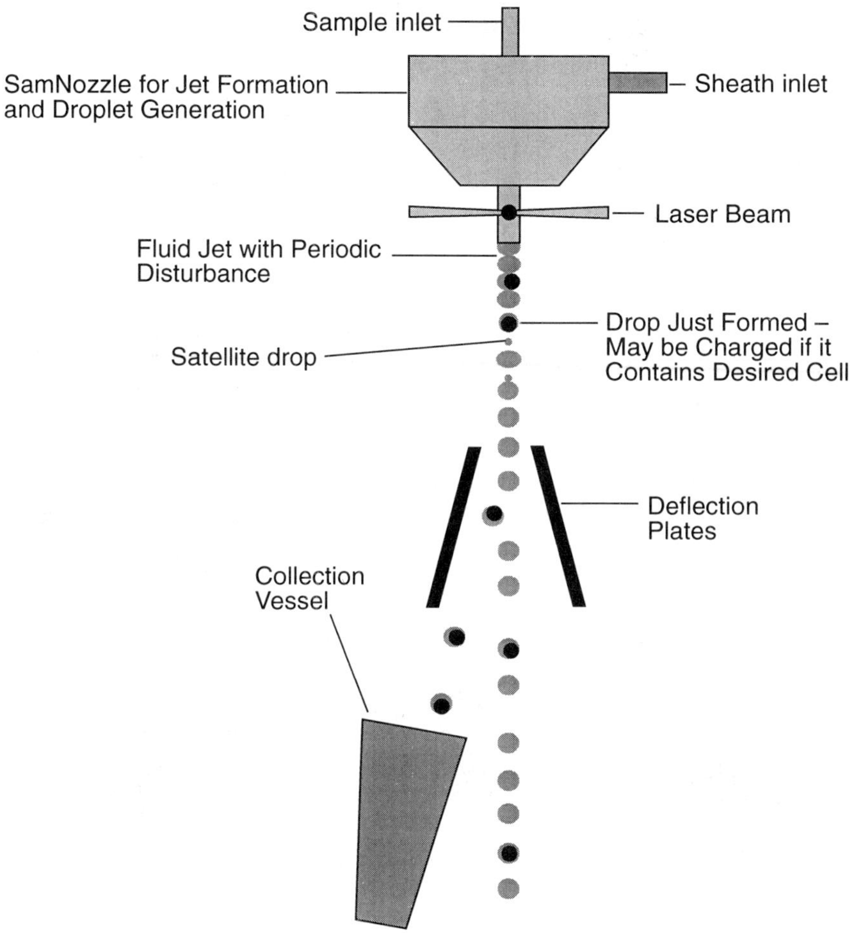

Figure 4 Schematic representation of the essential features of a droplet sorter. The drawing is not intended to be to scale. (By courtesy of Becton Dickinson Immunocytometry Systems, San Jose, CA.)

All of the information about a cell from the multiple detectors can be acquired within a few microseconds.

An example of multiparameter analysis (Fig. 5) shows data from a seven-parameter, five-color analysis of a human blood sample for the major types of lymphocytes. The blood sample was stained with five different monoclonal antibodies (mAbs) to human lymphocyte CD antigens (Ags), with

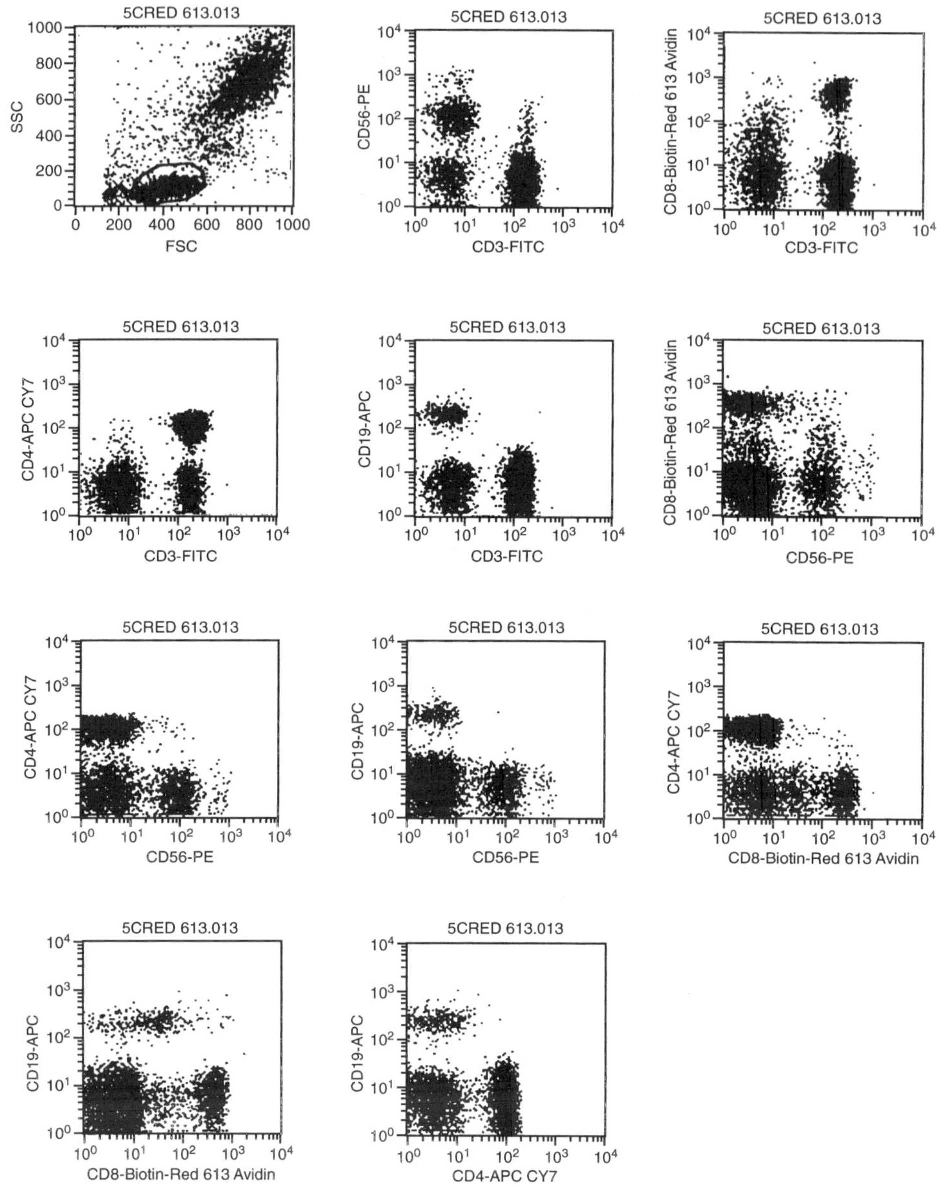

Figure 5 Example of multicolor analysis illustrating the large number of cell characteristics that can be simultaneously measured. A blood sample was stained with five

each Ab identified by a separate fluorescence color. A dual-laser cell sorter was used to analyze this five-color immunofluorescence sample. Each dot plot shows correlations between different CD Ags. Each fluorescence dot plot was gated on lymphocytes, which are identified in the encircled cluster of the SSC versus FSC dot plot in the upper left panel. The cells identified include total T cells (CD3), helper T cells ($CD3^+CD4^+$), suppressor T cells ($CD3^+CD8^+$), B cells ($CD19^+$), and natural killer (NK) cells ($CD3^-CD56^+$).

In addition to optical parameters, some flow cytometers and sorters can measure electronic cell volume or impedence. The first droplet sorter used this measurement (14). Electronic cell volume is measured by passing an electrical current through a small orifice. When a cell traverses the orifice, the electrical resistance within the orifice is changed and produces an electrical signal. If the plasma membrane of the cell is intact, the cell is a perfect insulator, and the cell produces a signal proportional to its volume. The signal also depends on the shape of the particle.

C. Advantages and Disadvantages of Flow Sorting

The major advantages to flow cytometric sorting are its ability to select cells or particles based on multiple, simultaneously measured characteristics; its ability to sort out a single cell for analysis or cloning: and the high purity of the sorted cells. The primary disadvantages are the relative complexity and expense of the instrumentation and the relatively long times required to obtain large numbers (millions) of sorted cells.

The ability to use multiparameter analysis to select sorted cells with high purity has been discussed earlier in this chapter. This would be one of the major reasons to choose flow sorting over a simpler, bulk-separation technique based on only one characteristic of the cells. Sorting of a single cell is accomplished with droplet sorters and can be a powerful tool for studying growth, differentiation, and development or for selecting clones of cells that have desirable characteristics. In theory, it may be possible to accomplish cell sorting by other methods to achieve the same result as flow cytometric sorting. For example,

---

different monoclonal antibodies (mAbs) to human lymphocyte CD antigens (Ags), with each Ab identified by a separate fluorescence color. A dual laser cell sorter was used to analyze this five-color immunofluorescence sample. Each dot plot shows correlations between different CD Ags. Each fluorescence dot plot was gated on lymphocytes, which are identified in the encircled cluster of the side scatter (SSC) versus forward scatter (FSC) dot plot in the upper left panel.

single cell sorting can be done using micromanipulators or laser tweezers, assuming that the desired cell can be identified in some way. Manually selecting the cells, however, is much slower than the flow cytometric approach.

If the number of selected cells required is very large, then flow cytometric sorting may be an impractically slow method. Flow cytometric sorting of viable cells can be performed at analysis rates of 10,000 cells/s in standard instruments with adequate purity and recovery of the sorted cells. The upper analysis rate for viable cell sorting is approximately 25,000 cells/s in the fastest sorters. Whether this sorting rate is adequate will depend on the fraction of desired cells in the sample and the number of sorted cells needed. A few hundred cells sorted onto a microscope slide may be enough to judge the morphology or staining pattern under a microscope. In contrast, sorting of bone marrow stem cells for experimental use could require at least 10,000,000 cells sorted from a starting sample containing less than 1% of desired stem cells. To accomplish this requires many hours of sorting even with the fastest achievable analysis speed (25,000 cells/s). Most applications require sorting times between these extremes.

III. SORTING MECHANISMS

A. Droplet Sorters

The most widely used flow sorting method is droplet sorting in which the fluid stream is broken into small droplets after cells pass through the sensing zone. Some droplets are empty and others contain a cell or cells. From the analysis in the sensing zone, the drops containing desired cells are known and can be given an electrical charge and physically sorted from the other drops by an electrostatic deflection. Several excellent texts provide detailed discussion of flow sorting technology (1–3), and only the essential elements will be considered here.

Figure 4 shows the important features of droplet formation, charging, and deflection. The fluid stream containing cells first passes through a laser beam where analysis takes place. The jet of fluid is then caused to break into drops in a stable and precisely timed way. Jets with diameters of 50–400 μm are used in commercial sorters, and droplets are formed at frequencies of 2000/s for the largest diameter jets to 100,000/s for the smallest diameter jets. Several methods can be used to stimulate drop formation. The most common method is to vibrate the entire nozzle assembly. In some designs the vibrating transducer acts directly on the liquid. Within well-defined frequency ranges, a stable oscillation is established on the fluid jet, and drops break off at a precise point below the laser beam. Small satellite drops form between the large drops and merge with the large drops a short distance from the break-off point.

The speed of particles traveling in the jet of fluid also is precisely controlled, so the time for a particle to pass from the laser beam to a drop that is about to break off is known. If the particle analyzed in the laser beam is to be sorted, then just as the drop is breaking off the jet, an electrical charge is applied to the stream. The stream is charged as the drop breaks off, and the drop carries an electrical charge after breaking free of the stream. The charge on the sample stream is then turned off and the following drops are not charged. The drops pass through an electric field formed between two plates across which a high voltage (> 3000 V) is applied. Charged drops are deflected in the electric field toward a collection vessel.

Most droplet sorters allow drops to be charged either positively or negatively, which allows drops to be deflected either to the left or to the right. Two different populations of cells can be sorted (left or right) during the same run.

Since the speed of the cells is well controlled, their position is always known after they pass through the laser beam. The electronics also can tell whether a nearby cell passed through the laser beam, but was not analyzed because the electronics were busy analyzing a previous cell. The electronics keep track of where each cell is in the fluid jet and "knows" whether any other cells are nearby. If any nearby cells are far enough apart to have been analyzed separately by the electronics, the system also can tell whether adjacent cells are both desired cells or if one is an unwanted cell. Thus the positions of all cells that pass through the laser beam are known, and the "identity" of cells are known, including cells whose identity is "not analyzed."

The electronics for sorting includes logic to determine whether a drop containing a desired cell is likely to have a coincident "passenger" cell and whether the passenger is a wanted cell or not. There is usually a choice of how many drops to deflect for each sort event. Deflecting a single drop has the advantage of the lowest likelihood of a passenger cell, but it is more difficult to maintain such a strict condition. Deflecting two or more drops for each sort event is common. The decision whether to sort a drop (or drops) containing passenger cells depends on the sorting mode.

B. Enclosed Sorters

Enclosed sorters are much slower cell separators than droplet sorters but have the advantages of lower cost and greater ease of use. Enclosed sorters can analyze cells at the same rate as droplet sorters, but the mechanisms in enclosed sorters cannot resopond to sort cells at rates greater than 300–1000/s.

Also, if cells are to be sorted in a high-purity mode, then the analysis rate with the enclosed sorter may need to be limited to approximately 1000 cells/s.

The major advantages of enclosed sorters is their ease of use and inherent containment of biohazardous samples. Enclosed sorters are no more complicated to use than the basic flow cytometry analyzer. There are no special adjustments that the user needs to make nor any set-up or monitoring of the sort mechanism. Since the cells are always enclosed and no aerosols are generated, there is less danger when sorting biohazardous samples, such as cells from patients infected with human immunodeficiency virus (HIV), than when a droplet sorter is used.

Catcher-Tube Sorting

Two types of sorting devices separate the cells within an enclosed chamber. The first, called catcher-tube sorting, is illustrated in Figure 6. This sorter uses a small tube that is normally positioned so the flowing sample stream passes to the side and is not collected. In this position, only sheath fluid enters the catcher tube. When a desired cell has been detected, the catcher tube rapidly moves into the sample stream at just the right time to capture the cell and then moves back to its resting position. The entire capture process takes less than 1 millisecond. At a maximum sorting rate of 300 cells/s, the catcher tube method is not efficient for sorting millions of cells. This simple sorting method

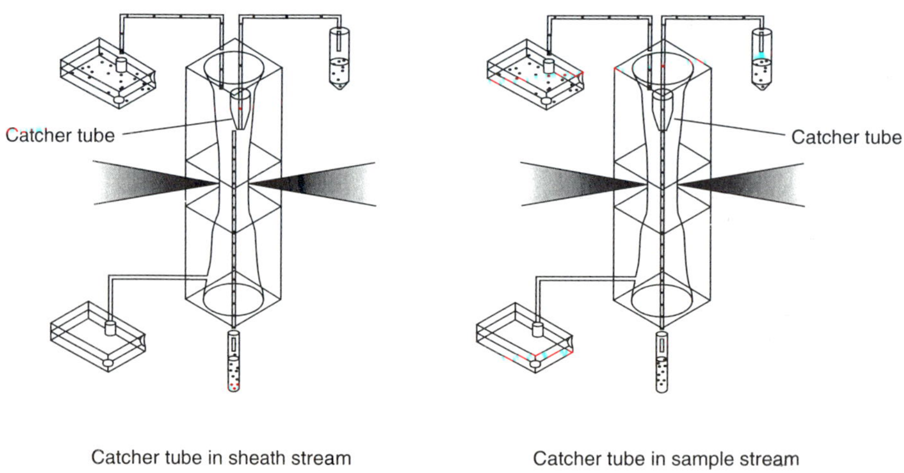

Figure 6 Schematic diagram illustrating operation of a catcher-tube sorter. (By courtesy of Becton Dickinson Immunocytometry Systems, San Jose, CA.)

has been very effective for a wide variety of applications including sorting cells for morphological examination, sorting transfected cells, sorting cells for subsequent staining by fluorescence in situ hybridization (FISH) with DNA probes or for PCR amplification of a specific gene, isolation of moderate numbers (400,000) of cells for functional studies, isolation of ^{14}C-labeled cells for isotopic analysis, and enrichment of parasites from water samples (15). The catcher-tube method also has been used to separate live cells for subsequent analysis for apoptosis (16).

Enrichment sorting of rare cells can be effectively performed even with the slow catcher-tube sorting mechanism. Rare cells that are present at frequencies of 1:100 to 1:100,000 can be rapidly enriched to purities of 20% or better by analyzing the cells at rates of 10,000 cells/s or faster. At an analysis rate of 25,000 cells/s, cells that are only 1% (1:100) of the total will arrive at a rate of 250/s.

Fluidic-Switching Sorting

A second approach to an enclosed sorter uses a branched fluid path after the sensing zone. Figure 7 illustrates the fluid paths and valving. A valve actuated by a piezoelectric transducer controls whether cells go down the normally open path to waste or through the separate path for sorted cells (17,18). Published descriptions of the fluidic-switching instruments thus far have used epi-illumination (i.e., the excitation light and emitted fluorescence both pass through the same lens but in opposite directions).

Fluidic-switching sorters have been used to isolate intact pancreatic is-

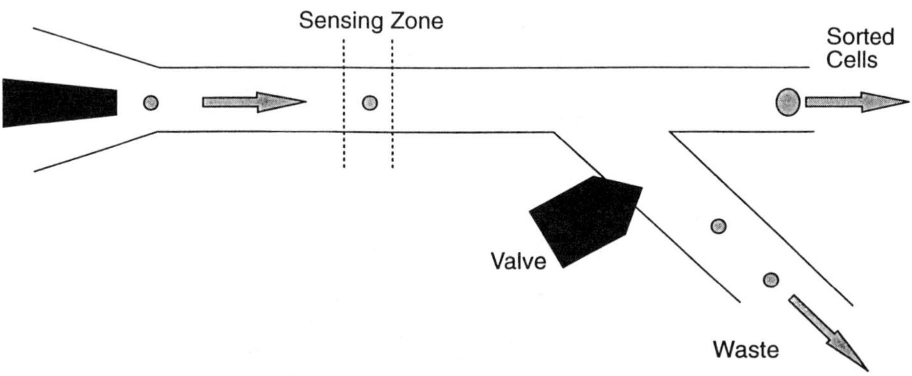

Figure 7 Schematic diagram of the essential features of a fluidic-switching sorter based on the description by Gray et al. (19).

lets (19) and to sort plant cell nuclei (20). Fluidic-switching sorting should be capable of the same applications described earlier for the catcher-tube sorter.

C. Cell Killing or Destructive Sorting

Although not used in commercial sorters, an alternative approach to cell selection is under investigation in several laboratories. Rather than select wanted cells, the alternative approach destroys unwanted cells or other biological particles. Cells are analyzed in a sensing zone and then pass through a region where they may be selectively killed. Most approaches to destructive sorting use an intense laser beam to cause photodamage to unwanted cells (21–23), but cell killing has also been accomplished with high electric fields that break down the plasma membrane (24).

D. Collection Devices

Many different containers and devices have been used to collect sorted cells. In the simplest case, the sorted cells are collected in a test tube. The cells may also be collected on a filter, and all or most of the liquid can be removed from the sorted suspension, leaving a very concentrated layer of cells. Cells may also be sorted directly into cytocentrifuge buckets. Two collection devices that work only with droplet sorting are microtiter wells and microscope slides. Many custom collection devices have also been made for specific applications.

Direct sorting of droplets onto specially prepared slides (discussed later) allows immediate examination of the sorted particles with a microscope. By moving the microscope slide slightly after each sorted drop, a pattern of individual cell droplets can be deposited on the slide (25). Single cells can also be sorted into microtiter wells containing appropriate medium. An automated mechanism for moving the tray to position wells for sorting is available as an option on commercial droplet sorters.

IV. CHARACTERIZING THE FLOW-SORTING PROCESS AND ITS LIMITATIONS

A. Sorting Modes

It would be ideal if a flow sorter could be set to simultaneously provide the highest purity of the selected cells and fastest sorting rate. Unfortunately, during the time that a wanted cell is being sorted, a second unwanted cell may be sorted with it. The sorter can be set to reject (abort) these potential sort events that include an unwanted cell, but the wanted cell will also not be sorted. The faster the cells are analyzed (i.e., the higher the cell throughput), the more frequent will be the coincidence of a wanted and unwanted cell during a sort event.

Cells arrive randomly to be analyzed and sorted, and the statistics of random processes determine the ultimate performance of flow sorters. Lindmo et al. (26) and Pinkel et al. (27) analyze sorting statistics in detail, and their results are reviewed here. Different sorting modes allow a trade-off between sorting rate, purity, and the fraction of desired cells that are actually sorted. The simplest mode is often called "enrichment" or "recovery." Any event that includes a wanted cell, whether or not an unwanted cell is also present, is sorted in this mode (Fig. 8B). The higher the arrival rate of cells in the sensing zone, the more frequent a coincidence of an unwanted cell with a wanted cell and the lower will be the purity of the selected cells.

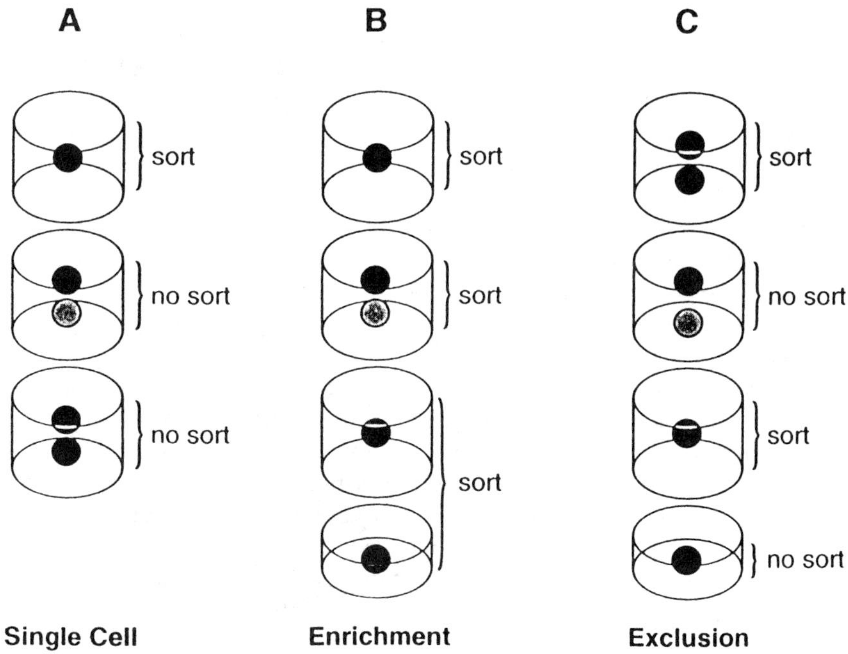

Figure 8 Illustration of the sorting modes for catcher tube sorting. (A) Single-cell mode; (B) enrichment mode; (C) exclusion mode. Details of the sort modes vary slightly with droplet sorting. (By courtesy of Becton Dickinson Immunocytometry Systems, San Jose, CA.)

If wanted cells are to be sorted at high purity, it is necessary to abort sort events that include a coincident, contaminating cell. If each sort event is to include one and only one desired cell, then even an event that includes two desired cells is aborted. This mode is called the "single-cell" (or "counting") mode (Fig. 8A). It is used when one wants to sort an exactly known number of desired cells.

If sort events can include two desired cells but must exclude unwanted cells, the mode is called "exclusion" (or "normal recovery") and is illustrated in Figure 8C. This mode gives high purity and the highest rate of sorting the desired cells. In practice, if the wanted cells are present at low frequency (< 5%), then the probability of having two desired cells in the same sort event is very low. In this case, there will be no significant difference in the sorting rate of the "exclusion" mode and the "single-cell" mode.

B. Purity and Efficiency

If a sorter is adjusted properly, purity in all but the enrichment mode can be nearly 100% of the desired cells, even at high event rates, and 98% purity would be a typical minimum expected. Achieving this high purity depends on electronic circuitry to recognize the presence of a "passenger" cell in the sort event and to abort events that will include an unwanted cell along with a wanted cell. Sorting in "enrich" mode does not abort unwanted cells, and purity is determined by the statistical frequency that unwanted cells are present in sort events. Simple probability theory using Poisson statistics can be used to predict the purity of enrichment sorting (26,27).

The statistical question can be asked as: What is the probability that no unwanted particle is present during the time a wanted cell is being sorted? The factors that affect sorting are the rate, r, at which particles arrive in the sensing zone; the fractions, s, of the subpopulation to be sorted; and the time, T, that a cell can be physically selected during a sort event. For enclosed sorters, T is the time during which the sort path is accessable to the sample stream during a sort event. For droplet sorters, $T = n/f$, where n is the number of droplets sorted per event and f is the droplet frequency.

The probability, P, that no other particle is coincident in the time interval T is given by the Poisson probability for zero events, or:

$$P = e^{-rT} \tag{1}$$

This is the probability of a pure sort of one and only one desired cell. If the desired cell is a fraction, s, of the total population, then the fraction of undesired cells is $(1 - s)$. The arrival rate of undesired cells is $(1 - s)r$ rather than r, and the probability that no undesired cell is present during the time a desired cell is sorted is:

$$P = e^{-(1-s)rT} \tag{2}$$

Equation (2) determines the purity of the sorted fraction in enrichment mode. Note that the higher the arrival rate of cells, the lower the purity. Likewise a longer time, T, for the sorting process reduces the purity. This is because a longer time is allowed for a second cell to be coincident with the one being sorted.

Because of coincidence of wanted and unwanted cells, all the wanted cells in a sample may not be sorted. Sorting efficiency is the fraction of wanted cells that pass through the sensing zone and are actually sorted. Efficiency is the theoretical maximum of this fraction for sorting in a particular mode. Efficiency does not account for any cells that may be lost or damaged during the sorting process. Another measure of the sorting process is "yield," which is the ratio of the actual number of cells collected divided by the number of wanted cells that passed through the sensing zone.

For enrichment sorting, the efficiency is 100%, since no sort events are aborted, but if high purity of the sorted cells is required, efficency will always be less than 100% due to aborted sort events that have coincidence of an unwanted cell. Since sort events that are not aborted are the same events that do not have a coincident unwanted cell, Eqs. (1) and (2) also apply to efficiency. For single cell sorting, the efficiency, E, is:

$$E = e^{-rT} \tag{3}$$

And for exclusion mode sorting the efficiency is:

$$E = e^{-(1-s)rT} \tag{4}$$

The same function [Eqs. (2) and (4)] describes the purity of enrichment sorting or the efficiency of exclusion mode sorting. Figure 9 shows a plot of purity of enrichment sorting or efficiency of exclusion sorting versus event rate for a case where the time, T, during which a cell can be physically selected in the sort event is 50 μs. Results for samples with desired cells representing 1, 10, and 50% of the total cells are shown. As the fraction, s, of the population to be sorted decreases, the purity or efficiency approaches the limit shown for the case where $s = 0.01$. In other words, when the wanted cells are a sufficiently rare subpopulation, a coincidence of a wanted cell will nearly always be with an unwanted cell.

Figure 10 shows the dependence of sorting efficiency on the interval, T, during which a sort event takes place. Sorting of a rare ($s < 0.01$) subpopula-

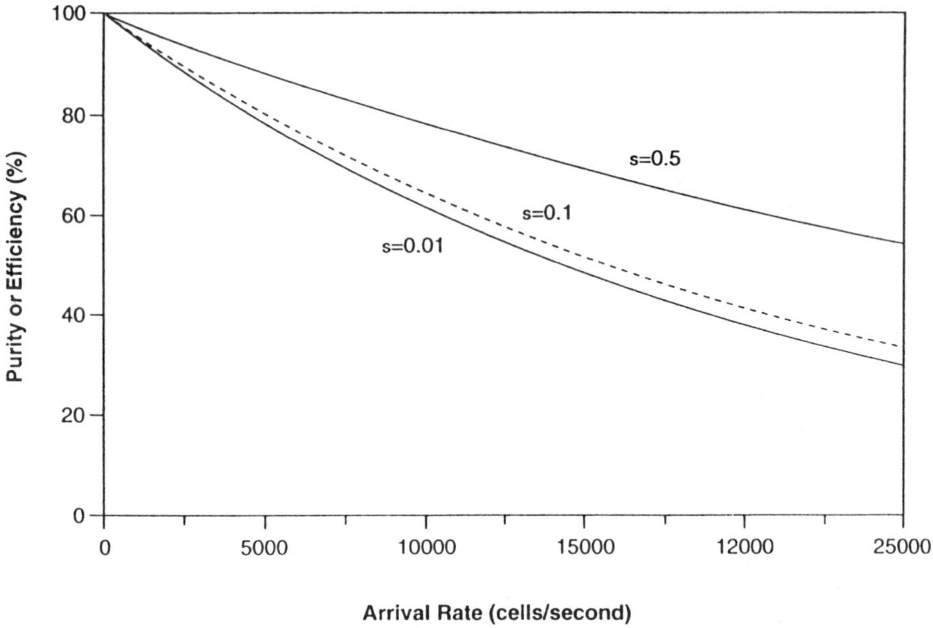

Figure 9 Plots of purity for enrichment sorting mode and efficiency for exclusion sorting mode based on Eqs. (2) and (4). Sort interval time, T, is assumed to be 50 μs. Results are shown for sorting subpopulations that represent the fraction, s, of the total sample equal to 0.5, 0.1, and 0.01.

tion is assumed. For droplet sorters, $T = 50$ μs is fairly typical for routine sorting, $T = 25$ μs is considered high-speed sorting, and $T = 200$ μs would be typical for sorting large particles using large-diameter nozzles. Enclosed sorters generally have $T > 200$ μs.

C. Sorted Cells/Second

As discussed, there is a trade-off between the rate at which cells are analyzed and the fraction of wanted cells that are sorted. As the analysis rate increases, a larger fraction of sort events are aborted to maintain purity and eliminate coincident cells. The rate at which cells are sorted is simply the efficiency, E, defined above multiplied by the arrival rate of wanted cells. For enrichment sorting, the sorting rate is essentially the same as the arrival rate of wanted

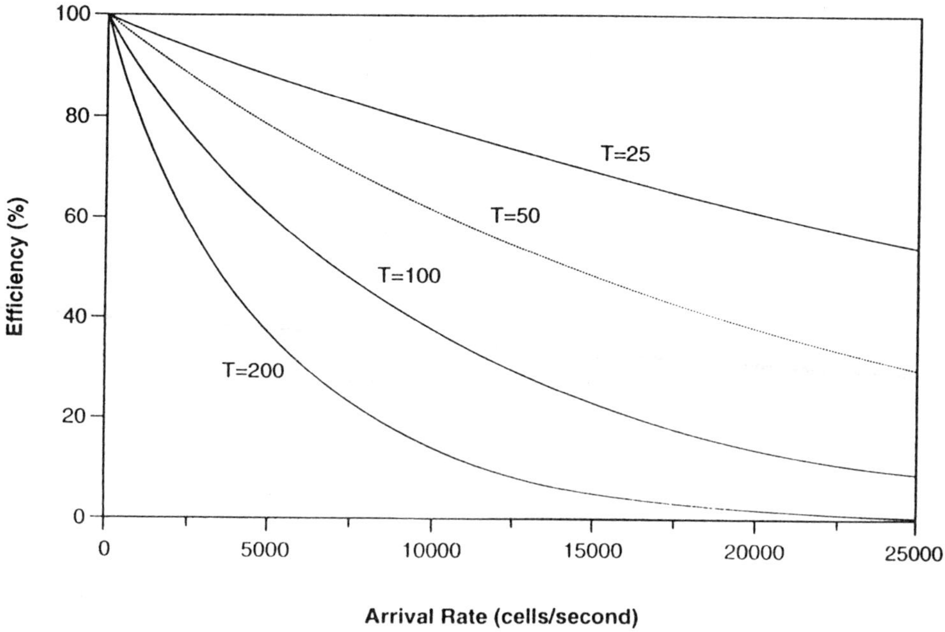

Figure 10 Dependence of sorting efficiency on the time, T (in μs), during which a cell can be physically sorted. Sorting of a rare event ($s < 0.01$) is assumed so that Eq. (3) in the text applies.

cells since no sort events are aborted. For single cell or exclusion sort modes the sorting rate, R, of wanted cells is:

$$R = sr \cdot e^{-rT} \quad \text{(for single-cell mode)} \tag{5}$$

$$R = sr \cdot e^{-(1-s)rt} \quad \text{(for exclusion mode)} \tag{6}$$

Figure 11 shows the sort rate for both exclusion mode and single-cell mode as a function of arrival rate, r, and at T = 50 μs. Exclusion mode is most beneficial when sorting a subpopulation that is a large fraction of the total. For a subpopulation that is 10% or less of the total population, there is little difference in sort rate with these two modes. There is a maximum sort rate, and if the cell arrival rate is too high, the number of sorted cells per second actually decreases. It can be shown that for single cell mode, the maximum rate of sorted cells occurs when

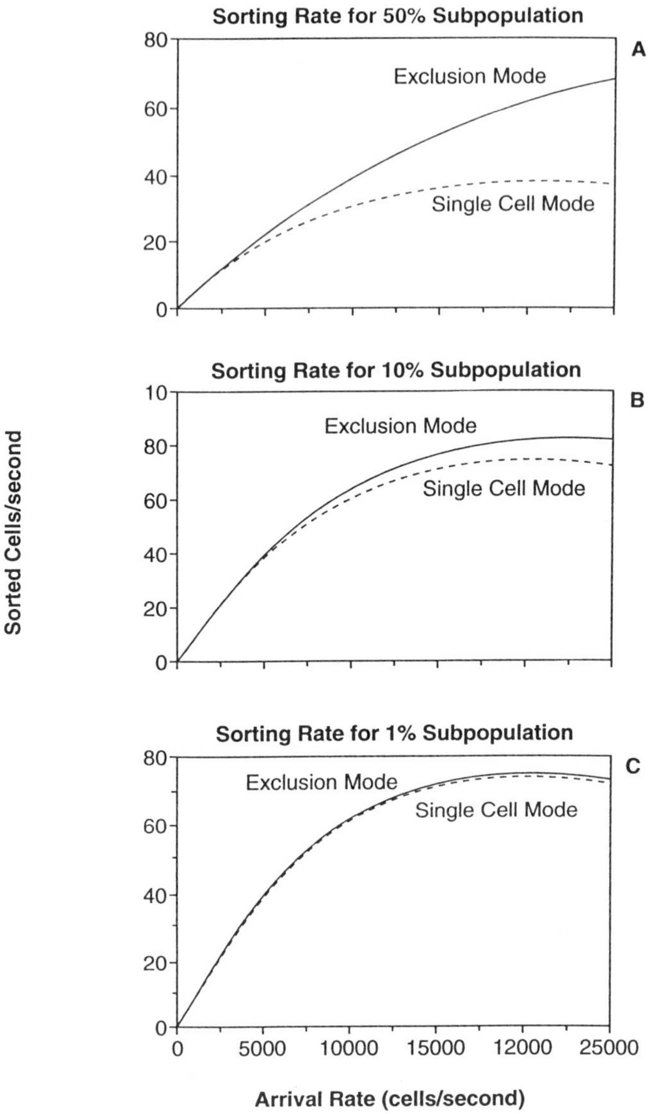

Figure 11 Theoretical sorting rates for single-cell mode [Eq. (5)] and exclusion mode [Eq. (6)] with T = 50 μs. (A) Sorted subpopulation is 50% of the total population. (B) Sorted subpopulation is 10% of total. (C) Sorted subpopulation is 1% of total.

$$r = \frac{1}{T} \tag{7}$$

and then

$$R_{max} = \frac{sr}{e} \tag{8}$$

The maximum sort rate occurs when the average time between cells arriving in the sensing zone is T, the time during which a particle is being physically sorted.

For exclusion mode, the equivalent formula is:

$$r = \frac{1}{(1 - s)T} \tag{9}$$

and again

$$R_{max} = \frac{sr}{e}$$

Exclusion mode is useful when the fraction, s, of cells to be sorted is large. When s is small (< 0.1), there is little or no difference between the maximum sort rates achievable in single cell and exclusion mode.

D. Recovery

The fraction of cells actually recovered in the sorted sample is the practical measure of flow sorting. The measurement of recovery takes into account the intentional loss of sort events required to maintain the desired purity level. Recovery is usually defined as the ratio of cells in the collection container divided by the number of sorted events. Recovery is usually less than 100% and depends on many factors including the type of container, the type of media the cells are sorted into, the adjustment of the sorter, and the types of cells. Fragile cells may simply be destroyed if the sorter is not adjusted properly. In some cases it may be necessary to slow down the sorting process to recover more cells. For example, it might be necessary to reduce the jet velocity and increase the jet diameter of a droplet sorter to reduce the forces on fragile cells.

V. COMBINING FLOW CYTOMETRIC SORTING WITH OTHER SEPARATION METHODS

Virtually any method of cell separation that maintains cells in single-cell suspensions can be used to enrich samples in a particular subpopulation before flow sorting. Examples include various types of centrifugal separation, immunomagetic separation, immunoabsorption, and selective lysis of unwanted cells. Since most other methods allow the sample to be efficiently processed in bulk, flow sorting is best used as a last step. A recent example of the use of immunomagnetic separation before sorting is described by Olweus (28). $CD34^+$ bone marrow prepartions were made 90% pure with 70–80% recoveries before sorting. It is also sometimes useful to do "two-pass" flow sorting, by first running the sample through at a high rate in enrichment mode and then resorting the enriched sample in a high-purity mode.

VI. SORTING APPLICATIONS

The analytical section of the sorter produces parameters that can be selected to become the basis for sort criteria, allowing selection of a subpopulation from a heterogeneous population for further study. For example, fluorescent-tagged Abs may be used to define lymphocyte subsets that may be isolated for functional characterization, or proliferating cells may be sorted for their position in the cell cycle based on DNA content and then further studied for drug response. There are many different applications. Rather than listing all of them, different types of sorting will be grouped and described with examples.

A. Verification Sorting

Often, the investigator will want to correlate a flow cytometric measurement with a more traditional observation or measurement such as microscopy. Cell type may be verified by morphology or artifacts may be investigated. Usually the sorting of a few cells will provide the means for verification. If the desired cells are not too rare, then a sort mode that emphasizes purity should be chosen.

A special device may be required for verification sorting. A good review of sorting collection techniques may be found in Shapiro's book on flow cytometry (1).

Methods for maintaining morphology for cells sorted directly onto microscope slides have been developed, including the use of poly-L-lysine by

Kanz et al. (29) and albumin or calf serum by Alberti et al. (30) for coating the slides. In the albumin procedure, it was noted that reduction of salt in the sheath fluid is important for preserving the morphology of the cells on the slide. It is also possible to sort a small number of cells into a tube and transfer them for cytospin preparation. Once the cells are on the slide, they may undergo further preparation including additional staining or potentially FISH analysis or autoradiography.

Methods have been developed for sorting and handling a small number of cells. Sorting onto a layer of calf serum over Ficoll Hypaque is described by Patrick and Keller (31). The cells are spun and removed from the interface between the Ficoll and calf serum. They also include the description of a simple device for drawing off the cells as they are sorted.

For concentrating the sample for microscopy or other biochemical analysis, Metezeau et al. (32) describe a method for sorting onto nitrocellulose filters with an applied vacuum. This keeps the local concentration of sorted material relatively high.

Some progress has been made toward eliminating the sorting step in verification by microscopy. The use of a flow cytometer and an imaging device in combination has been described (33,34). A charge-coupled device (CCD) camera was used to obtain images of selected particles after flow cytometric analysis. Instead of creating a deflection pulse for sorting, a light source was flashed at the appropriate time.

Special collection techniques have been developed for sorting cells for subsequent analysis by electron microscopy. It was found that small numbers of cells could be concentrated or "sandwiched" in an agarose case using microcentrifugation and then processed using conventional techniques for electron microscopy (35).

Other techniques in molecular biology are useful on small numbers of cells including PCR. The DNA from a few cells can be copied for further studies, including identification. Dunne et al. describe a method for collecting sorted cells on nitrocellulose filters that have been saturated with lysing buffer that preferentially immobilizes mRNA for detection and quantitation with specific probes (36).

A recent illustration of verification sorting comes from flow cytometric studies of apoptosis. In these studies various dyes and flow measurements were made that the investigators wanted to correlate with apoptotic cells. Chromatin condensation is characteristic of apoptosis and may be verified by light and electron microscopy (37). Another indicator of apoptosis is the fragmentation of nuclear DNA that results in a characteristic ladder when the DNA is isolated and analyzed by agarose gel electrophoresis (38). Sorting regions were determined for the apoptotic cells and controls. The cells were sorted and then prepared for microscopy or the sorted cells were lysed and the DNA was extracted for electrophoresis.

Omerod et al. (39) correlated apoptosis with propidium iodide staining and H33342 uptake in murine interleukin (IL)-3–dependent hemopoietic cells by sorting and subsequent microscopy and gel electrophoresis. Similar studies were done by Chiu et al. (40) to correlate apoptosis with different regions of the Hoechst 33342 emission spectrum. Schmid et al. (41) used sorting to verify the use of seven-amino actinomycin D for the detection of apoptotic cells. Vitale et al. (42) used sorting to characterize propidium iodide uptake in apoptotic thymocytes. Hardin et al. (43) validated methods for phenotyping lymphocytes undergoing apoptosis.

It is possible also to characterize apoptotic cells by nick translation assays or terminal deoxynucleotidal transferase (TdT). Frey (16) sorted cells to characterize various dyes for their ability to indicate apopotosis. The sorted fractions were processed for the TdT assay and then reanalyzed.

A final example is investigation of possible artifacts such as two cells stuck together. For example, a double positive event for two cell surface markers that does not make biological sense may be two cells attached together. The sorting of the events will allow direct observation and confirmation that the events are doublets or singlets.

B. Bulk Sorting into a Collection Tube

The most common type of cell sorting is bulk sorting into a collection tube. These may be very long sorts for large quantities of cells. Some care must be taken to aid the viability and recovery of sorted cells. Sterility of the sorter and the collection device may be essential for subsequent culturing or use of the sorted cells. Coating the collection tubes with albumin or calf serum will help prevent cells from sticking to the tube. It helps to sort directly into media and to mix the sorted sample. The choice of tube material is important. The charge on the deflected drops may build up on the surface of the tube, and this may result in the repulsion of incoming droplets. This usually does not happen with glass, but it may be a problem with polystyrene collection tubes. A wire appropriately connected to the tube and grounded may help remove charge build-up. If temperature control is necessary, a circulation water bath may be used to cool the cells or a peltier chiller (44) may be used.

The sort mode chosen for this type of sorting will depend on the purity desired. If the cells of interest are numerous enough to sort in a reasonable time in one pass, then a mode that emphasis purity should be used. This mode should allow for the sorting of events where passenger cells within the sort envelope are desired. If the desired population is rare, then a mode should be chosen to enrich for the population of interest. This will be discussed later under rare event sorting.

One of the many applications of bulk sorting into a tube has been the characterization of hematopoiesis and the isolation of progenitor cells. For the mouse, an assay was developed for progenitor cells by Till and McCulloch (45). The mouse was lethally irradiated to kill all of the blood-forming cells and then the bone marrow of a genetically identical donor transplanted. The inoculated cells migrated to the spleen and a colony formed that was believed to arise from a single stem cell. The colonies became visible to the eye and can be counted on the surface of the spleen. The clonigenic assay for progenitor cells was used to evaluate sorted subpopulations of cells for the presence of progenitor cells.

Subpopulations of bone marrow were sorted on the basis of light scatter, lectin binding, H33342 uptake, Rhodamine 123, and immunofluorescence. A review of murine stem cell purification by flow cytometry was written by Visser and DeVries (46).

Other types of bulk sorting include the selection of cells for a specific trait or variant cell types. In the case of hybridoma cells, this may be the selection of cells that secrete high levels of protein (47) or the selection for mutations where the isotype has switched (48). Other applications for strain improvement in microbiology have been investigated as well (49). Often the sorting is done in multiple rounds where the sorted cells may be grown in between sorts and purified further for the desired trait by more sorting.

Recent developments in molecular biology have led to the need to select cells that carry an introduced gene. To enhance the detection of cells with the gene of interest, a construct is made where the gene of interest is paired with another gene to mark the cell for flow detection. Genes for cell-surface Ags expression have been introduced and can be detected using mAbs (50). Another improved method for enhancing cell detection has been the introduction of genes that, when expressed, allow the production of enzymes that produce a fluorescent product when the appropriate substrate is added. The FACS-GAL method developed by the Herzenberg group is an example of this (51). More recently, the development of green fluorescent protein (GFP) from the jellyfish *Aequorea victoria* as a marker has been developed (52). Constructs with GFP will directly indicate that a cell has been transfected.

C. Single-Cell Deposition

If the desired endpoint of selection is a single cell with a selected trait, then the bulk sorted sample may be cloned by limiting dilution. A more direct method is single-cell deposition (25). Circuitry for sorting a single cell on demand has been developed. Combining single-cell sorting with an apparatus to move a microtiter plate under the sorted cell stream facilitates single-step phenotyping and cloning. Single-cell deposition also may be applied to verification sorting where a slide is moved under the sorted cell stream. This allows the observation of single events rather than a pooled event.

Another way to do single-cell sorting is to randomly deposit single cells into microtiter plate wells and then determine which sites are active or are of interest. One can then go back and look at the analysis data for the events of interest and see if the flow data for those events have common characteristics. This type of sorting has been called index sorting.

The ability to sort a single cell may be extended to the ability to accurately count and deposit cells for assays where the titration of the cells is important. It is possible to deposit any number of cells per well.

D. Large-Particle Sorting

In general, the design of the droplet sorters has emphasized the capability to sort lymphocyte size cells on the order of 10 μm with nozzles having orifice sizes of approximal 70 μm. Jett and Alexander investigated the relationship of particle size and orifice diameter on recovery and recommended using an orifice size approximately four times the particle size (53).

Galbraith and Harkins were motivated to sort plant protoplasts on the order of 50–100 μm and developed sort conditions where 80% recoveries could be obtained with particles that were about one-half the jet diameter (54). Freyer et al. determined conditions for sorting spheroids with diameters ranging from 41 to 96 μm and obtained reasonable recoveries (55).

The desire to sort pancreatic islets led Sweet and McAllister to develop a nozzle holder to sort particles 100–200 μm in diameter. Uniform droplet formation is achieved by applying energy directly to the sheath fluid via a piston rather than shaking the whole nozzle holder. Droplet sorting with orifice size up to 400 μm is possible (56).

The large particle droplet sorter was used at the University of Miami Medical School (57) to evaluate and sort pancreatic islet preparations. A gradient purified human islet prep was sorted into different size islets based on pulse width. In addition, rat islet cells were purified from a crude rat pancreatic digest. Islet volume yield was about 30% with a purity averaging 50%.

The fluid switching sorters have been used for large-particle sorting as well. A Partec sorter was modified by Gray and others to allow the passage of 300-μm particles. Successful sorting of rat islets was performed (19).

There are limitations to large-particle sorting. The optimal drop drive frequency for a 200-μm nozzle was found by Galbraith and Harkins to be about 6 kHz, which limits analysis rates to hundreds of events per second rather than thousands.

Another problem with large-particle sorting is sample particle suspension. Fulwyler used continuous mixing with a magnetic stirring wheel for islet sorting. The use of xanthum gum was proposed by Freyer et al. (58) as a suspension medium for large particles. The sedimentation velocity of 100- to 400-μm particles was nine times slower with 0.1% gum when compared with medium alone. The gum was found to be noncytotoxic and the shear viscosity low when the sample was made to flow.

E. High-Speed Sorting

The basic idea behind high-speed sorting is to increase the drop drive frequency to improve sort yield at higher sample supply rates. This has the additional requirement of reducing the instrument dead time and increasing the stream velocity. Most of the work in this area was done at the Lawrence Livermoore and Los Alamos National Laboratories primarily for the isolation of chromosomes. The drop-drive frequency often used for 70-μm nozzles is 27 kHZ and the stream velocity is 10 m/s. The high-speed sorters are routinely run for chromosome work at 50 m/s and the drop drive is about 220 kHz (59).

The use of sorters has been investigated as a means of purifying human hematopoietic progenitor cells for transplantation. The application of high-speed sorting has been investigated to improve the efficiency. SyStemix Inc. has developed a clinical high-speed sorter (CHSS). Cells have been sorted at 15,000–30,000 cells/s with mean purities of 90%. The drop-drive frequency was 60 kHz and the sheath pressure was reported to be 40 PSI (60–62). With a 70-μm nozzle this produces a stream velocity of about 20 m/s. Yields of 40% at 20,000 events/s have been described.

F. Rare Event Sorting

A major problem with rare event sorting is rare event detection. The use of multicolor staining, instrument cleanliness, and other techniques have allowed the detection of one in a million cells (63). Shapiro has a good discussion on the statistics and problems of rare event detection in his book (1). Once the particles of interest are detected, a sorting strategy can be determined. The yield, purity, amount of sample, abort rate, and time available for sorting become very important. Different approaches may need to be tried, and some care needs to be taken for the collection device or tube. Once the rare event is sorted, it must be retained. Enrichment sorting may help. If the rare event is determined by a unique positive marker, it may be possible to trigger on that parameter and have the sorter ignore all other events (64). The sample input rate may be increased, but the abort rate will stay low. Selecting the enrichment mode for sorting will help insure that the event is sorted, but there will be significant contamination from events that were not detected. If the purity is not adequate after enrichment sorting, the cells may be sorted again, choosing a sort mode to enhance purity. For more information on rare event sorting, see the review by Leary (65).

A current application of rare event sorting is in the detection and isolation of fetal cells in maternal circulation for prenatal genetic diagnosis. The ability to detect and isolate nucleated fetal erythrocytes has been demonstrated (66). It is estimated that there is one fetal cell per 20,000–50,000 maternal erythrocytes. A recent study described the combination of bulk separation methods and flow sorting for isolation of fetal cells (67). Magnetic beads and other bulk methods were evaluated. Normal multiparameter sorting was then used to provide preparations that were approximately 80% pure for FISH analysis.

G. Sex Selection

Another unique application of cell sorting is sex preselection based on separating X and Y chromosome–bearing sperm. The difference in DNA content is used as sort criteria for separating the sperm. Larry Johnson's Gerplasm and Gamete Physiology Lab at the U.S. Department of Agriculture has developed this into a practical technique for in vitro insemination. X and Y sperm populations can be made 85–90% pure. This technique has been applied to swine, rabbits, sheep, and cattle. The throughput is limited to approximately 400,000 sorted sperm cells/h. The technology is also being developed for application to humans for the prevention of sex-linked diseases. The sorting principles for sperm are the same for regular cells, but the signal detection and fluidics are modified. The sample injection tip is beveled to orient the sperm at the laser intersect, and fluorescence is measured at the forward as well as 90° angle. A recent review of the technology was written by Johnson (68).

H. Biohazard Testing for Droplet Sorters

There is a potential for droplet sorters to produce aerosals that may be hazardous with some biological samples. An assay was developed for evaluating aerosal containment by Merril et al. (69). T4 bacteriophage are used as a sample and plates with susceptible bacteria lawns are placed around the instrument to detect phage that may have escaped from the sorter. Plaques will form on the plates, indicating the presence of phages.

VII. CONCLUSIONS

Flow sorting is a unique method allowing separation of cells and other biological particles based on multiple, simultaneous characteristics. While inherently slower than bulk separation processes based on a single characteristic of cells, flow sorting allows separation based on multiple criteria that are difficult or impossible to use by other methods. Flow sorting can also complement other separation techniques, which can be used for initial enrichment of the cells of interest.

REFERENCES

1. Shapiro HM. *Practical Flow Cytometry*, 3rd ed. New York: Wiley-Liss, 1995.
2. Melamed MR, Lindmo T, Mendelsohn ML, eds. *Flow Cytometry and Sorting*, 2nd ed. New York: Wiley-Liss, 1990.
3. Van Dilla MA, Dean PN, Laerum OD, Melamed MR, eds. *Flow Cytometry: Instrumentation and Data Analysis.* Orlando, FL: Academic Press, 1985.
4. Bauer KD, Duque RE, Shankey TY, eds. *Clinical Flow Cytometry. Principles and Applications.* Baltimore, MD: Williams & Wilkins, 1993.
5. Ormerod MG, ed. *Flow Cytometry. A Practical Approach.* Oxford: IRL Press at Oxford University Press, 1990.
6. Given AL. *Flow Cytometry: First Principles.* New York: Wiley-Liss, 1992.
7. Darzynkiewicz Z, Robinson JP, Crissman HA, eds. *Methods in Cell Biology*, Vol. 33, *Flow Cytometry.* San Diego: Academic Press, 1990.

8. Darzynkiewicz Z, Robinson JP, Crissman HA, eds. *Methods in Cell Biology*, Vol. 42, *Flow Cytometry.* San Diego: Academic Press, 1994.
9. Robinson JP, ed. *Handbook of Flow Cytometry Methods.* New York: Wiley-Liss, 1993.
10. Radbruch A, ed. *Flow Cytometry and Cell Sorting.* Berlin: Springer, 1992.
11. Herzenberg LA, Sweet RG, Herzenberg LA. Fluorescence activated cell sorting. *Sci Am* 1976; 234:108.
12. Horan PK, Wheeless LL Jr. Quantitative single cell analysis and sorting. *Science* 1977; 198:149–157.
13. Arndt-Jovin DJ, Jovin TM. Automated cell sorting with flow systems. *Ann Rev Biophys Bioeng* 1978; 7:527–558.
14. Fulwyler MJ. Electronic separation of biological cells by volume. *Science* 1965; 150:910–911.
15. Sorting Application Notes 1-11 for FACSort/FACSCalibur. San Jose, CA: Becton Dickinson Immunocytometry Systems.
16. Frey T. Nucleic acid dyes for detection of apoptosis in live cells. *Cytometry* 1995; 21:265–274.
17. Duhnen J, Stegemann J, Wiezorek C, Mertens H. A new fluid switching flow sorter. *Histochemistry* 1983; 77:117–121.
18. Göhde H, Schumann J. Method and apparatus for sorting particles. US Patent 4,756,427 (1988).
19. Gray DWR, Gohde W, Carter N, Heiden T, Morris PJ. Separation of pancreatic islets by fluorescence-activated cell sorting. *Diabetes* 1989; 38(suppl 1):133–135.
20. Pfosser M, Amon A, Lelley T, Heberle-Bors E. Evaluation of sensitivity of flow cytometry in detecting aneuploidy in wheat using disomic and ditelosomic wheat-rye addition lines. *Cytometry* 1995; 21:387–393.
21. Herweijer H, Stokdijk W, Visser JW. High speed photodamage cell selection using bromodeoxyuridine/Hoechst 33342 photosensitized cell killing. *Cytometry* 1988; 9:143–149.
22. Martin JC, Jett JH. Photodamage, a basis for super high speed cell selection. *Cytometry* 1981; 2:114 (abstract).
23. Keij JF, Groenwegen AC, Dubelaar GB, Visser JW. High-speed photodamage cell selection using a frequency-doubled argon ion laser. *Cytometry* 1995; 19:209–216.
24. Bakker Schut TC, de Grooth BB, Greve J. A new principle of cell sorting by using selective electroporation in a modified flow cytometer. *Cytometry* 1990; 11:659–666.
25. Stovel RT, Sweet RG. Individual cell sorting. *J Histochem Cytochem* 1979; 27:284–288.
26. Lindmo T, Peters DC, Sweet R. Flow sorters for biological cells. In: Melamed MR, Lindmo T, Mendelsohn ML, eds. *Flow Cytometry and Sorting*, 2nd ed. New York: Wiley-Liss, 1990:153–161.
27. Pinkel D, Stovel R. Flow chambers and sample handling. In: Van Dilla MA, Dean

PN, Laerum OD, Melamed MR, eds. *Flow Cytometry: Instrumentation and Data Analysis*. Orlando, FL: Academic Press, 1985:108–114.

28. Olweus J, Terstappen LW, Thompson PA, Lund-Johansen F. Expression and function of receptors for stem cell factor and erythropoietin during lineage commitment of human hematopoietic progenitor cells. *Blood* 1996; 88:1594–1607.
29. Kanz L, Bross KJ, Mielke R, Lohr GW, Fauser AA. Fluorescence-activated sorting of individual cells onto poly-l-lysine-coated slide areas. *Cytometry* 1986; 7:491–494.
30. Alberti S, Stovel R, Herzenberg LA. Preservation of cells sorted individually onto microscope slides with a fluorescence-activated cell sorter. *Cytometry* 1984; 5:644–647.
31. Patrick CW, Keller RH. A simple device for the collection of cells sorted by flow cytometry. *Cytometry* 1984; 5:308–311.
32. Metezeau P, Berheim A, Berger R, Goldberg ME. A simple device to obtain high local concentrations of material sorted by flow cytometry for biochemical or morphological analysis. *Cytometry* 1984; 5:550–552.
33. Huller R, Glossner E, Schaub S, Weingartner J, Kachel V. The macro flow planktometer: a new device for volume and fluorescence analysis of macro plankton including triggered video imaging in flow. *Cytometry* 1994; 17:109–118.
34. Kubota F, Kusuzawa H, Kosaka T, Nakamoto H. Flow cytometer and imaging device used in combination. *Cytometry* 1995; 21:129–132.
35. Sebring RJ, Johnson NF, Spall WD. Transmission electron microscopy of small numbers of sorted cells. *Cytometry* 1988; 9:88–92.
36. Dunne JF, Thomas J, Lee S. Detection of mRNA in flow-sorted cells. *Cytometry* 1989; 10:199–204.
37. Wylie AH, Morris RE, Smith AL, Dunlop D, Currie AR. Chromatin cleavage in apoptosis: association with condensed chromatin morphology and dependence on macromolecular synthesis. *J Pathol* 1984; 142:67–77.
38. Wyllie AH. Glucocorticoid-induced thymocyte apoptosis is associated with endogenous endonuclease activation. *Nature* 1980; 284:555–556.
39. Omerod MG, Collins MK, Rodriguez-Tarduchy G, Robertson D. Apoptosis in interleukin-3-dependent haemopoietic cells: quantification by to cytometric methods. *J Immunol Methods* 1992; 153:57–65.
40. Chiu L, Cherwinski H, Ransom J, Dunne JF. Flow cytometric ratio analysis of the Hoechst 33342 emission spectrum: multiparametric characterization of apoptotic lymphocytes. *J Immunol Methods* 1996; 189:157–171.
41. Schmid I, Uittenbogaart CH, Keld B, Giorgi JV. A rapid method for measuring apoptosis and dual-color immunofluorescence by single laser flow cytometry. *J Immunol Methods* 1994; 170:145–157.
42. Vitale M, Zamai L, Mazzotti G, Cataldi A, Falcieri E. Differential kinetics of propidium iodide uptake in apoptotic and necrotic thymocytes. *Histochemistry* 1993; 100:223–229.

43. Hardin JA, Sherr DH, Demaria M, Lopez PA. A simple fluorescence method for surface antigen phenotyping of lymphocytes undergoing DNA fragmentation. *J Immunol Methods* 1992; 154:99–107.
44. Gottlinger C, Meyer KL, Weichel W, Muller W, Raftery B, Radbruch A. Cell-cooling in flow cytometry by Peltier elements. *Cytometry* 1986; 7:295–297.
45. Till JE, McCulloch EA. A direct measurement of the radiation sensitivity of normal mouse bone marrow cells. *Radiat Res* 1961; 14:213–219.
46. Visser JW, De Vries P. Identification and purification of murine hematopoietic stem cells by flow cytometry. In: Darzynkiewicz Z, Crissman HA, ed. *Methods in Cell Biology.* Vol. 33. *Flow Cytometry.* San Diego, CA: Academic Press, 1990:451–468.
47. Marder P, Maciak RS, Fouts RL, Baker RS, Starling JJ. Selective cloning of hybridoma cells for enhanced immunoglobulin production using flow cytometric sorting and automated laser nephelometry. *Cytometry* 1990; 11:498–505.
48. Kipps TJ, Herzenberg LA. Hybridoma immunoglobulin isotype switch variant selection using the fluorescence activated cell sorter. In: Weir DM, Blackwell CC, Herzenberg LA, eds. *Handbook of Experimental Immunology.* Edinburgh: Blackwell Publications, 1986:109.1–109.8.
49. Betz JW, Aretz W, Hartel W. Use of flow cytometry in industrial microbiology for strain improvement programs. *Cytometry* 1984; 5:145–150.
50. Kavathas P, Herzenberg LA. Stable transformation of mouse L cells for human membrane T-cell differentiation antigens, HLA and B2-microglobulin: selection by fluorescence activated cell sorting. *Proc Natl Acad Sci USA* 1983; 80:524–528.
51. Nolan GP, Fiering S, Nicolas JF, Herzenberg LA. Fluorescence activated cell analysis and sorting of viable mammalian cells based on B-D-galactosidase activity after transduction of Escherichia coli lacZ. *Proc Natl Acad Sci USA* 1988; 85:2603–2607.
52. Chalfie M, Tu Y, Euskirchen G, Ward WW, Prasher DC. Green fluorescent protein as a marker for gene expression. *Science* 1994; 263:802–805.
53. Jett JH, Alexander RG. Droplet sorting of large particles. *Cytometry* 1985; 6:484–486.
54. Harkins KR, Galbraith DW. Factors governing the flow cytometric analysis and sorting of large biological particles. *Cytometry* 1987; 8:60–70.
55. Freyer JP, Wilder ME, Jett JH. Viable sorting of intact multicellular spheroids by flow cytometry. *Cytometry* 1987; 8:427–436.
56. Schultz A, Houck D, McAllister W, Sweet R. Cell sorting: yields are improved using higher drop drive frequencies provided by a new method for drop generation. *Cytometry* 1991; 5(suppl):116.
57. Brendel MD, Houck D, Alejandro R, Mintz DH, Ricordi C. Monoclonal antibody mediated fluorescence-activated macrosorting of islets of Langerhans. Proceedings of the Artificial Insulin Delivery Systems and Islet and Pancreas Transplantation, Igls, Austria, 1995.

58. Freyer JP, Fillak D, Jett JH. Use of xanthum gum to suspend large particles during flow cytometric analysis and sorting. *Cytometry* 1989; 10:803–806.
59. Lindmo T, Peters DC, Sweet R. Flow sorters for biological cells. In: Melamed MR, Lindmo T, Mendelsohn ML, eds. *Flow Cytometry and Sorting*, 2nd ed. New York: Wiley-Liss, 1990:162–170.
60. Tichenor EH, Lopez F, Combs J, et al. Isolation and characterization of human hematopoietic stem cells using high speed flow cytometry. *Cytometry* 1994; 7(suppl):58.
61. Sasaki DT, Tichenor EH, Lopez F, et al. Development of a clinically applicable high-speed flow cytometer for the isolation of transplantable human hematopoietic stem cells. *J Hematother* 1995; 4:503.
62. Tsukamoto A, Sasaki D, Chen B, Hoffman R. Characterization and isolation of mobilized peripheral blood stem cells using a high speed cell sorter. In: Morstyn G, Sheridan WP, eds. *Cell Therapy.* Cambridge: Cambridge University Press, 1996:183–198.
63. Gross HJ, Verwer B, Houck D, Recktenwald D. Detection of rare cells at a frequency of one per million by flow cytometry. *Cytometry* 1993; 14:519–526.
64. McCoy JP, Chambers WH, Lakomy R, Campbell JA, Stewart CC. Sorting minor subpopulations of cells: use of fluorescence as the triggering signal. *Cytometry* 1991; 12:268–274.
65. Leary JF. Strategies for rare cell detection and isolation. In: Darzynkiewicz Z, Robinson JP, Crissman HA, eds. *Methods in Cell Biology.* Vol. 42. *Flow Cytometry.* San Diego: Academic Press, 1994:331–358.
66. Bianchi DW, Stewart JE, Garber MF, Lucotte G, Flint AF. Possible effect of gestational age on the detection of fetal nucleated erythrocytes in maternal blood. *Prenat Diag* 1991; 11:523–528.
67. Lewis D, Schober W, Murrell S, et al. Rare event selection of fetal nucleated erythrocytes in maternal blood by flow cytometry. *Cytometry* 1996; 23:218–227.
68. Johnson LA. Sex preselection by flow cytometric separation of X and Y chromosome-bearing sperm based on DNA difference: a review. *Reprod Fertil Dev* 1995; 7:893–903.
69. Merril JT. Evaluation of selected aerosol-control measures on flow sorters. *Cytometry* 1981; 1:342–345.

12

Employing Surface Markers for the Selection of Transfected Cells

Gregor Siebenkotten, Ute Behrens-Jung, and Stefan Miltenyi
Miltenyi Biotec GmbH, Bergisch Gladbach, Germany

Katja Petry
Institut für Genetik, Universität zu Köln, Köln, Germany

Andreas Radbruch
Deutsches Rheuma-Forschungszentrum, Berlin, Germany

I. INTRODUCTION

Transfection of DNA into eukaryotic cells is widely used to functionally analyze DNA, RNA, and proteins in a natural environment. For stable transfections, those rare cells, in which integration of a vector has led to its permanent persistence, usually are selected with a drug to which a vector-encoded protein confers resistance. In case of transient transfections, the time span of expression of a drug-resistance protein from the nonintegrated vector is too short for a selection. Thus, these cells are usually analyzed in bulk, which is often hampered by a huge background of untransfected cells. Now, the selection of transient and even stable transfectants using a surface marker has become practical, because of the development of rapid sorting of large numbers of cells without affecting their viability. In this chapter, we summarize experiments done with selectable surface markers and show that selection of transfectants by cell sorting strongly facilitates the analysis of transient transfectants. For stable transfection, this method clearly has certain advantages over the use of drug selection.

II. TRANSIENT TRANSFECTION

A. Enrichment and Analysis of Low-Frequency Transient Transfectants

Eukaryotic cells are transiently transfected to analyze promoters, enhancers, or other regulatory sequence elements of gene expression; to confirm the identity of a cloned gene; or to study functional regions of a protein by mutational analysis. After transfection, cells are analyzed by microscopy or by flow cytometry, or cell lysates are prepared to measure the activity of reporter enzymes with radioactive or fluorochrome substrates. Finally, RNA or protein can be specifically isolated for analysis.

These experiments often are influenced by an excess of untransfected cells. Drug selection of transient transfectants is not applicable because the period of expression of the transgenes is too short. It often is not reasonable to generate stable transfectants instead, because drug selection to obtain stable transfectants is time-consuming and expensive. In addition, the analysis of single-cell clones after selection bears the risk that the expression rate of the integrated constructs may be influenced by positional effects or that the drug treatment or mutations caused by the integration may change the characteristics of the cell. If the DNA of interest is transiently transfected along with DNA of a surface protein, this marker can be used for selection. The surface marker can be either encoded on one DNA fragment together with the gene of interest, or expressed from a cotransfected DNA. A cotransfected surface marker can be used for the selection because most of the cells that take up DNA incorporate both DNAs. Figure 1 shows the cotransfection rates of two expression vectors transfected with different ratios.

In experiments in which flow cytometry is used for the analysis of transfectants, intracellular markers such as the green fluorescent protein from the jellyfish *Aequorea victoria* (1) or the *Escherichia coli* β-galactosidase (2) can be employed for detection, and the cells do not necessarily have to be physically selected. A surface protein (human CD25) has been used as a transfection marker for human T- and B-lymphocyte lines in experiments in which a cotransfected second reporter gene (murine CD8α) was employed to perform single-cell promoter/enhancer analyses (3).

Physical isolation of transiently transfected cells is practical, especially with methods [e.g., panning or magnetic cell sorting (MACS)] that allow handling of larger cell numbers at one time than does flow cytometry. Padmanabhan and coworkers (4,5) used a viral glycoprotein (from vesicular

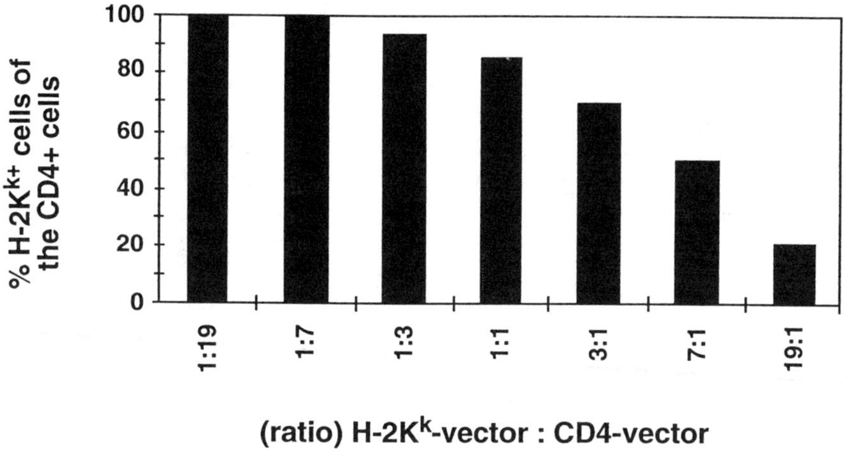

Figure 1 Efficiencies of transient cotransfections. Murine pre–B cells (1881A20 cell line) were electroporated with expression vectors at the indicated ratios. The vectors encoded murine major histocompatibility class (MHC) class I heavy chain (H-$2K^k$) and human CD4. After 17 hours, cells were stained with anti-CD4-phycoerythrin (PE) and anti–H-$2K^k$-fluorescein isothiocyanate (FITC) antibodies (Abs) and the percentage of H-$2K^{k+}$ cells among the $CD4^+$ cells was determined by fluorescence-activated cell analysis (FACScan) analysis.

stomatitis virus), the interleukin (IL)-2 receptor α-chain, or the multidrug resistance (*mdr*) gene as selectable surface markers and could enrich the cells expressing the cotransfected gene of interest to more than 90% purity in different human, monkey, and mouse cell lines by immunomagnetic sorting for the markers.

Selection of transiently transfected cells using surface markers allows analysis of cells that could hardly be used for transfection until now because transfection efficiencies were too low, like many primary cells. To test the advantage of immuno magnetic selection for transfection analysis of primary cells, Pilon and coworkers (6) transfected activated primary human B cells with a CAT-reporter vector along with a vector encoding CD2 as a selectable surface marker. After magnetic cell sorting, they detected 20 times more CAT activity in the positive than in the negative fraction.

We analyzed immunoglobulin class switch recombination in primary murine splenic B cells by cotransfection of a recombination substrate and a selectable surface marker. Expression of the murine major histocompatibility (MHC) class I heavy chain H-$2K^k$ on H-$2K^d$ background was taken as an indicator of transfection, while the substrate recombination was monitored by expression of human CD4, which appeared on the cell surface upon recombination. Two fragments of genomic CD4 on the substrate, interrupted by a transcriptional termination sequence, were reunited to one transcriptional unit by recombination. The H-2K^k–expressing cells were isolated by MACS and were monitored by a fluorescence-activated cell analysis (FACScan) scan analysis (K. Petry et al., in preparation). Polyclonally activated murine primary B cells were transfected by electroporation with an equimolar mixture of the recombination substrate vector, the H-2K^k–selection vector, and a third expression vector, encoding the polyomavirus large T antigen to facilitate replication of the substrate vector, bearing a polyoma origin of replication. One to three percent of the B cells could be transfected. By 15- to 50-fold enrichment of H-2K^k–expressing cells with magnetic cell sorting (MACS) and gating on the H-2K^{k+} cells in the FACScan analysis, recombination of $<4\%$ of the transfected B cells could be clearly determined (Fig. 2). Cotransfection of the H-2K^k selection vector and a CD4 expression vector revealed a cotransfection rate of approximately 80–90%. Analysis of additional cell surface markers

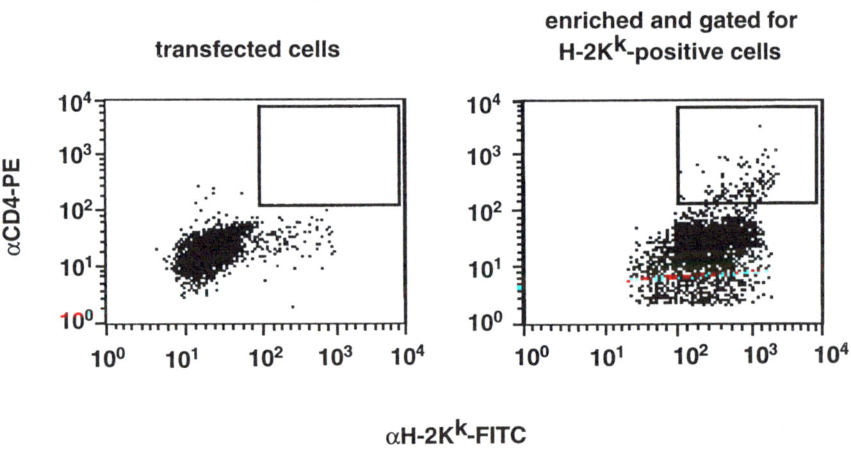

Figure 2 Detection of recombination of a substrate vector in transiently transfected primary cells. Polyclonally activated murine primary B lymphocytes were transfected with equimolar amounts of a recombination substrate (encoding CD4 after recombination), a selection vector (encoding H-2K^k), and a vector facilitating replication of the substrate vector. One day after transfection, H-2K^{k+} cells were enriched using magnetic cell sorting (MACS) and analyzed by fluorescence-activated cell analysis FACScan. By gating, only activated blastoid cells were analyzed and propidium iodide (PI)–stained dead cells were excluded. After magnetic enrichment and gating on H-2K^{k+} cells, an analyzable number of CD4$^+$ cells, harboring a recombined substrate vector, could be obtained.

(B220, IgM, and κ light chain) verified that the transfectants were B lymphocytes. Thus, using one marker as an internal standard for transfection, additional markers can serve as reporters for the activity of regulatory elements, transcription factors, recombination enzymes, etc. By selection of the transfected cells, the sensitivity of the detection of the reporter protein is strongly increased, which is especially important for experiments with cells that show low transfection efficiencies. Multiparameter analyses of living cells can be done on the single-cell level.

Surface marker expression can be detected very early. We transfected the murine pre–B-cell line 1881A20 with a CD4 expression vector and could clearly stain 25% of the CD4-expressing cells 2 hours after transfection, while 60% could be stained after 4 hours (data not shown). These cells divide approximately once every 12 hours. In a cell line with a doubling time of about 24 hours [e.g., Chinese hamster ovary (CHO) cells], 50% of the cells that had taken up DNA could be detected 6 hours after transfection. The selection of cells at such an early time point permits the analysis of events happening shortly after onset of expression of a given gene (e.g., to mimic a given signal) or of genes that cause growth inhibition or lead to cell death. Apart from that, transfection and analyses of the cells can be done in a single day.

B. Expression Cloning

Expression cloning, the isolation of genes using their expression in eukaryotic cells, is probably the most widely distributed application of selection of transfectants by surface markers. Complete cDNA libraries or aliquots of sheared genomic DNA are transfected, and each cell takes up only a few cDNA expression vectors or pieces of genomic DNA. Those rare cells that express the surface protein of interest upon transfection are isolated to recover the transfected DNA encoding the protein. In the first attempts, FACS or rosetting was applied to select cells expressing specific markers from cells stably transfected with fragments of total genomic DNA. These cells then were used as a source of target cDNA for a subtractive cDNA hybridization (7–10). Since 1987, direct cloning of a cDNA by transient expression in mammalian cells in combination with physical isolation of the positive cells by panning (11) or FACS (12) has become a more widely spread technique. Immunomagnetic selection instead of panning has been used for expression cloning (13–15). Since rare cells can be isolated to high purity from large cell numbers in a short time by magnetic cell sorting, this method has great advantages for expression cloning.

II. STABLE TRANSFECTION

A. Selection of Stable Transfectants in Cell Lines

Transfected DNA can either remain transiently in the cell or can be integrated into the host genome. Approximately 0.1–1% of the cells that take up DNA upon transfection will integrate it. These rare stably transfected cells must be isolated, which is usually done by drug selection, a time-consuming and expensive process. Apart from immunomagnetic selection in combination with (in support of) drug selection (16–19), isolation of stable transfectants solely by surface markers has been described. Stable supertransfection of cells already resistant to a drug was achieved by direct selection for a protein expressed on the cell surface (20). This procedure was described in more detail by Kappes et al. (21). After transfection of the two subunits of an MHC class II molecule, one step of magnetic immunoselection 2 days after transfection followed by three to four more steps in the following weeks led to an enrichment of stable transfectants to > 95%.

By using two selection steps, the first 1 day after transfection and the second 1 week later, we have selected stable transfectants faster and presumably at lower cost than by drug selection. A total of 5×10^7 cells were transfected with 50 μg of linearized vector carrying the murine MHC class I gene H-2K^k. Cells expressing H-2K^k were enriched by MACS after 20 hours. This selection should cover all cells that took up DNA, regardless whether or not the DNA is integrated into the genome. In the rapidly dividing cell line used (the murine pre–B-cell line 1881A20), transient expression of H-2K^k terminates after 3 days. Six days after transfection, 2×10^9 cells of the enriched population were magnetically enriched for H-2K^k a second time by passing them twice through MACS. The approximately 2% of the cells that still expressed H-2K^k could be enriched to 98% purity. This expression was stable in nearly all the cells; after one more week 96% of the cells could still be stained for H-2K^k (Fig. 3).

Another example of using a transfected surface marker for the selection of rare cells is the screening for rare recombination events. We stably transfected the murine pre–B cell line 1881A20 with the class switch recombination substrate described above. After stimulation with bacterial lipopolysaccharides (LPS), approximately 0.1% of the cells began to express human CD4. By two steps of MACS, we enriched these cells to 97% homogeneity (Fig. 4). We could then show by Southern blot analysis that all CD4$^+$ cells had recombined their substrate.

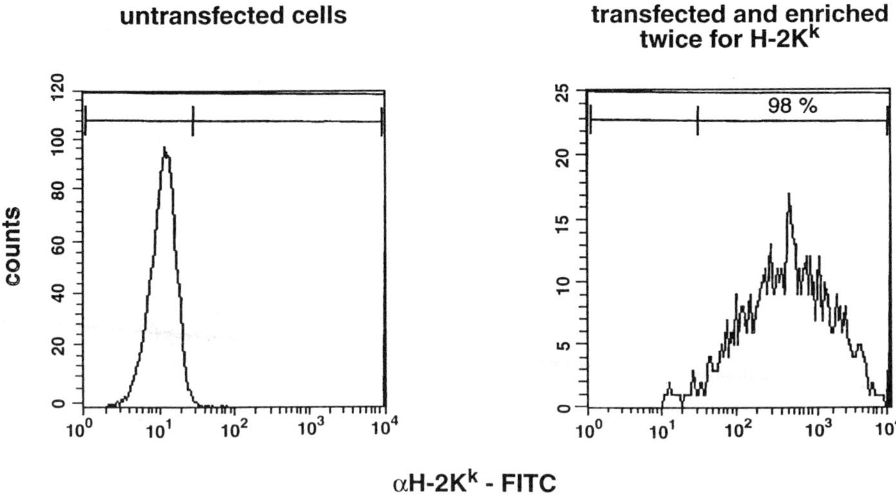

Figure 3 Selection of stable transfectants using a surface marker. Murine pre–B cells (1881A20 cell line) were transfected with an H-$2K^k$ expression vector. H-$2K^k$-expressing cells were selected by magnetic cell sorting (MACS) 1 day after transfection and a second time 1 week later, when only stable transfectants still expressed H-$2K^k$. The cells were stained with anti–H-$2K^k$-fluorescein isothiocyanate (FITC) antibodies (Abs) before transfection and after the second step of selection and analyzed by fluorescence-activated cell analysis (FACScan). Dead cells were excluded by PI staining and gating.

B. Selection of Stably Transfected Primary Cells for Gene Therapy

One further possible application for immunoselection of transfected cells is the isolation of surface reporter-positive cells after gene transfer for ex vivo gene therapy. Selection can take place as soon as nonintegrated vector DNA has vanished from the cells, which usually is a few days after transfection.

After retroviral transduction, cells with an integrated vector usually are selected by treatment with drugs against which the vector encodes a protein conferring resistance. The drug selection strongly prolongs the time the cells have to be handled ex vivo, and an influence of the drugs or the dying cells on the differentiation state of the resistant cells cannot be excluded. If the patient's immune system is not suppressed in the course of a gene therapy,

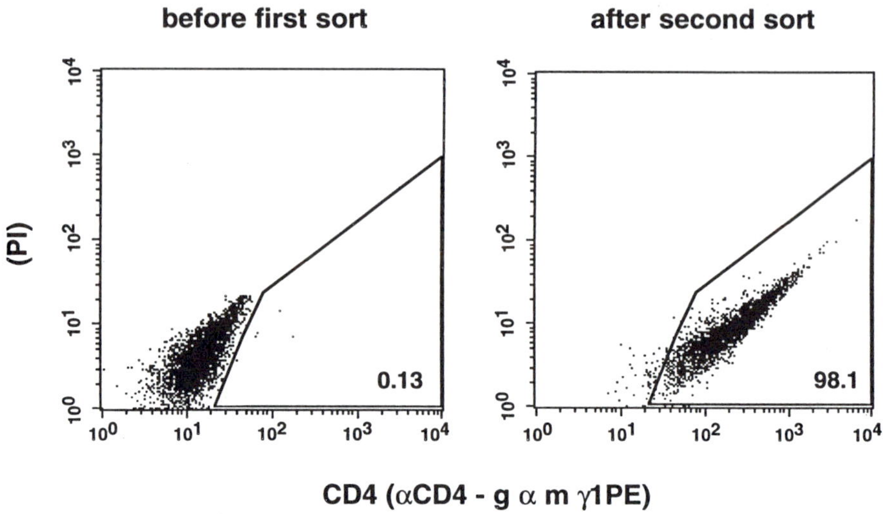

Figure 4 Detection of rare recombination events by using CD4 as a surface-selectable marker. Murine pre–B cells (1881A20 cell line) were stably transfected with a recombination substrate vector; a weak recombination activity was induced. Cells expressing CD4 upon recombination were isolated in two steps of magnetic cell sorting (MACS) and analyzed by fluorescence-activated cell analysis (FACScan). Dead cells are excluded by PI staining.

cells might soon be deleted from the body by immunoreactions caused by peptides derived from the foreign drug resistance protein and presented by MHC class I molecules. Recently, it was shown that cells expressing a viral *tk* gene in a gene therapeutic approach were rejected by $CD8^+$ host T cells specific for peptides derived from the *tk* product (22). If (retro)viral transduction procedures are rendered efficient enough to make selection unnecessary, the biological risk of effects of infectious viral particles is usually strongly increased.

If a surface selection marker is used in gene therapy, it would stay on the surface of the cells after their transfer into the body. Thus, it should be biologically completely nonfunctional, that is, it should not transduce any signal into the cell when cross-linked, e.g., by antibodies used for the selection; it should not lead to adhesion of the cell; it should not bind a soluble ligand and thus possibly influence its concentration in the body; and it should not interfere with the activity of other molecules on the cell surface by lateral interaction. Thus, such a marker must most likely be mutated or truncated. In addition, the marker must not be immunogenic or the recipient has to be tolerant.

Several different surface markers have been used for the detection and selection of primary cells transduced with a retroviral vector in the course of the development of gene therapy protocols. However, until now none of these proteins completely fulfilled the requirements for a marker that is neutral in gene therapy.

A mutated form of the endogenous murine prion protein (mPrP) or murine Thy-1 have been suggested as selectable surface markers for gene therapy (23,24). Human CD24 was used for the selection of retrovirally transduced murine bone marrow cells by FACS, which then successfully reconstituted lethally irradiated mice (25). Later the same group used murine CD24 [heat-stable antigen (Ag)] as a marker for flow cytometric selection of retrovirally transduced human CD34 cells (26). CD24 is a glycosyl phosphatidylinositol–anchored protein of only 31–35 amino acids. The most advanced experiments have been done with the human low-affinity nerve growth-factor receptor (LNGFR). Human peripheral blood mononuclear cells were infected by cocultivation with irradiated virus-producer cells, which resulted in a rate of 15% cells expressing virus-encoded LNGFR. By magnetic immunoselection, this population was enriched to > 98% purity (27). A mutated, nonfunctional form was used as a selectable marker of retrovirally transfected peripheral blood lymphocytes of six patients which were then reinfused into the patients (28). However, the result of this clinical trial has not been published to date.

Highly efficient transduction procedures such as cocultivation with cell lines producing the retroviral transfer vehicle are considered unsafe for gene therapy. Safer strategies such as infection with virus-containing supernatants often lead to transduction of < 1% of the cells.

The concept of magnetic selection of transfected cells using surface markers should be advantageous for retroviral-transfer approaches. In addition, this fast and gentle selection procedure might allow even less risky and less efficient gene-transfer approaches than transduction with supernatants, provided that there are enough cells to transfect.

ACKNOWLEDGMENT

We thank Andreas Dzionek for excellent technical help.

REFERENCES

1. Chalfie M, Tu Y, Euskirchen G, Ward WW, Prasher DC. Green fluorescent protein as a marker for gene expression. *Science* 1994; 263:802–805.
2. Nolan GP, Fiering S, Nicolas J-F, Herzenberg LA. Fluorescence-activated cell analysis and sorting of viable mammalian cells based on beta-D-galactosidase activity after transduction of *Escherichia coli lacZ. Proc Natl Acad Sci USA* 1988; 85:2603–2607.
3. Mazda O, Teshigawara K, Fujimoto S, Hattori N, Dou YM, Katsura Y. A reporter system using a flow cytometer to detect promoter/enhancer activity in lymphoid cell lines. *J Immunol Methods* 1994; 169:53–61.
4. Padmanabhan R, Corsico CD, Howard TH, et al. Purification of transiently transfected cells by magnetic affinity cell sorting. *Anal Biochem* 1988; 170:341–348.
5. Padmanabhan R, Howard T, Gottesman MM, Howard BH. Magnetic affinity cell sorting to isolate transiently transfected cells, multidrug-resistant cells, somatic cell hybrids, and virally infected cells. *Methods Enzymol* 1993; 218:637–651.
6. Pilon M, Gullberg M, Lundgren E. Transient expression of the CD2 cell surface antigen as a sortable marker to monitor high frequency transfection of human primary B cells. *J Immunol* 1991; 146:1047–1051.
7. Kavathas P, Herzenberg LA. Stable transformation of mouse L cells for human membrane T-cell differentiation antigens, HLA and beta 2-microglobulin: selection by fluorescence-activated cell sorting. *Proc Natl Acad Sci USA* 1983; 80:524–528.
8. Kavathas P, Sukhatme VP, Herzenberg LA, Parnes JR. Isolation of the gene encoding the human T-lymphocyte differentiation antigen Leu-2 (T8) by gene transfer and cDNA subtraction. *Proc Natl Acad Sci USA* 1984; 81:7688–7692.
9. Littman DR, Thomas Y, Maddon PJ, Chess L, Axel R. The isolation and sequence of the gene encoding T8: a molecule defining functional classes of T lymphocytes. *Cell* 1985; 40:237–246.
10. Maddon PJ, Littman DR, Godfrey M, Maddon DE, Chess L, Axel R. The isolation and nucleotide sequence of a cDNA encoding the T cell surface protein T4: a new member of the immunglobulin gene family. *Cell* 1985; 42:93–104.
11. Seed B, Aruffo A. Molecular cloning of the CD2 antigen, the T-cell erythrocyte receptor, by a rapid immunoselection procedure. *Proc Natl Acad Sci USA* 1987; 84:3365–3369.
12. Yamasaki K, Taga T, Hirata Y, et al. Cloning and expression of the human interleukin-6 (BSF-2/IFN beta 2) receptor. *Science* 1988; 241:825–828.
13. Mallett S, Fossum S, Barclay AN. Characterization of the MRC OX40 antigen of activated CD4 positive T lymphocytes-a molecule related to nerve growth factor receptor. *EMBO J* 1990; 9:1063–1068.
14. Hamann J, Fiebig H, Strauss M. Expression cloning of the early activation antigen

CD69, a type II integral membrane protein with a C-type lectin domain. *J Immunol* 1993; 150:4920–4927.

15. Zannettino AC, Rayner JR, Ashman LK, Gonda TJ, Simmons PJ. A powerful new technique for isolating genes encoding cell surface antigens using retroviral expression cloning. *J Immunol* 1996; 156:611–620.
16. Kieffer N, Fitzgerald LA, Wolf D, Cheresh DA, Phillips DR. Adhesive properties of the beta 3 integrins: comparison of GP IIb-IIIa and the vitronectin receptor individually expressed in human melanoma cells. *J Cell Biol* 1991; 113:451–461.
17. Deckert M, Kubar J, Bernard A. CD58 and CD59 molecules exhibit potentializing effects in T cell adhesion and activation. *J Immunol* 1992; 148:672–677.
18. König R, Huang L-Y, Germain RN. MHC class II interaction with CD4 mediated by a region analogous to the MHC class I binding site for CD8. *Nature* 1992; 356:796–798.
19. Roche PA, Teletski CL, Karp DR, Pinet V, Bakke O, Long EO. Stable surface expression of invariant chain prevents peptide presentation by HLA-DR. *EMBO J* 1992; 11:2841–2847.
20. Bonneville M, Ito K, Krecko EG, et al. Recognition of a self major histocompatibility complex TL region product by gamma delta T-cell receptors. *Proc Natl Acad Sci USA* 1989; 86:5928–5932.
21. Kappes DJ, Strominger JL, Monos DS. A novel method for generating stable high-level tranfectants involving direct immunomagnetic selection for cell-surface epitopes: expression of HLA class-II genes in HeLa cells. *Gene* 1991; 108:245–252.
22. Riddell SR, Elliott M, Lewinsohn DA, et al. T-cell mediated rejection of gene-modified HIV-specific cytotoxic T lymphocytes in HIV-infected patients. *Nat Med* 1996; 2:216–223.
23. Tumas DB, Spangrude GJ, Brooks DM, Williams CD, Chesebro B. High-frequency cell surface expression of a foreign protein in murine hematopoietic stem cells using a new retroviral vector. *Blood* 1996; 87:509–517.
24. Planelles V, Haislip A, Withers-Ward ES, et al. A new reporter system for detection of retroviral infection. *Gene Ther* 1995; 2:369–376.
25. Pawliuk R, Kay R, Lansdorp P, Humphries RK. Selection of retrovirally transduced hematopoietic cells using CD24 as a marker of gene transfer. *Blood* 1994; 84:2868–2877.
26. Conneally E, Bardy P, Eaves CJ, et al. Rapid and efficient selection of human hematopoietic cells expressing murine heat-stable antigen as an indicator of retroviral-mediated gene transfer. *Blood* 1996; 87:456–464.
27. Mavilio F, Ferrari G, Rossini S, et al. Peripheral blood lymphocytes as target cells of retroviral vector-mediated gene transfer. *Blood* 1994; 83:1988–1997.
28. Bonini C, Verzeletti S, Servida P, et al. Transfer of the HSV-TK gene into donor peripheral blood lymphocytes for in vivo immunomodulation of donor antitumor immunity after allo-BMT. *Blood* 1994; 84:110a (abstract 427).

13

CD34+ Cell Sorting and Enrichment: Applications in Blood Banking and Transplantation

Hassan T. Hassan, K. Gutensohn, A. R. Zander, and P. Kuehnl
Hamburg University Hospital Eppendorf, Hamburg, Germany

I. HUMAN HEMATOPOIETIC CD34+ CELLS

The human CD34 cell-surface antigen (Ag) is a highly glycosylated 115 kDa type I integral membrane protein expressed in only 1–4% of normal adult bone marrow mononuclear cells (1). All hematopoietic progenitor cells including the early multipotential long-term culture initiating cells (LTC-IC), blast colony-forming cell (Bl-CFC), mixed colony-forming cell (MIX-CFC), granulocyte-macrophage colony-forming cell (GM-CFC), erythroid blast-forming cell (E-BFC), and megakaryocytic colony-forming cell (Meg-CFC) are CD34+ (2). As the hematopoietic progenitor cells differentiate and become committed to particular lineages, CD34 expression is progressively reduced and all precursor and mature hematopoietic cells are CD34− (3).

Seven anti-CD34 monoclonal antibodies (mAbs) [My10 (HPCA-1), BI.3C5, 12.8, p115.2, ICH-3, TüK3, and QBEND10] have been divided according to the binding epitopes into three classes (4): Class I Abs (My1O, BI.3C5, 12.8, and ICH-3) directed against epitopes that are sensitive to neuraminidase, chymopapain, and glycoprotease; Class II Abs (QBEND10) directed against epitopes that are sensitive to chymopapain and glycoprotease, but not neuraminidase; and Class III Abs (p115.2 and TüK3) directed against epitopes that are insensitive to all these enzymes.

The human CD34 gene has been localized to band 32 on the long arm of chromosome 1, a region containing a cluster of genes encoding hematopoietic regulatory and signaling molecules and eight exons spanning 26 kb of DNA (5). The CD34 protein is a member of the sialomucin family of surface molecules and shows no homology to other proteins currently entered into the database (6).

II. SUBSETS OF HUMAN HEMATOPOIETIC CD34$^+$ CELLS

The small population of human hematopoietic CD34$^+$ cells represents, both in molecular and functional terms, a heterogeneous collection of of stem and progenitor cells. The myeloid committed CD34$^+$ cells coexpressing CD33 Ag represent two thirds of the population in all three sources of hematopoietic cells: bone marrow, peripheral blood after mobilization, and cord blood (Table 1). The B lymphoid–committed CD34$^+$ cells coexpressing CD19 Ag represent about 20% of the population in marrow versus only $\leq 3\%$ in both mobilized peripheral blood and cord blood (Table 1). The T lymphoid–committed CD34$^+$ cells coexpressing CD2 Ag represent $\leq 5\%$ of the population in all three sources of hematopoietic cells (Table 1).

The primitive small subset of HLA-DR$^-$ or CD38$^-$CD34$^+$ cells varies among the three sources: 3–29% in marrow, 14–46% in mobilized peripheral blood, and 2–12% in cord blood (Table 1). Both CD34$^+$CD33$^+$ myeloid-committed cells and CD34$^+$HLA-DR$^-$ long-term reconstituting hematopoietic stem cells were significantly more in the CD34^{+-} mobilized peripheral blood cells compared with the CD34$^+$ marrow cells ($p < 0.01$) (Table 1). The rapid hematological recovery after peripheral blood progenitor cell (PBPC)

Table 1 Comparison of CD34$^+$ Cell Subsets in Normal Human Bone Marrow Harvests (BM), rHuG-CSF–Mobilized Peripheral Blood Leukapheresis Products (PB), and Cord Blood (CB)

CD34$^+$ cells	BM	PB	CB
CD34$^+$CD33$^+$ cells	61% (19–77)	73% (48–93)	54% (26–82)
CD34$^+$HLA-DR$^-$ cells	5% (3–29)	29% (14–46)	3% (2–12)
CD34$^+$CD19$^+$ cells	20% (14–30)	1% (1–3)	1% (0–2)
CD34$^+$CD2$^+$ cells	4% (3–6)	1% (1–3)	1% (0–2)

Numbers represent median (range) of results from 26 bone marrow harvests, 31 peripheral blood progenitor cell leukapheresis products, and 35 cord blood collections.

transplantation could be due to the increased number of CD34$^+$CD33$^+$ myeloid-committed cells and of the CD34$^+$HLA-DR$^-$ long-term reconstituting hemopoietic stem cells in comparison with bone marrow transplantation.

Most human CD34$^+$ cells from all three sources of hematopoietic cells express the adhesion molecules of the (CD29/CD49d,e) β1 and (CD11a/CD18) β2 integrin family (3). In addition, other prominent adhesion molecules including CD31 (PECAM-1), CD43 (sialomucin), CD44 (H-CAM), CD54 (ICAM-1), CD55 (DAF), and CD58 (LFA-3) are expressed on most human hematopoietic CD34$^+$ cells (3).

Although the exact mechanism of CD34$^+$ cells homing to the bone marrow stroma remains to be determined, a homing mechanism that mimics the multistep process identified for leukocyte-endothelial cell interactions has been suggested (6). The first step, initiating cell-to-cell interactions, is envisioned to require interaction between CD34 and an L-selectin–like molecule, and the second step to involve stronger binding of an integrin molecule (e.g., VLA-4) on the CD34$^+$ cell to its ligand (e.g., VCAM-1) on the marrow stroma. This might create a stable interaction between circulating hematopoietic stem/progenitor cells and the marrow capillary endothelium. These stem/progenitor cells, now halted in their circulation, may then pass through the endothelial layer, adhere to the marrow extracellular stromal matrix, and undergo hematopoietic proliferation and differentiation.

III. CYTOKINE RECEPTORS ON HUMAN HEMATOPOIETIC CD34$^+$ CELLS

The receptors of several hematopoietic cytokines including stem cell factor receptor (*c-kit*), *flk-2/flt3/stk*-1 receptor, both units of interleukin (IL)-3 receptor, granylocyte-macrophage colony-stimulating factor (GM-CSF) receptor, and thrombopoietin *mpl* receptor are expressed in a substantial proportion of the human hematopoietic CD34$^+$ cells (7). The expression of *c-kit*, GM-CSF, and *mpl* receptors are associated mainly with the more immature CD34$^+$CD33$^-$ cells (7).

IV. CD34$^+$ CELLS SORTING FOR LEUKEMIA DIAGNOSIS AND PROGNOSIS

The CD34 is expressed in about 40–60% of cells from adult patients with acute myeloid leukemia (AML) with all French-American-British classification

(FAB) subtypes except acute promyelocytic leukemia (APL; M3), myelodysplastic features, and/or secondary AML, chronic myeloid leukemia (CML) cells especially from patients with accelerated and blastic phases, 70% of children with B-acute lymphocytic leukemia (ALL), and most Philadelphia chromosome $(Ph)^+$ ALL cells (8). Adult patients with $CD34^+$ AML cells were shown to have low total leukocytic count, high frequency of chromosomal abnormalities, and significantly low remission and survival rates (9). Also, in patients with myelodysplasia, CD34 expression was associated with transformation to AML and poor survival rate (10). Moreover, in vivo transplantation studies showed the expansion of AML cells in severe chronic immunodeficiency (SCID) mice to be restricted to the $CD34^+$ population (11). On the other hand, CD34 is not expressed on the predominant malignant cells in patients with chronic Iymphocytic leukemia or hairy cell leukemia.

V. $CD34^+$ CELLS SORTING TO QUANTITATE HEMATOPOIETIC STEM/PROGENITOR CELLS IN ALLOGRAFTS AND AUTOGRAFTS BEFORE TRANSPLANTATION

Flow cytometric counts of CD34 immunostained mononuclear cells from harvested bone marrow or collected pheriphal blood leukapheresis products are widely used to predict the yield of progenitor cells in allografts and autografts before transplantation. In the steady state, normal adult blood contains nearly undetectable levels of $CD34^+$ cells, but after mobilization with recombinant human hematopoietic growth factors (e.g., granulocyte colony-stimulating factor, rHuG-CSF, rHuGM-CSF, and/or rHuIL-3) with or without chemotherapy, the percentage of $CD34^+$ cells may increase to 1–5%.

The $CD34^+$ cell count predicts the yield of progenitor cells that can be collected by leukapheresis (12). Although difficulties with interlaboratory standardization of $CD34^+$ cell enumeration do not permit an exact stipulation, there is a general agreement about a threshold number between 1 and 2.5×10^6 $CD34^+$ cells/kg patient body weight, above which prompt recovery of neutrophil and platelets within 2 weeks after myeloeblative therapy is essentially assured (12).

However, there is no standardized way to enumerate $CD34^+$ cells by flow cytometry since using single-parameter (CD34 staining only) or dual-parameter (CD34 staining together with either lymphocyte gating or side-scatter gating) analysis rarely correlates with estimates obtained by fluorescence microscope. Also, isotype-matched control Ab staining of minor populations of cells, nonspecifically stained adherent cells, and cellular debris, all of which may be recorded by flow cytometry, can often mask or inflate the true number of the small $CD34^+$ cell population. Therefore, the clinical threshold of progenitor cells required for rapid engraftment after transplantation will become more precise by careful analyses of the subsets of $CD34^+$ cells and correlating each with the rapidity and durability of neutrophil and platelet recovery.

Using double-simultaneous staining with CD34 phycoerythrin (PE) and CD45 fluorescein (FITC), CD33 FITC, CD38 FITC, or HLA-DR FITC and collecting four data parameters (linear forward scatter, linear side scatter, log FITC and log PE fluorescence) represents the most sensitive, specific, and reproducible available means of enumerating $CD34^+$ cells in allografts or autografts before transplantation (13).

VI. $CD34^+$ CELL ENRICHMENT TO PURIFY HEMATOPOIETIC STEM/PROGENITOR CELLS IN ALLOGRAFTS AND AUTOGRAFTS BEFORE TRANSPLANTATION

There is increasing evidence that marrow and peripheral blood frequently contain occult tumor cells that can contribute to malignant relapse after autotransplantation (14,15). Since several tumor cells, including breast cancer, ovarian cancer, neuroblastoma, Iymphomas, myeloma, and $CD34^-$ leukemias, do not express CD34, positive selection of $CD34^+$ hematopoietic stem/progenitor cells offers an effective method of passively purging autografts of the occult malignant cells before transplantation that could reduce the relapse rate.

Autologous transplantation with selected $CD34^+$ cells rather than the whole leukapheresis product or bone marrow harvest has two important advantages: reducing the tumor-cell contamination by as much as 4 logs and reducing thc volume of frozen autograft, facilitating storage and avoiding side effects related to the infusion of large volumes of cryoprotectant and debris of lysed cells (16). To achieve 3- to 4-log depletion of tumor cells during the clinical-scale selection of $CD34^+$ cells, a purity of 90% or more should be obtained (Table 2). Unfortunately, the current $CD34^+$ cell clinical-scale selection technologies have not achieved consistent autografts of 90% or more pure $CD34^+$ cells (17). Therefore, it seems unlikely that $CD34^+$ cell selection alone will fully deplete the autograft of tumor cells (Table 2). A maximally effective purging of tumor cells often requires a combination of positive selection of CD34 cells and negative selection of tumor cell methodologies.

Table 2 Mathematical Model Showing That Selection of CD34$^+$ Cells Alone Does Not Resolve the Problem of Tumor-Cell Contamination of Autografts Before Transplantation

	Starting population		
	CD34$^+$ cells 5×10^7 (1%)	Autograft cells 5×10^9	Tumor cells 5×10^7 (1%)
Efficiency of CD34$^+$ cell selection (%)	CD34$^+$ recovered cells	Contaminated with: Autograft cells	Contaminated with: Tumor cells
50	2.5×10^7	2.5×10^7	2.5×10^5
60	3.0×10^7	2.0×10^7	2.0×10^5
70	3.8×10^7	1.2×10^7	1.2×10^5
80	4.0×10^7	1.0×10^7	1.0×10^5
90	4.5×10^7	0.5×10^7	5.0×10^4
99	4.95×10^7	0.05×10^7	5.0×10^3

The assumption has been made that 5×10^9 mononuclear bone marrow cells were harvested that are contaminated with 5×10^7 tumor cells (1%) and contain 5×10^7 CD34$^+$ cells (1%) of the total harvested mononuclear cells.

In the allogeneic transplantation, the biological effects depend upon the composition of the allograft. The goal is to achieve engraftment and rapid hematopoietic recovery and to avoid the development of graft-versus-host disease (GVHD) without loss of potential graft-versus-leukemia (GVL) effect. The development of GVHD is dependent on the presence of T lymphocytes. Recently, successful allogeneic PBPC transplantation has been shown by several groups to result in shorter hematological recovery and hospitalization times and fewer financial costs while sparing the donors the discomfort and risks of the marrow harvest procedure (18). Peripheral blood progenitor cells from healthy donors contain 1 log excess of T lymphocytes representing a potential risk for GVHD; allogeneic PBPC transplantations have shown no severe acute GVHD except in very few cases (18). Therefore, the risk of inducing severe acute GVHD is not as high as was expected during the early period of allogeneic PBPC transplantation; however, the incidence and severity of chronic GVHD after allogeneic PBPC transplantation remains to be determined due to the short-term follow-up to date of these pilot clinical studies. The main advantage for allogeneic transplantation of selected CD34$^+$ cells has been shown to be a reduction of the high T-lymphocyte content that contributes to GVHD while retaining enough cells for immune reconstitution and a potential GVL effect, without T-cell–depletion methods such as complement-mediated lysis or counterflow centrifugal elutriation, which have been complicated by graft failure and/or increased rate of relapse (19).

In addition the demonstration of donor-derived long-term hematopoiesis after allogeneic transplantation of selected CD34$^+$ cells in cancer patients treated with myeloablative therapy has provided the final proof that human hematopoietic stem cells are contained in the CD34$^+$ cell population (19).

The selection of CD34$^+$ cells is currently being used to concentrate hematopoietic stem/progenitor cells for gene-therapy trials (14,15). The CD34$^+$ cell enrichment is performed to increase the retrovirus:cell ratio in the subsequent ex vivo retroviral infection protocols. This is currently neccessary because high titers of the appropriate retroviruses cannot yet be obtained and retroviral infection of unselected hematopoietic cells is usually inefficient.

VII. CD34$^+$ CELL ENRICHMENT TO EXPAND HEMATOPOIETIC STEM/PROGENITOR CELLS IN ALLOGRAFTS AND AUTOGRAFTS BEFORE TRANSPLANTATION

After myeloablative therapy followed by hematopoietic stem/progenitor cell transplantation, an obligatory period of 2 weeks of pancytopenia, during which the patients are susceptible to febrile episodes, infections, and bleeding, exists despite the use of post-transplant hematopoietic growth factors such as rHuGM-CSF, rHuG-CSF, and/or rHuIL-3 (20). With the availability of many hematopoietic growth factors, it has become reasonable to consider ex vivo culture expansion of clinical transplant grafts to provide a larger number of hematopoietic stem/progenitor cells, which could shorten the critical period of severe pancytopenia and consequently decrease initiation risk, hospitalization time, and financial costs. Other additional advantages of ex vivo hematopoietic stem/progenitor cell expansion include a concomitant depletion (passive purging) of tumor cells in autografts and also a reduction of T lymphocytes (indirect depletion) in allografts during the culture before transplantation.

In vivo studies of long-term reconstitution have shown ex vivo expanded hematopoietic stem/progenitor cells to successfully engraft lethally irradiated mice (21,22). Also, clinical-scale systems of ex vivo expansion of human CD34$^+$ hematopoietic progenitor cells have been successfully established (23). More significantly, two recent pilot clinical studies have shown ex vivo expanded hematopoietic stem/progenitor cells to successfully engraft 18 patients with cancer (9 breast cancer, 5 lung cancer, and 4 other cancers) after high-dose chemotherapy (24,25). The hematopoietic engraftment in both studies was prompt with rapid and sustained neutrophil and platelet recovery within 10 days and no infusional toxic effects of ex vivo expanded hematopoietic stem/progenitor cells (24,25). These encouraging results need to be reproduced in a large number of cancer patients in further clinical studies using myeloablative therapy instead of only high-dose chemotherapy. The appropriate dosing of these ex vivo expanded hematopoietic stem/progenitor cells shall be determined in these further studies.

Of concern, current ex vivo culture systems result in significant increases in the number of total and progenitor hematopoietic cells, but the frequencies of $CD34^+$ cells decline after 1 week of culture to very low levels. There is both in vivo (21,22) and in vitro (23) evidence that sufficient hematopoietic long-term reconstituting stem cells are still maintained in the ex vivo cultured grafts. The optimal ex vivo expansion culture system remains to be identified and several preclinical studies curently are being done to optimize the conditions of the ex vivo expansion culture system using several different combinations of hematopoietic growth factors aiming at both a shorter period of culture and a fewer number of hematopoietic growth factors for an effective increase in hematopoietic stem and progenitor cells.

Clearly from a clinician's point of view, the best ex vivo expansion technique would be one in which the three lineages of hematopoietic progenitor cells (erythroid, myeloid, and megakaryocytic) are expanded and available for patient's benefit (26). The ability to efficiently expand all three lineages' progenitor cell numbers in vitro would limit the number of stem and progenitor cells that must be actually harvested from the patient or donor, lessening the expense, hazards, inconvenience, and potential morbidity incurred during the bone marrow harvest and cryopreservation. Control of the lineage obtained during ex vivo expansion of marrow progenitor cells would allow the creation of components capable of treating specific deficits such as refractory and aplastic anemias, thalassemias, or congenital and acquired neutropenias or thrombocytopenias.

Identifying the effective, particular combination of hematopoietic cytokines that could expand each lineage in a sustained long-term manner would insure that all three lineages are adequately provided, promoting a rapid, sustained, complete, trilineage hematopoietic reconstitution after myeloablative therapy (26). The combination of stem cell factor (rHuSCF), rHuIL-3, rHuIL-11, erythropoietin (rHuEPO), and thrombopoietin (*mpl* ligand) seems to be an ideal candidate that can expand ex vivo all three lineages' progenitor-cell numbers in human long-term bone marrow (26) and cord blood (27) cultures.

REFERENCES

1. Silvestri F, Banavali S, Baccarani M, Preisler HD. The CD34 haemopoietic progenitor cell associated antigen: biology and clinical applications. *Haematologica* 1992; 77(3):265–273.
2. Civin CI, Gore SD. Antigenic analysis of hematopoiesis: a review. *J Hematother* 1993; 2(2):137–144.
3. Knapp W, Strobl H, Scheinecker C, Bello-Fernandez C, Majdic O. Molecular characterization of $CD34^+$ human hematopoietic progenitor cells. *Ann Hematol* 1995; 70(6):281–296.
4. Schlossman S, Boumsell S, Gilks W, et al. *Leukocyte Typing V: White Cell Differentiation Antigens.* New York: Oxford University Press, 1995:840.
5. Molgaard H, Spurr N, Greaves M. The hematopoietic stem cell antigen, CD34, is encoded by a gene located on chromosome 1. *Leuk Res* 1989; 3:773.
6. Krause DS, Fackler MJ, Civin CI, May WS. CD34: structure, biology, and clinical utility. *Blood* 1996; 87(1):1–13.
7. McClanahan T, Dalrymple S, Barkett M, Lee F. Hematopoietic growth factor receptor genes as markers of lineage commitment during in vitro development of hematopoietic cells. *Blood* 1994; 84(11):3981–3982.
8. Hassan HT, Freund M. Recombinant human stem cell factor protects CD34 positive human myeloid leukemia cells from chemotherapy-induced apoptosis. In: Pieters R, ed. *Drug Resistence in Leukemia & Lymphoma.* Reading: Harwood Academic Publishers; 1996:123–126.
9. Myint H, Lucie NP. The prognostic significance of the CD34 antigen in AML. *Leuk Lymphoma* 1992; 7(5–6):425–429.
10. Sullivan SA, Marsden KA, Lowenthal RM, et al. Circulating $CD34^+$ cells: an adverse prognostic factor in myelodysplastic syndromes. *Am J Hematol* 1992; 39(2):96–101.
11. Lapidot T, Sirard G, Vormoor J, et al. A cell initiating human AML after transplantation into SCID mice. *Nature* 1994; 367:645.
12. Juttner CA, To LB, Haylock DN, et al. Blood progenitor cells: biologic and clinical issues. *Bone Marrow Transplant* 1995; 15:S15.
13. Sutherland DR, Keating A, Nayar R, Anania S, Stewart AK. Sensitive detection and enumeration of CD34 positive cells in peripheral and cord blood by flow cytometry. *Exp Hematol* 1994; 22(10):1003.–1010.
14. Rill DR, Santana VM, Roberts WM, et al. Direct demonstration that autologous bone marrow transplantation for solid tumors can return a multiplicity of tumorigenic cells. *Blood* 1994; 84(2):380–383.
15. Deisseroth A, Zu S, Claxton D. Genetic marking shows that Ph^+ cells represent in autologous transplants of CML contribute to relapse after autologous bone marrow transplantation in CML. *Blood* 1994; 83(10):3068–3076.
16. Hassan HT, Stockschläder M, Krüger W, Hoffknecht MM, Zander AR. G-CSF administration following peripheral-blood selected CD34 positive cell autologous

transplantation accelerates hematological recovery after myeloablative therapy. *Br J Haematol* 1996; 92(4):1027–1029.
17. Hassan HT, Zeller W, Schleimer B, et al. CD34 positive cell selection using Cellpro Ceprate in 55 G-CSF–mobilized peripheral blood grafts collected for transplantation. *Bone Marrow Transplant* 1996; 1:S50.
18. Tanaka J, Kesai M, Imamura M, Asaka M. Clinical application of allogeneic peripheral blood stem cells transplantation. *Ann Hematol* 1995; 71(6):265–269.
19. Hassan HT, Zeller W, Stockschläder M, Krüger W, Hoffknecht MM, Zander AR. Comparison between bone marrow and G-CSF–mobilised peripheral blood allografts undergoing clinical scale CD34 positive cell selection. *Stem Cells* 14:419–429.
20. Hassan HT, Zander AR. Thrombocytopenia after autologous peripheral blood stem/progenitor cell transplantation: unresolved problem and possible approaches to resolve it. *J Hematother* 5:407–414.
21. Muench MO, Firpo MT, Moore MA. Bone marrow transplantation of interleukin-1 plus kit ligand ex-vivo expanded bone marrow accelerates hematopoietic reconstitution in mice without the loss of stem cell lineage and proliferative potential. *Blood* 1993; 81(12):3463–3473.
22. Knobel KM, McNally MA, Berson AE, et al. Long-term reconstitution of mice after ex-vivo expansion of bone marrow cells: differential activity of cultured bone marrow and enriched stem cell populations. *Exp Hematol* 1994; 22:1227.
23. Purdy MH, Hogan CJ, Hami L, et al. Large volume ex vivo expansion of CD34 positive hematopoietic progenitor cells for transplantation. *J Hematother* 1995; 4:515.
24. Brugger W, Heimfeld S, Berenson RJ, Mertelsmann R, Kanz L. Reconstitution of hematopoiesis after high-dose chemotherapy by autologous progenitor cells generated ex vivo. *N Engl J Med* 1995; 333(5):283–287.
25. Williams SF, Lee WJ, Bender JG, et al. Selection and expansion of peripheral blood CD34 positive cells in autologous stem cell transplantation for breast cancer. *Blood* 1996; 87(5):1687–1691.
26. Hassan HT, Biermann B, Zander AR. Maintenance and expansion of erythropoiesis in human long-term bone marrow cultures in presence of erythropoietin plus stem cell factor and interleukin-3 or interleukin-11. *Eur Cytokine Network* 1996; 7:129.
27. Hassan HT, Plantholz D, Zander AR. Ex vivo expansion of pre-CFUs and CFU-GM from CD34 positive human umbilical cord blood cells. *Exp Hematol* 24:1025.

Appendix A

Cellular Properties for Cell Separation

Table A1 Cell Densities

Cell type	Density [g/mL]
T cells	1.077
B cells	1.077
Monocytes	1.064
Granulocytes	1.09
NK cells	1.06
Erythrocytes	1.115
Osteoblasts	1.055

Compiled by Dr. Peter Van Vlasselaer, Activated Cell Therapy, Mountain View, California.

Table A2 Cell Surface Antigen Levels (estimated number of antigens per cell; some numbers are averages from multiple references)

Antigen	Lymphocytes	Monocytes	Granulocytes
CD2	22,000		
CD3	80,750		
CD4	59,000	18,000	
CD5	60,675		
CD7	25,000		
CD8	144,325		
CD11a	26,500	54,000	17,000
CD11b	12,000	52,000	45,000
CD11c	<1,000	21,000	11,000
CD13	<1,000	20,500	17,000
CD14	<1,000	123,000	<3,000
CD16	79,000	12,000	329,000
CD18	30,500	101,500	42,000
CD19	27,000		
CD20	123,750		
CD21	14,000		
CD22	17,000		
CD23	4,400		
CD24	39,000		
CD25	68,000		
CD28	12,200		
CD33	<1,000	19,500	10,000
CD37	99,000		
CD38	10,100		
CD44	102,000	201,000	73,000
CD45	191,750	79,300	30,300
CD45RA	56,000		
CD45RO	14,000		
CD46	11,000	21,000	21,000
CD54	<1,000	19,000	12,000
CD55	7,000	26,000	15,000
CD57	398,900		
CD58	5,000	17,000	12,000
CD69	165,000		
HLA-DR	151,300		
HLAClass 1	121,000	174,000	53,000

Compiled by Dr. Robert Hoffman, Becton Dickinson Immunocytometry Systems, San Jose, California, from

1. Bikoue A, George F, Poncelet P, Mutin M, Janossy G, Sampol J. Quantitative analysis of leukocyte membrane antigen expression: normal adult values. *Cytometry* 1996; 26:137–147.

2. Denny TN, Stein D, Garcia A, Scolpino A, Holland B, Connor E. Fluorescence intensity of CD4 and CD8 cells from healthy adult donors over time. *Cytometry* 1993; 6(suppl):58. Abstract 55C.
3. Lenkei R, Anderssen B. Determination of the antibody binding capacity of lymphocyte membrane antigens by flow cytometry in 58 blood donors. *J Immunol Meth* 1995; 183:267–277.
4. Poncelet P, Bikque A, Rostaing O, George F, Sampol J. Quantitative assessment of human leucocyte membrane antigens: definition of normal densities measured in immunocytometry with the QiF1 assay. *Cytometry* 1993; 6(suppl):60. Abstract 83C.
5. Poncelet P, Bikque A, Lavabre T, Poinas G, Parant M, Duperray C, Sampol J. Quantitative assessment of human lymphocytes membrane antigens: definition of normal densities measured in immuno-cytometry with the QIFI assay. *Cytometry* 1991 5(suppl):82. Abstract 418A.
6. van Wauwe JP, De Mey JR, Goossens JG. OKT3: a monoclonal anti-human T lymphocyte antibody with potent mitogenic properties. *J Immunol* 1980; 124:2708–2713.
7. Warnes G, Biggerstaff JP, Savidge GF. Quantification of lymphocyte activation markers by three colour subpopulation gating. *Cytometry* 1994; 7(suppl):32. Abstract 91C.

Appendix B

CD Antigen Designations*
(as of February 1996)

*As of February 1996. An up-to-date list is available on the Internet at http://histo.cryst.bbk.ac.uk/wwwfiles/cd_table.html. Reviewed by David Buck, AmCell Corporation, Sunnyvale, California.

CD antigen	Cellular expression	Molecular* weight (kD)	Functions	Other names	Family relationships
CD1a,b,c,d	Cortical thymocytes, Langerhans cells, dendritic cells, B cells (CD1c), intestinal epithelium (CD1d)	43–49	MHC class I-like molecule, associated with β2-microglobulin. May have specialized role in antigen presentation		IgSF
CD2	T cells, thymocytes, NK cells	45–48	Adhesion molecule, binding CD58 (LFA-3), can activate T cells	T11, LFA-2	IgSF
CD2R	T cells	45–58	Activation-sensitive conformational form of CD2	T11-3	IgSF
CD3	Thymocytes, T cells	g: 25–28 d: 20 e: 20 z: 16 h: 22	Associated with the T-cell antigen receptor, required for cell surface expression of and signal transduction by TCR	T3	IgSF (gde) z/h related to FcR γ chain
CD4	Thymocyte subsets, helper and inflammatory T cells (about two-thirds of peripheral T cells), monocytes, macrophages	55	Coreceptor for MHC class II molecules, binds lck on cytoplasmic face of membrane, receptor for HIV-I and HIV-2 gp120	T4, L3T4	IgSf

CD5	Thymocytes, T cells, subset of B cells	67	Binds to CD72	T1, Ly1	Scavenger receptor
CD6	Thymocytes, T cells, B cell CLL	100–130	CD166	T12	Scavenger receptor
CD7	Pluripotential hematopoietic cells, thymocytes, T cells	40	Unknown, marker for T cell ALL and pluripotential stem cell leukemias		IgSF
CD8	Thymocyte subsets, cytotoxic T cells (about one-third of peripheral T cells)	a: 32–34 b: 32–34	Coreceptor for MHC class I molecules, binds Ick on cytoplasmic face of membrane	T8, Lyt2,3	IgSF
CD9	Pre-B cells, eosinophils, basophils, platelets, monocyte EC	22–27	Possible role in platelet aggregation and activation		Tetraspanning membrane protein
CD10	B- and T-cell precursors, bone marrow stromal cells	100	Zinc metalloproteinase, marker for pre-B ALL	Neutral endopeptidase, common acute lymphocytic leukemia antigen (CALLA)	

CD antigen	Cellular expression	Molecular* weight (kD)	Functions	Other names	Family relationships
CD11a	Lymphocytes, granulocytes, monocytes, and macrophages	180	αL subunit of integrin LFA-1 (associated with CD18); binds to CD54 (ICAM-1), ICAM-2, and ICAM-3	LFA-1	
CD11b	Myeloid and NK cells	165	αM subunit of integrin CR3 (associated with CD18); binds CD54, complement component iC3b, and extracellular matrix proteins	Mac-1	
CD11c	Myeloid cells	150	αX subunit of integrin CR4 (associated with CD18); binds fibrinogen	CR4, p150, 95	
CDw12	Monocytes, granulocytes, platelets	90–120	Unknown		
CD13	Myelomonocytic cells	150–170	Zinc metalloproteinase	Aminopeptidase N	
CD14	Myelomonocytic cells	53–55	Receptor for complex of LPS and LPS binding protein (LBP)		

CD15	Neutrophils, eosinophils, monocytes		Branched penta-saccharide, expressed on glycolipids and many cell surface glycoproteins; the sialylated form is a ligand for CD62E (ELAM)	Lewis-x (Lex)	
CD16	Neutrophils, NK cells, macrophages	50–80	Component of low-affinity Fc receptor, FcγRIII, mediates phagocytosis and ADCC	FcgRIII	IgSF
CDw17	Neutrophils, monocytes, platelets		Lactosyl ceramide, a cell surface glycosphingolipid		
CD18	Leukocytes	95	Integrin β2 subunit, associates with CD11a, b, and c		
CD19	B cells	95	Forms complex with CD21 (CR2) and CD81 (TAPA-1), coreceptor for B cells		IgSF
CD20	B cells	33–37	Possible role in regulating B-cell activation		Tetraspanning membrane protein

CD antigen	Cellular expression	Molecular* weight (kD)	Functions	Other names	Family relationships
CD21	Mature B cells, FDC	145	Receptor for complement component C3d, EBV; with CD19 and CD81 forms coreceptor for B cells	CR2	CCP superfamily
CD22	Mature B cells	α: 130 β: 140	Adhesion of B cells to monocytes, T cells	BL-CAM	IgSF
CD23	Mature B cells, activated macrophages, eosinophils, follicular dendritic cells, platelets	45	Low-affinity receptor for IgE, ligand for CD19:CD21:CD81 coreceptor	FcγRII	C-type lectin
CD24	B cells, granulocytes	35–45	Unknown	Possible human homologue of mouse heat-stable antigen or J11d.	
CD25	Activated T cells, B cells, monocytes	α: 55 β: 75	IL-2 receptor	Tac	a: CCP b: cytokine receptor superfamily
CD26	Activated B and T cells, macrophages	110	Protease	Dipeptidyl peptidase IV	
CD27	Medullary thymocytes, T cells	50–55	Unknown		NGF receptor superfamily

CD28	T-cell subsets, activated B cells	44	Activation of naive T cells, receptor for costimulatory signal (signal 2) binds CD80 (B7-1) and B7-2	Tp44	IgSF
CD29	Leukocytes	130	Integrin β1 subunit, associates with CD49a–f in VLA-1 integrin		
CD30	Activated B and T cells	105–120	Unknown	Ki-1	NGF receptor superfamily
CD31	Monocytes, platelets, granulocytes, B cells, endothelial cells	130–140	Possibly an adhesion molecule	PECAM-1	IgSF
CD32	Monocytes, granulocytes, B cells, eosinophils	40	Low-affinity Fc receptor for aggregated Ig/immune complexes	FcγRII	IgSF
CD33	Myeloid progenitor cells, monocytes	67	Unknown		IgSF
CD34	Hematopoietic precursors, capillary endothelium	105–120	Ligand for CD62 (L-selectin)		
CD35	Erythrocytes, B cells, monocytes, neutrophils, eosinophils, FDC	250	Complement receptor 1, binds C3B and C4b, mediates phagocytosis	CR1	CCP superfamily
CD36	Platelets, monocytes	88	Unknown	Platelet GPIV	
CD37	Mature B cells, mature T cells, myeloid cells	40–52	Unknown		Tetraspanning membrane protein

CD antigen	Cellular expression	Molecular* weight (kD)	Functions	Other names	Family relationships
CD38	Early B and T cells, activated T cells, germinal center B cells, plasma cells	45	Unknown		
CD39	Activated B cells, activated NK cells, macrophages, dendritic cells	78	Unknown, may mediate adhesion of B cells		
CD40	B cells, monocytes, dendritic cells	50	Receptor for co-stimulatory signal for B cells, binds CD40 ligand(T-BAM)		NGF receptor superfamily
CD40L	Activated CD4 T cells	39	Ligand for CD40	T-BAM, CD154	TNF-like
CD41	Platelets, megakaryocytes	125/22 dimer	αIIb integrin, associates with CD61 to form GPIIb; binds fibrinogen, fibronectin, von Willebrand factor, and thrombospondin		
CD42a,b,c,d	Platelets, megakaryocytes	a: 23 b: 135, 23 c: 22 d: 85	Binds von Willebrand factor, thrombin; essential for platelet adhesion at sites of injury	a: GPIX b: GPIba c: GPIbb d: GPV	

CD43	Leukocytes, except resting B cells; $CD5^+$ B cells	115–135 (neutrophils) 95–115 (T cells)	Binds CD54 (ICAM-1)	Leukosialin, sialophorin
CD44	Leukocytes, erythrocytes	80–95	Binds hyaluronic acid, mediates adhesion of leukocytes	Hermes antigen, Pgp-1
CD45	Leukocytes	180–240	Tyrosine phosphatase, augments signaling through antigen receptor of B and T cells, multiple isoforms result from alternative splicing (see below)	LCA, T200, B220
CD45RO	T-cell subsets, B-cell subsets, monocytes, macrophages	180	Isoform of CD45 containing none of the A, B, and C exons	
CD45RA	B cells, T-cell subsets (naive T cells), monocytes	205–220	Isoforms of CD45 containing the A exon	
CD45RB	T-cell subsets, B cells, monocytes, macrophages, granulocytes	190–220	Isoforms of CD45 containing the B exon	T200
CD45RC	—	—	—	—

CD antigen	Cellular expression	Molecular* weight (kD)	Functions	Other names	Family relationships
CD46	Hematopoietic and nonhematopoietic nucleated cells	56/66 (splice variants)	Membrane cofactor protein; binds to C3b and C4b to permit their degradation by Factor I	MCP	CCP superfamily
CD47	All cells	47–52	Unknown, associated with Rh blood group		
CD48	Leukocytes	40–47	Unknown	Blast-1	IgSF
CD49a	Activated T cells, monocytes	210	α1 integrin, associates with CD29, binds collagen, laminin	VLA-1	
CD49b	B cells, monocytes, platelets	165	α2 integrin, associates with CD29, binds collagen, laminin	VLA-2	
CD49c	B cells	125	α3 integrin, associates with CD29, binds laminin, fibronectin	VLA-3	
CD49d	B cells, thymocytes	150	α4 integrin, associates with CD29, binds fibronectin, Peyer's patch HEV, VCAM-1	VLA-4	
CD49e	Memory T cells, monocytes, platelets	135, 25 dimer	α5 integrin, associates with CD29, binds fibronectin	VLA-5	

CD49f	Memory T cells, thymocytes, monocytes	120, 25 dimer	α6 integrin, associates with CD29, binds laminin	VLA-6	
CDw50	Thymocytes, T cells, B cells, monocytes, granulocytes	130	ICAM-3 Countes Rec for LFA-1 (CD11a)		
CD51	Platelets, megakaryocytes	125, 24 dimer	αV integrin associates with CD61; binds vitronectin, von Willebrand factor, fibrinogen, and thrombospondin	Vitronectin receptor	
CDw52	Thymocytes, T cells, B cells (not plasma cells), monocytes, granulocytes	21–28	Unknown, target for antibodies used therapeutically to deplete T cells	CAMPATH-1	
CD53	Leukocytes	35–42	Unknown	MRC OX44	Tetraspanning membrane protein
CD54	Hematopoietic and nonhematopoietic cells	85–110	ICAM-1, binds CD11a/CD18 (LFA-1) and CD11b/CD18 (Mac-1) integrins, receptor for rhinovirus	ICAM-1	IgSF
CD55	Hematopoietic and nonhematopoietic cells	60–70	DAF, binds C3b, disassembles C3/C5 convertase	DAF	CCP superfamily

CD antigen	Cellular expression	Molecular* weight (kD)	Functions	Other names	Family relationships
CD56	NK cells	175–185	Isoform of NCAM, adhesion molecule	NKH-1	IgSF
CD57	NK cells subsets of T cells, B cells, and monocytes		Oligosaccharide, found on many cell surface glycoproteins	HNK-1, Leu-7	
CD58	Hematopoietic and nonhematopoietic cells	55–70	LFA-3, binds CD2, adhesion molecule	LFA-3	IgSF
CD59	Hematopoietic and nonhematopoietic cells	19	Binds complement components C8 and C9, blocks assembly of membrane attack complex	Protectin, Mac inhibitor	
CD60	T-cell subsets, platelets, monocytes		Oligosaccharide present on gangliosides		
CD61	Platelets, megakaryocytes, macrophages	105	Integrin β3 subunit, associates with CD41 (GPIIb/IIIa) or CD51 (vitronectin receptor)		
CD62E	Endothelium	110	ELAM, binds sialyl-Lewis x, mediates rolling interaction of neutrophils on endothelium	ELAM-1, E-selectin	C-type lectin

CD62L	B cells, T cells, monocytes, NK cells	76	LAM, binds CD34, Gly CAM, mediates rolling interactions with endothelium	LAM-1, L-selectin, LECAM-1	C-type lectin
CD62P	Platelets, megakaryocytes, endothelium	140	Adhesion molecule; binds sialyl Lewis x; mediates interaction of platelets with neutrophils and monocytes and rolling interaction of neutrophils on endothelium	P-selectin, PADGEM	C-type lectin
CD63	Activated platelets, monocytes, macrophages	53	Unknown, is lysosomal membrane protein translocated to cell surface after activation		Tetraspanning membrane protein
CD64	Monocytes, macrophages	72	High-affinity receptor for IgG	FcγRI	IgSF
CDw65	Myeloid cells		Oligosaccharide component of a ceramide dodecasaccharide		
CD66a	Neutrophils	160–180	Unknown, member of CEA family	BGP-1	IgSF
CD66b	Granulocytes	95–100	Unknown, member of CEA family	Previously CD67	IgSF

CD antigen	Cellular expression	Molecular* weight (kD)	Functions	Other names	Family relationships
CD66c	Neutrophils, colon carcinoma	90	Unknown, member of CEA family	NCA	IgSF
CD66d	Neutrophils	30	Unknown, member of CEA family		IgSF
CD66e	Adult colon epithelium, colon carcinoma	180–200	Unknown, member of CEA family	CEA	IgSF
CD66f			New cluster, pregnancy-specific glycoprotein		
CD68	Monocytes, macrophages, neutrophils, basophils, large lymphocytes	110	Unknown	Macrosialin	
CD69	Activated B cells, activated T cells, activated macrophages, activated NK cells	28, 32 homodimer	Unknown, early activation antigen	AIM	
CD70	Activated B cells, activated T cells, macrophages	75, 95, 170	CD27 ligand	Ki-24	
CD71	Activated leukocytes	90–95 homodimer	Transferrin receptor	T9	
CD72	B cells	86 homodimer	Unknown, ligand for CD5	Lyb-2	C-type lectin
CD73	B-cell subsets, T-cell subsets	69	Ecto-5′-nucleotidase, dephosphorylates nucleotides to allow nucleoside uptake		

CD74	B cells, macrophages, monocytes, MHC class II positive cells	33,35,41,43 (alternate initiation and splicing)	MHC class II-associated invariant chain	Ii, Ig	
CDw75	Mature B cells, T-cell subsets		Unknown, possibly oligosaccharide, dependent on sialylation		
CDw76	Mature B cells, T-cell subsets		Unknown, possibly oligosaccharide, dependent on sialylation		
CD77	Germinal center B cells		Unknown	Globotriaocylce ramide (Gb3), Pk blood group	
CD78	B cells		Unknown	Ba	
CD79a,b	B cells	a: 32–33 b: 37–39	Components of B-cell antigen receptor analogous to CD3, required for cell-surface expression and signal transduction	Iga, Igb	IgSF
CD80	B-cell subsets	60	Costimulator, ligand for CD28 and CTLAA-4	B7 (now B7-1), BB1	IgSF
CD81	Lymphocytes	26	Associates with CD19, CD21 to form B-cell coreceptor	Target of antiproliferative antibody	Tetraspanning membrane protein
CD82	Leukocytes	50–53	Unknown	R2	Tetraspanning membrane protein

CD antigen	Cellular expression	Molecular* weight (kD)	Functions	Other names	Family relationships
CD83	Activated B cells, activated T cells, circulating dendritic cells (veil cells)	43		HB15	
CDw84	Monocytes, platelets, circulating B cells	73		GR6	
CD85	Monocytes, circulating B cells	120, 83		GR4	
CD86	Monocytes, activated B cells	80		FUN-1,GR65, B7-2	
CD87	Granulocytes, monocytes, macrophages, activated T cells	50–65	Urokinase plasminogen activator receptor	UPA-R	
CD88	Polymorphonuclear leukocytes, macrophages, mast cells	40	Receptor for complement component C5a	C5aR	Rhodopsin superfamily
CD89	Monocytes, macrophages, granulocytes, neutrophils, B-cell subsets, T-cell subsets	50–70	IgA receptor	FcaR	IgSF
CDw90	CD34+ prothymocytes (human) thymocytes, T cells (mouse)	18	Unknown	Thy-1	IgSF
CD91	Monocytes	600	α2 macroglobulin receptor	Unknown	

CDw92	Neutrophils, monocytes, platelets, endothelium	70	Unknown	GR9	
CD93	Neutrophils, monocytes, endothelium	120	Unknown	GR11	
CD94	T-cell subsets, NK cells	43	Unknown	KP43	
CD95	Wide variety of cell lines; in vivo distribution uncertain	43	Binds TNF-like ligand, induces apoptosis	Apo-1, Fas	NGF receptor superfamily
CD96	Activated T cells	160	Unknown	T-cell activation Increased late expression (TACTILE)	
CD97	Activated cells	74, 80, 90	CD55 ligand	GR1	
CD98	T cells, B cells, NK cells, granulocytes, all human cell lines	80, 40 heterodimer	Unknown	4F2	
CD99	Peripheral blood lymphocytes, thymocytes	32	Unknown	MIC2, E2	
CD100	Broad expression on hematopoietic cells	150	Unknown	GR3	
CDw101	Granulocytes, macrophages	140	Unknown	BPC#4	
CD102	Resting lymphocytes, monocytes, strongest on vascular endothelial cells	55–65	Binds CD11a/CD18 (LFA-1 but not CD11b/CD18 (Mac-1)	ICAM-2	IgSF

CD antigen	Cellular expression	Molecular* weight (kD)	Functions	Other names	Family relationships
CD103	Intraepithelial lymphocytes, 2–6% of peripheral blood lymphocytes	150,25	aE integrin	HML-1, aE integrin	
CD104	Epithelia, Schwann cells, some tumor cells	220	β4 integrin	β4, β4 integrin	
CD105	Endothelial cells, bone marrow cell subset, in vitro activated macrophages	95 homodimer	Unknown, possibly ligand for an integrin	Endoglin	
CD106	Endothelial cells	100,110	Adhesion molecule, ligand for VLA-4	VCAM-1	IgSF
CD107a	Activated platelets	110	Unknown; lysosomal membrane protein translocated to the cell surface after activation	LAMP-1	
CD107b	Activated platelets	120	Unknown; lysosomal membrane protein translocated to the cell surface after activation	LAMP-2	
CDw108	Activated T cells in spleen, some stromal cells	80	Unknown	GR2	

CD109	Activated T cells, platelets, endothelial cells	170/150	Unknown	PAF, GR56	
CD110-113			Not yet assigned		
CD114	Monocytes, macrophages, neutrophils	95/130	G-CSF receptor chain		
CD115	Monocytes, macrophages	150	M-CSF receptor	M-CSFR, c-fms	IgSF
CD116	Monocytes, neutrophils, eosinophils, endothelium	75–85	GM-CSF receptor a chain	GM-CSFRα cytokine receptor superfamily	
CD117	Hematopoietic progenitors	145	SCF receptor	*c-kit*	IgSF tyrosine kinase
CD118	Broad cellular expression		Interferon α, β receptor	IFNa,bR	
CD119	Macrophages, monocytes, B cells, endothelium	90–100	Interferon γ receptor	IFNgR	
CD120a	Hematopoietic and nonhematopoietic cells, highest on epithelial cells	50–60	TNF receptor; binds both TNFα and TNFβ	TNFR-I	NgF receptor superfamily
CD120b	Hematopoietic and nonhematopoietic cells, highest on myeloid cells	75–85	TNF receptor, binds both TNFα and TNFβ	TNFR-II	NgF receptor superfamily
CD121a	Thymocytes, T cells, fibroblasts	80	Type I interleukin 1 receptor; binds IL-1α and IL-1β	IL-1R type I	IgSF

CD antigen	Cellular expression	Molecular* weight (kD)	Functions	Other names	Family relationships
CD121b	B cells, macrophages, monocytes	60–70	Type II IL-1R; binds IL-1α and IL-1β	IL-1R, type II	IgSF
CD122	NK cells, resting T-cell subpopulation, some B-cell lines	75	IL-2 receptor β chain	IL-2Rβ	
CD123	Bone marrow stem cells, granulocytes, monocytes, megakaryocytes	70	IL-3 receptor α chain	IL-3R	Cytokine receptor superfamily, fibronectin type III superfamily
CD124	Mature B and T cells, hematopoietic precursor cells	140	IL-4, IL-13 receptor	IL-4R	Cytokine receptor superfamily, fibronectin type III superfamily
CDw125	Eosinophils, basophils	60	IL-5 receptor	IL-5R	Cytokine receptor superfamily, fibronectin type III superfamily
CD126	Activated B cells and plasma cells, weak on most leukocytes	80	IL-6 receptor α subunit	IL-6R	IgSF cytokine receptor superfamily, fibronectin type III superfamily

CD127	Bone marrow lymphoid precursors, pro-B cells, mature T cells, monocytes	65–75	IL-7 receptor	IL-7R	Fibronectin type III superfamily
CDw128	Neutrophils, basophils, T-cell subset	58–67	IL-8 receptor	IL-8R	Rhodopsin superfamily
CD129					
CD130	Activated B cells and plasma cells, weak on most leukocytes, endothelial cells	130	Common receptor subunit for IL-6, IL-11, LIF, CT-1	IL-6Rβ	IgSF cytokine receptor superfamily, fibronectin type III superfamily
CDw131		120	Common β chain for IL-3, IL-5, GM-CSF		
CD132		64	Common γ chain for IL-2, IL-4, IL-7 IL-9, IL-15		Cytokine receptor superfamily
CD134	Activated CD4 T cells	50	OX-40 human homolog		
CD135	Bone marrow subset	130–150	Flt-3 receptor		
CDw136		180	MSPR RON.RTK		
CDw137	Activated CD8 T cells	30	Costimulatory molecule		TNF/NGF receptor family
CD138	Pre-B cells, plasma, EC	63	Syndecan-1		
CD139	B cells	209/228			
CD140a	Broad tissue expression	180			
CD140b	Broad tissue expression				
CD141	Myelocytes, EC, smooth muscle	100	Thrombin receptor		

CD antigen	Cellular expression	Molecular* weight (kD)	Functions	Other names	Family relationships
CD142	EC	45	Tissue factor, cofactor for factor 7		
CD143	EC, epithelial cells, macrocytes	160–170	Angiotensin converting enzyme		
EC144	EC	135	Vascular EC cadherin; cell–cell adhesion		
CDw145	Pan EC stromal cells	25,90,125			
CD146	EC, melanoma, T cells	118–125			IG superfamily
CD147	EC, lymphoid, neutrophils, monocytes	50–60	Neurothelim-adhesion molecule		
CD148			DEP-1, HPTP, chromosome 11		
CDw149	Broadly reactive lymphocytes		MEM133		
CDw150	B cells, T cells, thymocytes, dendritic cells	75.95 glycosyolated	SLAM costimulatory activity marker		
CD151	Lysosomes, monocytes, neutrophils, platelets, EC	28	PETA-3, 4-TM		
CD152	Activated T cells	44	Alternative receptor for CD80, CD86	CTLA-4	
CD153	Activated T cells	40	CD30 ligand		TNF family
CD154	B cells	32–39	CD40 ligand		TNF family
CD155	Monocytes, hematopoietic stem cells, thymocytes, neurons	80–90	Polio virus receptor		

CD156	Monocytes, macrophages, neutrophils	60	ADAM 8	
CD157	Monocytes, neutrophils, bone marrow stroma, EC, FDC.E60/61	42–45		
CD158a	NK inhibitory receptor	p58.1	Cw2, Cw4, Cw5, Cw6 restricted	
CD158b	NK inhibitory receptor	p58.2	Cw1, Cw3, Cw5, Cw6 restricted	
CD161	NK cells, T cells, monocytes	58, 70, 140	Regulates NK-mediated cytolytic activity	
CD162	EC, monocytes, bone marrow, spleen		Homodimer receptor, leukocyte rolling	
CD163	Monocytes, neutrophils (cytoplasm and surface)	130		
CD164	Epithelial cells, bone marrow stroma	80	Mucinlike molecule, on some carcinomas	
CD165	Thymocytes, NK, T cells	37	T-cell leukemia association	
CD166	Activated T and B cells, EC, fibroblasts	100	ALCAM, CD6 ligand	Ig-like domain

ADCC, antibody-dependent cellular cytotoxicity; AIM, activator inducer molecule; ALL, acute lymphocytic leukemia; BGP, biliary glycoprotein; CCP, complement control protein; CEA, carconoembryonic antigen; CLL, chronic lymphocytic leukemia; DAF, decay accelerating factor; EC, endothelial cells; ELAM, endothelium leukocyte adhesion molecule; GM-CSF, granulocyte-macrophage colony-stimulating factor; ICAM-I, intercellular adhesion molecule-1; IGSF, Ig superfamily; IL, interleukin; LAM, leukocyte adhesion molecule; LAMP, lysomal-associated membrane protein; LCA, leukocyte, common antigen; LFA, leukocyte function–associated antigen; M-CSF, macrophage colony–stimulating factor; MHC, major histocompatibility complexs; NCA, nonspecific cross-reacting antigen; NCAM, neural cell adhesion molecule; NGF, nerve growth factor; NK, natural killer; PA, platelet-activating factor; SCF, stem cell factor; TCR, T-cell receptor; TNF, tumor necrosis factor.

Glossary

A	ampere
AAV	adeno-associated virus
Ab	antibody
ADA	adenosine deaminase
Ag	antigen
AIDS	acquired immune deficiency syndrome
ALL	acute lymphocytic leukemia
AML	acute myeloid leukemia
APL	acute promyelocytic leukemia
Bl-CFC	blast colony-forming cell
BSA	bovine serum albumin
C′	serum complement
CACE	computer-assisted centrifugal elutriation
CFU	colony-forming unit
CFU-blast	blast colony-forming unit
CFU-GM	granulocyte-macrophage colony-forming unit
CFU-meg	megakaryocytic colony-forming unit
CFU-mix	multilineage colony-forming unit
CHO	Chinese hamster ovary
CHSS	clinical high-speed sorter
CIMS	continuous immunomagnetic sorter
cm	centimeter
CML	chronic myelogenous leukemia
cpm	counts per minute
DACS	density-adjusted cell sorting
DAF	decay-accelerating factor
DMEM	Dulbecco's Modified Eagle Medium
DMSO	dimethyl sulfoxide
DPBSE	Dulbecco's phosphate buffered saline with EDTA
E-BFC	erythroid burst-forming cell
EDTA	ethylenediaminetetraacetic acid
EPO	erythropoietin

FAB	French-American-British
FACS	fluorescence-activated cell sorting
FBS	fetal bovine serum
FISH	fluorescence in situ hybridization
FITC	fluorescein isothiocynate
FL	fluorescence
FSC	forward light scatter
GC	glucocerebroside
G-CSF	granulocyte colony-stimulating factor
GFP	green fluorescent protein
GM-CFC	granulocyte-macrophage colony-forming cell
GM-CSF	granulocyte-macrophage colony-stimulating factor
GMP	Good Manufacturing Procedures
GVHD	graft-versus-host disease
GVL	graft-versus-leukemia
HDP	high-density particles
HEPES	*N*-2-hydroxyethylpiperazine-*N'*-2-ethanesulfonic acid
HIV	human immunodeficiency virus
HPP-CFC	high proliferative potential colony-forming cell
HSA	human serum albumin
Hu	human
IDE	Investigational Device Exemption
IFN	interferon
IL	interleukin
IND	Investigational New Drug (application)
IU	International Unit
kb	kilobase
kD	kilodalton
kHz	kilohertz
LNGFR	low-affinity nerve growth factor receptor
LPS	lipopolysaccharide
LTC-IC	long-term culture-initiating cell
MAb	monoclonal antibody
MAC	membrane attack complex
MACS	magnetic cell sorting
Meg-CFC	megakaryocyte colony-forming cell
MHC	major histocompatibility class
MIX-CFC	mixed colony-forming cell
MDR	multidrug resistant
mL	milliliter
mPrP	murine prion protein
N	newton
NGF	nerve growth factor

NGFR	nerve growth factor receptor
NK	natural killer
PBMC	peripheral blood mononuclear cell
PBPC	peripheral blood progenitor cell
PBS	phosphate-buffered saline
PCR	polymerase chain reaction
PE	phycoerythrin
PEG	polyethylene glycol
Ph	Philadelphia
PHA	Phytohemagglutinin
PI	propidium iodide
pm	picometer
PMNL	polymorphonuclear leukocytes
PMP	peroxisomal membrane protein
psi	pounds per square inch
PVP	polyvinylpyrrolidone
r	recombinant
rpm	rotations (revolutions) per minute
SBA	soy bean agglutinin
SCF	stem cell factor
SCID	severe combined immunodeficiency disease
SEM	scanning electron microscopy
SSC	side light scatter
T	tesla (unit of measurement)
TCR	T-cell receptor
TdT	terminal deoxynucleotidyl transferase
TIL	tumor-infiltrating lymphocyte
USP	United States Pharmacopeia
v	volume
VBM	vertebral body marrow
w	weight

Index